AF564566

Abiotic & Biotic Stress Management in Plants

About the Editors

Dr. Bhav Kumar Sinha was born on 12th September 1975 at Haiderchak, Nalanda, Bihar. He did his Master's Degree in Plant Physiology from Banaras Hindu University, Varanasi in 1999 and Ph.D. (Plant Physiology) from Chaudhary Charan Singh Haryana Agricultural University, Hisar in 2004. In July 2008, he joined Sher-e-Kashmir University of Agricultural Sciences and Technology of Jammu, Jammu & Kashmir as Asst. Professor cum-Junior Scientist. Since then he is serving this University as Plant Physiologist and has given a direction in teaching, Plant Physiological research and extension. Dr. Sinha's research areas are Stress Physiology and Hormonal Physiology. He has published more than 23 original research paper in Indian and international journals, seven book chapters in different book and two practical manual. He is life member of serveral professional society and has handled one externally final project.

Reena (D.O.B – 01/12/1975) working as Senior Scientist (Entomology) at ACRA, Dhiansar, SKUAST-J, has done her B.Sc. (Agriculture) from Banaras Hindu University, Varanasi, M.Sc. (Agril. Entomology) from University of Agricultural Sciences, Dharwad and secured Ph.D. (Entomology) degree from C.C.S. Haryana Agricultural University, Hisar. She has qualified NET conducted by A.S.R.B., New Delhi and CSIR, New Delhi. She is also the recipient of Junior Research Fellowship (ICAR) during MSc. (Ag) degree program and Department of Science and Technology (DST), Government of India, Young Scientist Project Award (2009- 2012) under fast track scheme for young scientists. She is life member of several professional societies and has handled two externally funded project as PI and two as Co-PI. She has delivered several expert lectures and has published 25 research papers in journals of national and international repute.

Abiotic & Biotic Stress Management in Plants

Volume-I: Abiotic Stress

Bhav Kumar Sinha
Reena

NEW INDIA PUBLISHING AGENCY
New Delhi – 110 034

NEW INDIA PUBLISHING AGENCY
101, Vikas Surya Plaza, CU Block, LSC Market
Pitam Pura, New Delhi 110 034, India
Phone: + 91 (11) 27 34 17 17 Fax: + 91 (11) 27 34 16 16
Email: info@nipabooks.com
Web: www.nipabooks.com

Feedback at feedbacks@nipabooks.com

ISBN: 978-93-86546-57-9

Composed and Designed by NIPA

Sher-e-Kashmir University of Agricultural Sciences and Technology - Jammu

Dr Jag Paul Sharma
Director Research

Foreword

Plants are often exposed to various abiotic and biotic stresses. They have developed specific mechanisms to adapt, survive and reproduce under these stresses. Together, these stresses constitute the primary cause of crop losses worldwide, reducing average yields of most major crop plants. Current climate change scenarios predict an increase in mean temperatures and drought that will drastically affect global agriculture in the near future. In agriculture abiotic and biotic stress not only cause huge reduction in crop yields but also increase cost of cultivation, reduce input use efficiency, impair quality of produce. A complete understanding on physiological and molecular mechanisms especially signaling cascades in response to abiotic and biotic stresses in tolerant plants will help to manipulate susceptible crop plants and increase agricultural productivity in the near future. The biology of plant cell is more complicated with any foreign stimulus from the environment; multiple pathways of cellular signaling and their interactions are activated. These interactions mainly evolved as mechanism to enable the plant systems to respond to stress with minimum and appropriate physio- biochemical processes. Advanced aagricultural approaches are also required for sustainable solutions to the huge global problem of 'hidden hunger' and it may performed by the bio-fortification for increased micronutrient intakes and improved micronutrient status in the food. Management of cultural practices in conjunction with use of plant bio-regulators and chemicals will give a long way in management of various abiotic and biotic stresses.

Detailed discussion regarding cultural and agronomical management during stress condition will help not only to the researchers but also to the field functionaries and farmers. In this regard, an attempt has been made by Dr. Bhav Kumar Sinha and his team, Division of Plant Physiology, Faculty of Basic Sciences, Chatha, SKUAST-Jammu to come up with a volume entitled "Abiotic and Biotic Stress Management in Plants: Volume- I: Abiotic Stress" containing about 12 chapters.

This publication shall be of immense use to researchers, undergraduate and post graduate students along with other stake holders who is dealing with green ecosystem.

I congratulate the authors for their painstaking efforts in bringing out this publication.

Dr. Jag Paul Sharma

Main Campus Chatha, Jammu-180 009, J&K, INDIA
Tel: 0191-2263973
Mob: 09419134737
e-mail: jpsdr2015@gmail.com

Preface

Plants encounter a wide range of environmental insults during a typical life cycle and have evolved mechanisms by which to increase their tolerance of these through both physical adaptations, biochemical changes molecular and cellular changes that begin after the onset of stress. Environmental rudeness faces by the plants in the form of abiotic and biotic stress that seriously reduces their production and productivity. Approximately 70% of crops could have been lost due to both abiotic and biotic factors. Variety of distinct abiotic stresses, such as availability of water (drought, flooding), extreme temperature (chilling, freezing, heat), salinity, heavy metals (ion toxicity), photon irradiance (UV-B), nutrients availability, and soil structure are the most important features of and has a huge impact on growth and development and it is responsible for severe losses in the field and the biotic stress is an additional challenge inducing a negative pressure on plants and adding to the damage through herbivore attack or pathogen. Multiple stress exposure gives a possible outcome that Plant system develops tolerance to one environmental stress may affects the tolerance to another stress, for example, after exposure of plants to abiotic stress leading to enhanced biotic stress tolerance, wounding increases salt tolerance in tomato plants. In tomato plants, localized infection by *Pseudomonas syringae* pv. tomato (*Pst*) induces systemic resistance to the herbivore insect *Helicoverpa zea*.

Therefore, the subject of *Abiotic and Biotic Stress Management* is gaining considerable significance in the contemporary world. This book entitled "Abiotic and Biotic Stress Management in Plants: Volume- I: Abiotic Stress" deals with an array of topics in the broad area of abiotic stress responses in plants focusing "*problems and their management*" by selecting some of the widely investigated themes. Chapter 1: Cell signalling in plants during abiotic and biotic stress, Chapter 2: Salinity stress induced metabolic changes and its management, Chapter 3: High temperature stress: responses, mechanism and management, Chapter 4: Low temperature stress induced changes in plants and their management, Chapter 5: Biotechnological approaches to improve abiotic stress tolerance-I Chapter 6: Biotechnological approaches for improving abiotic stress tolerance-II, Chapter 7: Nutritional poverty in wheat under abiotic stress scenario, Chapter 8: Strategies for improving soil health under current climate change scenario Chapter 9: Abiotic stress management in Pulse crops, Chapter 10: Mitigation Strategies of Abiotic Stress in Fruit Crops, Chapter 11: Impacts of Abiotic Stress and Possible Management Options in Vegetable Crops, Chapter 12: Abiotic stress: impact and management in ornamental crops. We fervently believe that this book will provide good information and understanding of abiotic stress problems and their management in plants.

I would like to extend my gratitude to all contributors for their authoritative and up to date scientific information organized in a befitting manner. We thank the supporting staff of Division of Plant Physiology who have helped us in coming up with publication. The cooperation extended by Dr. J.P. Sharma, Director Research of the University is duly acknowledged. Valuable cooperation extended by Dr. S.A. Mallick, Dean, Faculty of Basic Sciences, in multifarious ways is gratefully acknowledged.

Last but not the least, I owe thanks to my son Krishna Sinha and Tanmay Sinha for taking care of me during this project.

Bhav Kumar Sinha
Reena

Contents

List of Contributors

Asha Rani
Department of Molecular Biology, Biotechnology & Bioinformatics, CCS Haryana Agricultural University, Hisar 125004

Ashok
Department of Crop Improvement and Biotechnology, College of Horticulture, University of Horticultural Sciences, Sirsi – 581 401 Karnataka

Amit Jasrotia
Division of Fruit Sciences, Sher-e-Kashmir University of Agricultural Sciences & Technology of Jammu, Main Campus Chatha, Jammu – 180 009

Arvinder Singh
Division of Vegetable Science and Floriculture, Sher-e-Kashmir University of Agricultural Science and Technology of Jammu, Chatha, Jammu-180009

Athani, S.I.
Department of Fruit Science, College of Horticulture, University of Horticultural Sciences Sirsi – 581 401, Karnataka

Bhav Kumar Sinha
Division of Plant Physiology, SKUAST- J, Faculty of Basic Sciences, Main Campus, Chatha Jammu – 180 009

G.K. Rai
School of Biotechnology, SKUAST-J, Chatha, Jammu, J&K 180009

Gurdev Chand
Division of Plant Physiology, SKUAST- J, Faculty of Basic Sciences, Main Campus, Chatha Jammu – 180 009

Jyoti Taunk
Department of Molecular Biology, Biotechnology & Bioinformatics, CCS Haryana Agricultural University, Hisar 125004

Jitender Singh
Centre of Excellence in Agri-Biotechnology, Department of Biochemistry & Physiology, College of Biotechnology SVPUAT, Meerut

Madhuri Gupta
Centre of Excellence in Agri-Biotechnology, Department of Biochemistry & Physiology, College of Biotechnology SVPUAT, Meerut

Mallikarjun Awati
Department of Biotechnology and Crop Improvement and Dept. of Soil Science and Agricultural Chemistry

Magdeshwar Sharma
Mega Seed project, SKUAST, Jammu-180009

Monika
Department of Molecular Biology, Biotechnology & Bioinformatics, CCS Haryana Agricultural University, Hisar 125004

Neelam R Yadav
Department of Molecular Biology, Biotechnology & Bioinformatics, CCS Haryana Agricultural University, Hisar 125004

Naveen Kumar
Wheat and Barley section, Department of Genetics and Plant Breeding, Chaudhary Charan Singh Haryana Agricultural University, Hisar- 125004 (Haryana), India

Pradeep K. Rai
Advanced Centre for Horticulture Research, SKUAST-J Udheywalla, Jammu, J&K 180 018

Ratnakar M. Shet
Department of Crop Improvement and Biotechnology, College of Horticulture, University of Horticultural Sciences, Sirsi – 581 401 Karnataka

Renu Munjal
Wheat and Barley section, Department of Genetics and Plant Breeding, Chaudhary Charan Singh Haryana Agricultural University, Hisar- 125004 (Haryana), India

Reena
ACRA, SKUAST- J, Rakh Dhiansar, Bari Brahmna, Jammu-181133

Ram C. Yadav
Department of Molecular Biology, Biotechnology & Bioinformatics, CCS Haryana Agricultural University, Hisar 125004

S. M. Prasanna
University of Horticultural Sciences, Udyanagiri, Bagalkot-587104, Karnataka state, India

Sapalika Dogra
Division of Plant Physiology FBSc. SKUAST-J, Jammu-180009

Shantappa, T
Department of Crop Improvement and Biotechnology, College of Horticulture, University of Horticultural Sciences, Sirsi – 581 401 Karnataka

Shivanand Hongal
Department of Vegetable, College of Horticulture, University of Horticultural Sciences, Sirsi – 581 401, Karnataka

Pankaj Kumar
Centre of Excellence in Agri-Biotechnology, Department of Biochemistry & Physiology, College of Biotechnology SVPUAT, Meerut

Shivani Khanna
Centre of Excellence in Agri-Biotechnology, Department of Biochemistry & Physiology, College of Biotechnology SVPUAT, Meerut

Mini Sharma
Centre of Excellence in Agri-Biotechnology, Department of Biochemistry & Physiology, College of Biotechnology SVPUAT, Meerut

Parshant Bakshi
Division of Fruit Sciences, Sher-e-Kashmir University of Agricultural Sciences & Technology of Jammu, Main Campus Chatha, Jammu – 180 009

S K Maurya
Department of Vegetable Science, College of Agriculture, G B Pant University of Agriculture & Technology Pantnagar Uttarakhand

V.K. Wali
Division of Fruit Sciences, Sher-e-Kashmir University of Agricultural Sciences & Technology of Jammu, Main Campus Chatha, Jammu – 180 009

V C Dhyani
Department of Agronomy, College of Agriculture, G B Pant University of Agriculture & Technology Pantnagar, Uttarakhand

Vithal Navi
Department of Agricultural Microbiology, College of Agriculture,University of Agricultural Sciences, Dharwad-580 005

Nomita Laishram
Division of Vegetable Science and Floriculture, Sher-e-Kashmir University of Agricultural Science and Technology of Jammu, Chatha, Jammu-180009

1

Cell Signaling in Plants During Abiotic and Biotic Stress

Bhav Kumar Sinha, Reena and Gurdev Chand

Plants are often exposed to various abiotic and biotic stresses and have developed specific mechanisms to adapt, survive and reproduce under these stresses (Pieterse *et al*., 2009). Abiotic stresses include drought, water logging, high temperature, cold(low temperature), salinity, chemical pollution (xenobiotics), uv radiation, heavy metal and oxidative and plants are also challenged by biotic stresses through microbial pathogens such as myco-plasma, nematodes, fungi bacteria (Tippmann *et al.*, 2006). The biology of plant cell is more complicated with any foreign stimulus from the environment; multiple pathways of cellular signaling and their interactions are activated. These interactions mainly evolved as mechanism to enable the plant systems to respond to stress with minimum and appropriate physio- biochemical processes. Abiotic and biotic stress induces signals and theses signals are recognized by receptors, followed by generation of secondary messengers *e.g.* activation of ion channels, production of reactive oxygen species (ROS) Xiong *et al*., 2002, accumulation of hormones (Bari and Jones, 2009; Peleg and Blumwald, 2011) such as salicylic acid (SA), ethylene (ET), jasmonic acid (JA) and abscisic acid (ABA). Secondary messengers are responsible for modulating intracellular level of calcium, often initiating protein phosphorylation cascade, which may leads to the activation of various proteins directly involved in cellular protection (Tippmann *et al.*, 2006) Fig.1.

Plants can perceive biotic and abiotic stresses via specific receptors. The stress recognition is usually followed by production of messengers such as reactive oxygen species (ROS) or Inositol-1,4,5-triphosphate (IP3), and modulatation of intracellular calcium(Ca2+). Production of these early messengers can be receptor mediated, however, in many abiotic stresses (*e.g.*, drought, cold, salt stress) membrane overexcitation can directly lead to ROS generation. The messengers also seem to influence each other, *e.g.*, via IP3 gated calcium channels. Receptors, ROS or Ca2+ dependent signals initiate specific phosphorylation cascades (*e.g.*, mitogen activated protein kinases - MAPK

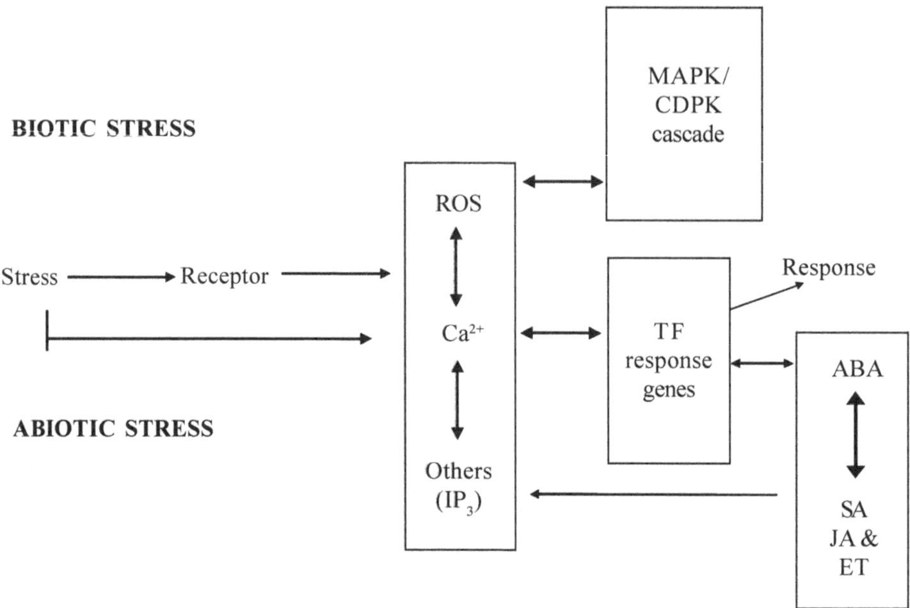

Fig. 1: Signalling network in plants under biotic and abiotic stress.

calcium dependent protein kinases - CDPK) and finally interact with the promotor regions of transcription factors (TF) and response genes. The regulation of gene transcription can additionally depend on the activity of plants hormones and other regulatory molecules (*e.g.*, abscidic acid - ABA, salicylic acid - SA, jasmonic acid - JA, ethylene - ET). The plant hormones can also influence the ROS and Ca2+ levels and initiate a second round of signalling (modified from Tippmann *et al.*, 2006).

Interestingly, multiple stress exposure gives a possible outcome that Plant system develops tolerance to one environmental stress may affects the tolerance to another stress (Bowler and Fluhr, 2000) (Table 1). This phenomenon is also known as cross- tolerance, showing that plants possess a powerful regulatory system that allows them to adapt quickly to a changing environment (Bowler and Fluhr, 2000, Capiati *et al.*, 2006 and Suzuki *et al.*, 2012) for example, after exposure of plants to abiotic stress leading to enhanced biotic stress tolerance, wounding increases salt tolerance in tomato plants (Capiati *et al.*, 2006). Stout *et al.*, 1999 reported that, in tomato plants, localized infection by *Pseudomonas syringae* pv. tomato (*Pst*) induces systemic resistance to the herbivore insect *Helicoverpa zea*. Positive relationship between abiotic and biotic stress may also possible (Abouqamar *et al.*, 2009), as demonstrated by the reduced infection of tomato by *Botrytis cinerea* and *Oidium neolycopersici* following the application of drought stress (Achuo *et al.*, 2006) and exposure of ozone may

Table 1: Cross tolerance between biotic and abiotic stresses (Tippmann *et. al.*, 2006)

Plant species	Abiotic Stress	Biotic Stress	Possible common signal compound	Reference
Positive influene of abiotic stress adaptation towards biotic stress tolerance				
Tobacco	ozone, UV	Tobacco mosaic Virus	SA	Yalpani *et.al.*, 1994
Arabidopsis	ozone	*Pseudomonas syringae* (bacterium)	SA, ROS	Sharma *et. al.*, 1996
Bell pepper	Ionic stress, hyperosmolarity,oxidation	*Pseudomonas syringae* (bacterium)	SA, JA	Hong and Hwang, 2005
Medicago	drought	*Aphanomyces euteiches* (Fungus)	ABA	Colditz *et. al.*, 2004
Arabidopsis	drought	*Pernospora parasitica*	SA	Chini *et. al.* 2004
Barley	Salinity	*Botrytis graminis*(fungus)	ABA	Wiese 2004
Arabidopsis	Salinity, drought	*Botrytis cinerea* (fungus)	JA, ROS	Mengiste *et al*. 2003
Rye	cold	*Microdochium nivale* (fungus)	ET,SA	Yu *et al*. 2001
Positive influence of biotic stress adaptation towards abiotic stress tolerance				
Arabidopsis	drought	*Paenibacillus polymyxa* (Rhizobacteria)	SA, ABA?	Timmusk and Wagner, 1999
Banana	salinity	*Glomus mosseae* (Mycorrhizal fungi)		Yano-Melo *et al*. 2003
Tomato	salinity	*Glomus mosseae* (Mycorrhizal fungi)		Al-Karaki 2000
Arabidopsis	drought (-) salinity, heat	*Peronospora parasitica* (fungus)	SA	Chini *et al*. 2004
Negative influence of abiotic stress on biotic stress				
Citrus	salinity	*Phytophthora* spp. (fungus), nematodes	?	Syvertsen and Levy , 2005
Red pine	drought	*Sphaeropsis sapinea* (fungus)	?	Blodgett *et al*. 1997, Stanosz *et al*. 2001
Oak	drought	*Armillaria spp.* (fungus)	?	Boyer, 1995

induce resistance to virulent *Pseudomonas syringae* strains in *Arabidopsis* (Sharma *et al*., 1996). On the contrary, biotic stress can also interfere to increase the resistance to abiotic stress. This effect is easily visible when plants are under pathogen attack. Goel *et al*., 2008 revealed that, the Infection due to pathogen attack may cause closure of stomata to obstruct pathogen entry and as a result water loss is reduced and leads to an enhanced plant resistance under abiotic stress. Xu *et al*., 2008 shows that, the viral infection protects plants against drought stress. Verticillium infection in Arabidopsis plants induced the expression of the Vascular-Related No Apical meristem ATAF and Cup-Shaped Cotyledon (NAC) domain (VND) transcription factor *VND7*. VND7 induced *de novo* xylem formation ensuring the water storage capacity results increases plant drought tolerance (Reusche *et al*., 2012).

Here we will elaborate on various cell signaling in abiotic and biotic stress that share signal components or have common stress responses for physiological adaptation. We focus on signaling compounds (reactive oxygen species (ROS), calcium, protein kinases, abscisic acid and salicylic acid) which usually work under both abiotic and biotic stress condition.

Signalling components

Most imperative components of stress-induced signalling pathways (reactive oxygen species, calcium, protein kinases, abscisic acid and salicylic acid) and signaling cascades seem to have equal importance in response to environmental changes and pathogen challenges. At this juncture, we want to discuss some examples of interacting pathways and involved key signals, which have relevance as broad stress responses for abiotic and biotic stress. A range of signal transducers exist in plants, we will concentrate on Ca2+, ROS, ABA, SA and protein kinases involved in phosphorylation cascades.

Calcium (Ca2+) as second messengers

In plants, Calcium (Ca2+) is a major component, which is elicited by different stress exposure. Cytosolic Ca2+ levels increase in plant cells in response to various environmental stresses, including pathogen challenge, osmotic stress, water stress, cold and wounding (Takahashi *et al*., 2011). In addition, several stresses induce the biosynthesis of second messengers that require Ca2+ for signal transduction, *e.g.*, ROS, ABA and IP3 (Tippmann *et al*., 2006). Plant Ca2+ signals are involved in an array of intracellular signaling pathways after pest invasion. After herbivore feeding dramatic increase in Ca2+ influx, followed by the activation of Ca2+-dependent signal transduction pathways (Arimura and Maffei, 2010). Transient increases in Ca2+ of varying duration are initiated by oxidative stress, pathogen elicitors and hyperosmotic treatments, often in conjunction with an oxidative burst (Rentel and Knight, 2004). During drought and any other water limiting stresses, oscillations of Ca2+ are important for the regulation of guard cell-mediated stomatal pore closure (Allen *et al.*, 2001). Rapid increase in cytosolic Ca2+ influx through membrane Ca2+ ion channels, carriers and fluxes from internal vacuolar store (Errakhi *et al.*, 2008). These channels are activated by membrane depolarization, reported in response to different stresses such as light intensity and fungal elicitors. ROS induction also activates a hyper polarization dependent current in guard cells (Kluesner e*t al*., 2002). Moreover, both components are highly interconnected: Ca2+ signaling components such as calmodulins CaMs) and calcium-dependent protein kinases (CDPKs) regulate ROS production by NADPH-oxidases (Takahashi *et al.*, 2011). As high levels of Ca2+ in the cytoplasm are toxic for the cell, therefore the influx of calcium into the cell has to be controlled (Tippmann, *et al.*, 2006). Ca2+ plays a very important role in abiotic stress responses. In transgenic *Arabidopsis* plants, calcineurin B-like protein function as a differential regulator of salt, drought and cold responses in plants and leads to enhanced tolerance to salt and drought but reduced tolerance to freezing (Cheong *et al.*, 2003).

Whole genome expression meta-analysis experiments under different abiotic and biotic stress treatments revealed a substantial number of genes that are commonly regulated under abiotic and biotic stress conditions (Shaik and Ramakrishna, 2013 and 2014). Ma and Bohnert, 2007, reported that, 197 commonly regulated genes identified include response to ABA, SA, jasmonic acid (JA), and ethylene (ET), major stress hormones controlling adaptation to abiotic and biotic stress.

Thus, it is very much clear that Ca2+ play major role in tolerance against abiotic and biotic stress and it will quit interesting to see how the Ca2+ stress signals puzzle can be solved. In future, the manipulation of cellular Ca2+ signaling opens new avenues in developing more resistant crops.

Reactive Oxygen Species (ROS)

Universal truth of abiotic and biotic stress is changes in cytosolic Ca2+ concentration and of the intercellular redox state. Minor alterations in normal condition of plants may responsible for the production of ROS. ROS are partially reduced forms of atmospheric oxygen and include H_2O_2, OH and O_2^- (Mittler 2002). Levels of ROS rapidly increase in the cells of local tissue soon after exposure of stresses, which includes drought, cold, ozone high temperature, heavy metal, flooding, salinity, insect attack and pathogen (Dey *et al.*, 2010, Takahashi, *et al.*, 2011). Cell wall peroxidases and plasma membrane localized NADPH oxidases provide a system, which produces H_2O_2 in response to abiotic stress and pathogen attack. Gonzalez *et al.*, 2011 repoted that, mitochondria also serve as the site of ROS production in marine algae (*Ulva compressa*) after abiotic stress exerted by copper. Smirnoff, 1995 reported that, ROS causing lipid peroxidation and damage to the Photosystem II reaction centre. Consequently, in non-stressed plant, a state of redox homeostasis is kept in a fine tuned balance between ROS production and scavenging by a range of antioxidant systems. Superoxide dismutase (SOD), ascorbate peroxidase (APX), non-specific peroxidase (POX), glutathione reductase (GR) and catalase (CAT) are the foremost antioxidant enzymes of the plants. SOD reacts with O_2^- to produce H_2O_2, which is scavenged by POXs and CATs. Among the different peroxidases APX, which uses ascorbate as an electron donor, has an important role in H_2O_2 detoxification and is well described in plants. The production of ROS is fine-modulated by the plant to avoid tissue damage (Kissoudis *et al*., 2014).

After pathogen challenge increased ROS production results in a stronger hypersensitivity response, favoring resistance against biotrophic pathogens but compromising resistance against necrotrophic fungi (Kobayashi *et al*., 2012). The oxidative burst is one of the earliest responses to pathogen recognition in pathogen defence mechanisms. Plants synthesize ROS to mediate cell death as part of the host resistance mechanisms (Zhou *et al.,* 2000). Acknowledgment of an invading pathogen results in the activation of plasma membrane-associated NAD(P)H oxidases and peroxidases, significantly enhanced production of O_2. Dismutation of O_2^- by SOD generates H_2O_2, which accumulated in the cell and activate defence mechanisms, ultimately leading to PCD (Jabs *et al*., 1997). ROS act as antimicrobial toxic compounds against invading pathogens, strengthen the cell wall by cross-linking proteins and may induce PCD in plant cells. However, it has been shown that, high level of ROS to cell death and lower level are mostly responsible to regulate the plant's stress responses (Choudhury, et al., 2013). To achieve higher local concentrations, the plant suppresses its own detoxification machinery. In tobacco, for instance, tobacco mosaic virus infection leads to suppression of the activity of ROS detoxifying enzymes like cytosolic APX and CAT (Mittler *et al*., 1998).

ROS are responsible for stress-induced tolerance in *Arabidopsis thaliana* after infection with the vascular pathogen *Verticillium* spp. by increasing drought tolerance due to *de novo* xylem formation and the resulting enhanced water flow. The production of ROS can help in cell-to-cell communication by signal transduction through the *Respiratory Burst Oxidase Homologue D* (*RBOHD*) and can act as a secondary messenger by modifying protein structures and activating defense genes. Davletova *et al.,* 2005 reported that the transcription factor *Zat12* was involved in both abiotic and biotic stress and that *Zat12* could be a regulator in ROS scavenging.

In Arabidopsis, ROS-sensitive transcription factors activated by ROS production and it lead to the induction of genes participating in the stress responses. Gechev *et al.*, 2006 proposed that ROS were inducers of tolerance by activating stress response-related factors like mitogen-activated protein kinases (MAPKs), transcription factors, antioxidant enzymes, dehydrins, and low-temperature-induced-, heat shock-, and pathogenesis-related proteins.

Priming with specific chemicals for stress tolerance is responsible for certain modifications in ROS signaling. Treatment of cucumber plants with brassinosteroids lead to arise in H_2O_2 levels and primed the plants for both biotic and abiotic stress tolerance. H_2O_2 priming for salt tolerance in citrus moderately increased the abundance of oxidized and *S*-nitrosylated proteins, and the level remained the same after stress application, however, non-treated plants were more sensitive to the stress (Tanou *et al*., 2009).

It has been observed that, defects in the ROS production pathways impair plant defences significantly because, ROS itself can constitute a stress component and higher tolerance to oxidative stress can increase plant stress tolerance. Thus, a picture unfolds where the nature, timing and amount of ROS is important for the effect on the plant: low levels of ROS are involved in stress signalling inducing protective genes, while high concentrations usually lead to the induction of PCD.

Phytohormones

Plant hormones were first defined as the organic substances which, being produced in one part of the plant and is transferred to another part and there persuade a specific physiological process. It is also called as phytohormones. The five classical phytohormones: auxin, cytokinin, ethylene (ET), gibberellins and ABA are the chemical messengers synthesize in trace quantities and recently identified brassinosteroids, JA and SA are also potent phytohormones act as chemical messengers. Phytohormones move throughout the plant body via the xylem or phloem transport stream, move short distances between cells or are maintained in their site of synthesis to exert their influence on target cells where they bind transmembrane receptors located at the plasma membrane or endoplasmic reticulum or interact with intracellular receptors. Hormonal signaling includes alterations in gene expression patterns and in some cases non genomic responses. Varied concentration of plant hormones and tissue sensitivity to them regulate physiological process that has profound effects on growth and development. Existence of phytohormomes affects all phases of the plant life cycle and their responses to environmental stresses, both biotic and abiotic. Hormonal signalling is significant for plant defenses against abiotic and biotic stresses (Taiz and Zeiger, 2010; Williams, 2010).

Phytohormones regulate the responses against adverse environments are grouped into two types: those that play a major role in response to biotic stress (ET, JA and SA) and those that have central roles in regulating the abiotic stress responses (ABA). Commonly the biotic defense signaling networks mediated by phytohormones are dependent on the nature of the pathogen and its mode of pathogenicity. SA plays a central role in the activation of defense responses against biotrophic and hemi-biotrophic pathogens as well as the establishment of systemic acquired resistance. By contrast, JA and ET are generally associated with defense against necrotrophic pathogens and herbivorous insects. Concerning to abiotic stress, ABA is the most studied stress-responsive hormone; it is involved in the responses to drought, osmotic and cold stress (Peleg and Blumwald, 2011; Bari and Jones, 2009; Vlot, *et al*., 2009; Wasilewska, *et al*., 2008).

Abscisic Acid

ABA is considered as major hormone involved in the regulation of various aspects of plant growth and development and act as the central regulator of abiotic stress resistance in plants and coordinates an array of functions (Wasilewska *et al*., 2008, Finkelstein, 2013 and Wani and Kumar, 2015), enabling plants to cope with different stresses. Well established regulatory function of ABA is water balance and osmotic stress tolerance (Zhu 2002) and even under non-stress conditions, plants retain a low level of ABA, which increases in response to abiotic stress conditions. Drought and salinity activate *de novo* ABA synthesis to prevent further loss of water by transpiration through stomata (Levitt, 1980), mediated by turgor pressure change of the guard cell. ABA is responsible for fast changes in intracellular Ca2+ concentration and stimulates further signalling in the cell. At the time of osmotic stress, ABA induces the proteins accumulation involved in the biosynthesis of osmolytes (*e.g.*, proline, trehalose), which increases the stress tolerance of plants (Nayyar *et al*., 2005). Drought stress induced accumulation of ABA and direct exogenous application of ABA has been reported to enhance freezing tolerance and accumulate protective AFP proteins, which might confer additional resistance (Mantyla *et al*., 1995). ABA is also needed for protection against the oxidative damage of heat stress (Larkindale and Knight, 2002). Whereas, plant mutants, which are defective in the biosynthesis of ABA, have been shown to be less tolerant to environmental stresses such as low temperatures and heat (Xiong *et al*,. 2002). Evidently, the intracellular concentration of ABA has to be kept well balanced, as ABA biosynthesis impaired mutants show reduced fitness and often less stress tolerance. In contrast, ABA signalling in abiotic stress is quite complex, it is evidenced that, salt tolerant (*sto*) mutant has reduced osmotic stress tolerance and is deficient in ABA accumulation (Ruggiero *et. al*., 2004).

ABA also plays a vital role in biotic stress and it has been observed that, exogeneous ABA treatment increases the susceptibility of various plant species to bacterial and fungal pathogens (Thaler and Bostock, 2004 and Mohr and Cahill, 2007). Edwards and Allen, 1983 noticed that, ABA reduced penetration resistance of powdery mildew in barley leaves after floating leaves in ABA solution by an unknown mechanism. ABA induces susceptibility of plants to pathogens by reducing the gene expression of defence-related transcripts in potato and soybean. ABA-deficient tomato mutants show a reduction in susceptibility to the necrotroph *Botrytis cinerea* (Asselbergh *et. al.,* 2007) and ABA-deficient *Arabidopsis* has reduced susceptibility to the oomycete *Hyaloperonospora parasitica* (Mohr and Cahill, 2003). In general, ABA is involved in the negative regulation of plant defenses against various biotrophic and necrotrophic pathogens.

ABA also acts as a positive regulator of defense is reported thereof (Mauch-Mani and Mauch, 2005). However, increased concentration of ABA under the effect of abiotic stress induces stomatal closure, as a "secondary effect", the entry of biotic assailants through these passive ports of the plant is prevented. Consequently, under such circumstances, the plant is protected from abiotic as well as from biotic stress (Melotto *et al.,* 2006). ABA protects plants against *Alternaria brassicicola* and *Plectosphaerella cucumerina* indicating that ABA acts as a positive signal for defense against some necrotrophs (Ton and Mauch-Mani, 2004). Moreover, mutants which are deficient in ABA are more sensitive to infection by the fungal pathogens *Pythium irregulare* (Adie *et al.*, 2007) and *Leptosphaeria maculans* (Kaliff *et al.*, 2007). ABA is required for the synthesis of â-aminobutyric acid (BABA) is a non-protein amino acid, which enhances resistance through restriction of pathogen growth (Jakab *et al.*, 2001). BABA "primed" plants enhanced callose deposition in the interaction with two necrotrophic fungi, *Alteraria brassicola* and *Plectosphaerella cucumerina* and effectively stopping the disease progression that leads to higher resistance against these pathogens. Whereas callose deposition priming is absent in *abi4-1* and *aba1-5* mutants, hence ABA signalling regulates BABA-mediated priming for callose deposition (Ton and Mauch- Mani, 2004). Transcriptome and meta analyses of expression profiles tainted by necrotroph *Pythium irregulare* infection, recognized many JA-induced genes, although, the ABA responsive element (ABRE) appears in the promoters of many of the defense genes (Adie *et al.*, 2007; Wasilewska *et al.*, 2008). Therefore, it indicates that ABA plays an important role in the activation of plant defense through transcriptional reprogramming of plant cell metabolism. Furthermore, ABA is obligatory for JA biosynthesis and the expression of JA responsive genes after *Phytium irregular* infection (Adie *et al.*, 2007).

Apparently, the role of ABA as a regulator during the conversation between abiotic and biotic stress strongly depends on the timing of the stress perception: does the infection hit a plant that had already been exposed previously to abiotic stress or does an infected plant become additionally exposed to abiotic stress (Yasuda *et al.,* 2008).

Salicylic acid

Salicylic acid (SA) is important hormone which activates defense responses against hemibiotrophic and biotrophic pathogens, it is also important for establishment of systemic acquired resistance (SAR) (Vlot *et al*., 2009) and exogenous application of SA results in induction of *PR* genes increasing resistance to a broad range of pathogens. Mutants of Arabidopsis and tobacco in which endogenous SA levels are less, fail to develop SAR or express ion of *PR* genes and when these plants are treated with the SA synthetic analog, 2,6-dichloro-isonicotinic acid, resistance and *PR* genes expression are resumed (Vlot *et al*., 2009). The SAR is induced systemically by a signal generated in the inoculated leaf which is transmitted via the phloem to the uninfected portions of the plant (Parker, 2009). JA signaling directly regulated by GA signaling, mediated through direct binding of the GA repressor protein DELLA to JAZ proteins and relieving JA signaling repression (Hou *et al*., 2010). DELLA proteins appear to be central nodes in abiotic and biotic stress cross-talk and positively affect ROS detoxification (beneficial for acclimation to abiotic stress) through higher expression of ROS detoxification genes (Achard *et al*., 2008). DELLAs also sensitize JA signaling (through binding of DELLAs to JAZ) at the expense of SA signaling, enhancing resistance to necrotrophic pathogens (Navarro *et al*., 2008). SA levels rise coincidently with or just prior to SAR and systemic *PR* gene expression or peroxidase activation in pathogen-infected tobacco or cucumber and it was also detected in the phloem of pathogen-infected cucumber and tobacco, because a significant amount of SA in the systemic leaves of pathogen-infected tobacco and cucumber is transported from the inoculated leaf (Vlot *et al*., 2009). Treatment of plants with JA results in enhanced resistance to herbivore challenge. Mutants defective in the biosynthesis or perception of JA show compromised resistance to herbivore attackers (Bari and Jones, 2009). Attack of herbivores such as *Manduca sexta* in tobacco induces the JA signaling activity (Paschold *et. al*., 2007). Similarly, JA signaling is induced in tomato and *Arabidopsis* by *Tetranychus urticae* and *Pieris rapae*, respectably (Reymond *et al*., 2004 and De Vos *et al*., 2005). However, not all herbivores activate JA signaling in plants (Bari and Jones, 2009). The production of proteinase inhibitors (PIs) and other anti-nutritive compounds such as polyphenol oxidase (PPO), threonine deaminase (TD), leucine amino peptidase and acid phosphatase (VSP2) are mediated by JA in order to deter, sicken or kill the attacking insect (Howe and Jander, 2008).

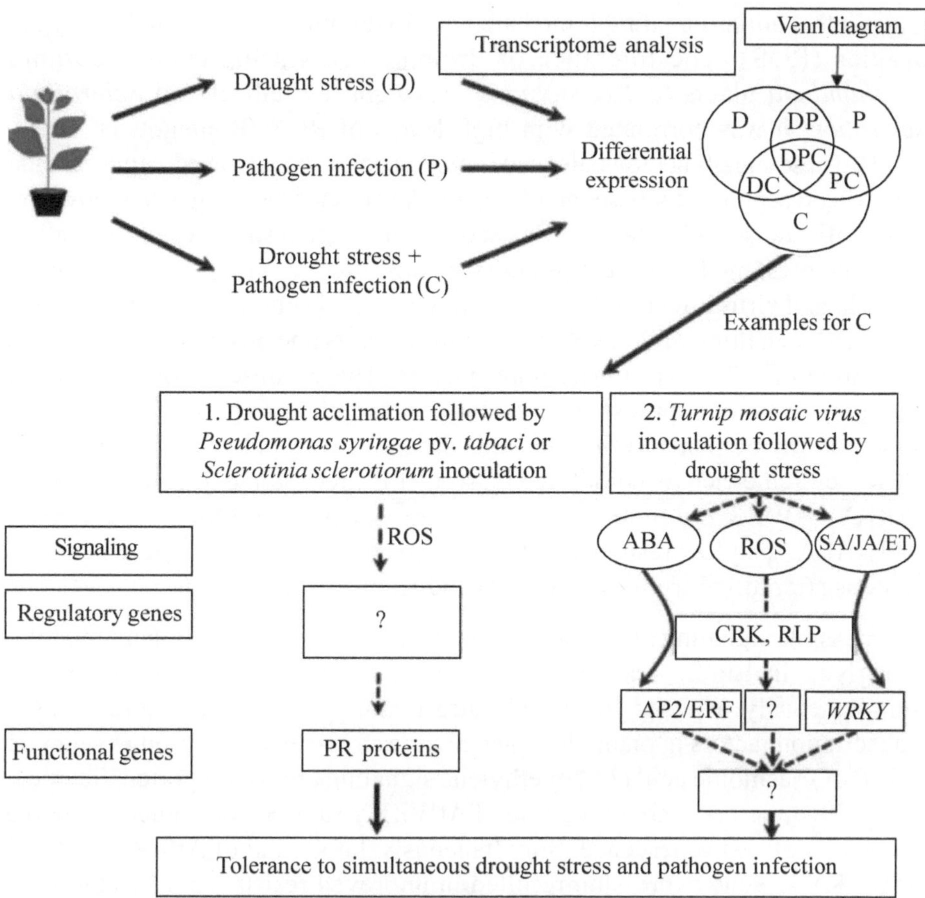

Fig. 2: Model for expected molecular responses of plants exposed to a combination of drought stress and pathogen infection

This representation compares the transcriptome profile of plants under respective individual stresses. Venn diagram is shown to indicate possible gene regulation scenarios. In addition to several shared genes, certain number of genes can be specifically regulated under each stress condition. D – genes unique to drought stress; P – genes unique to pathogen; C – genes unique to drought stress and pathogen combination (tailored response); DP –genes shared between drought stress and pathogen; DC –genes shared between drought stress and combination of drought and pathogen; PC –genes shared between pathogen and combination of drought and pathogen; and DPC – genes shared between drought stress, pathogen and combination of drought and pathogen. Genes specifically induced under stress combination "C" reflects the tailored molecular mechanisms regulated in plants simultaneously exposed to drought and pathogen. Individual stress responses of some of these unique genes induced under stress combination suggest that initial signaling is mediated by phytohormones and reactive oxygen species (ROS). The tolerance of drought stress acclimated *Nicotiana benthamiana* plants to *Pseudomonas syringae* pv. tabaci and *Sclerotinia sclerotiorum* was correlated with high levels of ROS (Ramegowda, *et al.*, 2013). These signaling molecules can initiate specific signal transduction cascade involving receptor like kinases and receptor like proteins resulting in the activation of specific transcription factors. Based on the individual stress response studies, it can be presumed that the regulatory events after recognitions of combined drought and virus infection involves salicylic acid, jasmonic acid and ethylene mediated regualtion of WRKY transcription factors and abscisic acid mediated regulation of AP2/ERF transcription factors. These transcription factors can further activate or suppress functional genes thereby bringing in tolerance or susceptibility of plants to simultaneous drought stress and pathogen infection. CRK – cysteine-rich receptor like kinases; RLP – receptor-like protein kinases; WRKY – WRKY domain containing transcription factors; AP2/ERF – APETALA2/ethylene response factors; dotted arrows indicate possible signaling response (modified from Ramegowda and Kumar, 2015).

The possible signaling events under combined drought and pathogen infection are given in Fig.2. Transcriptome and metabolome changes in plants simultaneously exposed to drought stress and pathogen infection. WRKY transcription factors in plant biotic and abiotic stress response through salicylic acid (SA), jasmonic acid (JA) or ethylene signaling are well reported. Scarpeci *et al.*, 2013, describe about the role of AtWRKY30 in SA dependent negative regulation of leaf senescence. Simultaneously, knock-out of AtWRKY18 and AtWRKY40 gene expression resulted in improved resistance of Arabidopsis plants to biotrophic powdery mildew fungus *Golovinomyces oronti* which was accompanied by altered SA and JA signaling, EDS1 genes expression and accumulation of phytoalexin camalexin (Schön *et al.*, 2013). Furthermore, individual stress response of AP2/ERF transcription factors in mediating biotic and abiotic stress response through ABA was studied. AtERF11 has been shown to negatively regulate ABA-mediated control of ethylene synthesis thereby averting the negative effect of ethylene on plant growth and development (Li *et al.*, 2011). Another AP2/ERF transcription factor, RAP2.6 granted resistance against beet cyst nematode *Heterodera schachtiiin* Arabidopsis roots by enhanced callose deposition in syncytia (Ali *et al.*, 2013) and showed hypersensitivity to exogenous ABA and abiotic stresses during seed germination and early seedling growth in Arabidopsis (Zhu *et al.*, 2010). CRK7, a cysteine-rich receptor like kinase, has been shown to mediate oxidative signaling induced by apoplastic ROS (Idänheimo *et al.*, 2014).

Jasmonic acid

The JA pathway regulates response to abiotic stress, defenses against insect herbivores and necrotrophic fungal pathogens and also defenses against biotrophic pathogens such as the powdery mildews (Ellis and Turner, 2001). In *Arabidopsis* leaves, jasmonates control the expression of an estimated 67-85% of wound- and insect-regulated genes. Treatment of plants with JA results in enhanced resistance to herbivore challenge. Mutants are defective in biosynthesis or perception of JA shows compromised resistance to herbivore attackers (Bari and Jones, 2009). JA regulates wound responses and defense against insect pests, and is also implicated in drought responses. However, microarray analysis of gene expression in wild-type and *coi1* Arabidopsis plants that were wounded, attacked by insects, or exposed to water stress reveals a surprisingly large overlap of *COI1*-dependent genes regulated by wounding and by water stress, and an unexpectedly different profile of genes regulated by wounding and by herbivory (Reymond *et al*., 2000). Attack of herbivores such as *Manduca sexta* in tobacco induces the JA signaling activity (Paschold *et al*., 2007). Similarly, JA signaling is induced in tomato and *Arabidopsis* by *Tetranychus urticae* and *Pieris rapae*, respectably (Reymond *et al*., 2004 and De Vos *et al*., 2005). However, not all herbivores activate JA signaling in plants (Bari and Jones, 2009). The production of proteinase inhibitors (PIs) and other anti-nutritive compounds such as polyphenol oxidase (PPO), threonine deaminase (TD), leucine amino peptidase and acid phosphatase (VSP2) are mediated by JA is responsible for discouraging, repelling or killing of the insects (Howe and Jander, 2008). Terpenoids and other volatile compounds produced by an herbivore-attacked plant are recognized by other carnivorous and parasitoid insects (Williams, 2011).

Protein/MAP Kinases are signal transmitteres

Signal transductions between all eukaryotes are dependent on phosphorylation of proteins by protein kinases. Ca2+ dependent protein kinases (CDPK) and a group of serine/threonine kinases, mitogen activated kinases (MAPK) have a special importance for the signal transduction of diverse cellular processes under various abiotic and biotic stress responses (Fujita *et al*., 2006). CDPKs contain a carboxyterminal calmodulin-like domain containing EF-hand calcium-binding sites plus a N-terminal protein kinase domain (Cheng *et al*., 2002). Thus, the signaling pathways activated in response to abiotic and biotic stresses situated in part on CDPKs. The *Arabidopsis* genome encodes 34 CDPKs, but only few substrates of these enzymes have been identified (Uno *et al*., 2009). Since MAPK cascades are activated by a range of stimuli including abiotic stress (cold, drought, salinity, ROS and heat), biotic stress (wounding pathogens and pathogen elicitors) ethylene, ABA and SA, they could have a role in the combination of both the stresses (Tena *et al.* 2001 and Samajova *et al.*, 2013). MAPKs are central for the transduction of cellular signals by activation and repression of downstream target protein activity (Fig. 1). Environmental factors stimulate a MAP kinase kinase kinase (MAPKKK), which phosphorylates kinase kinase (MAPKK). MAPKK activates MAPK, which in turn activates downstream cellular proteins by phosphorylation(Jonak *et al.*, 2002). MAPK signaling is of equal importance in biotic and abiotic signaling, defects in MAPK mediated signaling affect not only pathogen related responses. For instance *Arabidopsis MAPK3* responds to a variety of abiotic stress including drought, cold, salinity and is sensitive to mechanical stress (Mizoguchi *et al.*, 1996) and *MKK2* is hypersensitive to salt and cold stress through impaired action for *MAPK4*and *MAPK6* (Teige *et al.*, 2004 and Gudesblat *et al.*, 2007). In Alfalfa, *MMK4*, an orthologue of *AtMPK3*, responds to drought and cold (Jonak *et al.*, 1996). MAPK pathways activated by pathogen attack are mediated by SA, and the resulting expression of *PR* genes induces defense reactions (Xiong *et al.*, 2003). The Arabidopsis protein VIP1 is translocated into the nucleus after phosphorylation by MPK3 and acts as an indirect inducer of *PR1* (Pitzschke *et al.*, 2009). Chinchilla *et al.,* 2007 observed that pathogen associated molecular patterns (PAMPs) like flagellin trigger MAPK cascades in order to establish pathogen response signaling. Over-expression of the *OsMPK5* gene and also kinase activity of OsMPK5 induced by ABA responsible to increased abiotic and biotic stress tolerance. *OsMPK5* seems to play a double role in the rice stress response, one as a positive regulator of resistance to the necrotrophic brown spot pathogen *Cochliobolus miyabeanus* and the second as a mediator of abiotic stress tolerance (Teige *et al.*, 2004 and Brader *et al.*, 2007). Tomato plants also activate MPK1 and MPK2 against UV-B, wounding, and pathogens in order to enhance their defense mechanisms (Holley *et al.*, 2003).

In summary, MAP kinase cascades comprise a foremost mechanism for activation of abiotic and biotic defence responses and in mediating hormones and other signals.

Transcription factors

During abiotic and biotic stress plant responses to stimuli involve a network of molecular mechanisms that vary depending on the nature of signals. In the signal transduction pathway that leads from the perception of stress signals to the expression of stress-responsive genes, transcription factors play an indispensable role. TFs are a group of master proteins that interact with *cis*-elements present in promoter regions upstream of genes and regulate their expression. TFs involved in stress crosstalk comprise a diverse collection of TF families in plants, such as NAC, MYB, AP2/ERF, WRKY, and others, viewing the complexity of the genetic regulatory networks under abiotic and biotic stress condition (Atkin- son and Urwin, 2012 ; Shaik and Ramakrishna, 2014). Many members of TFs families are involved in regulation of multiple physiological processes such as metabolism, cell cycle progression, growth, development, reproduction and leaf senescence, an vital component of both abiotic and biotic stresses (Zhou *et al*., 2010; Hussain *et al*., 2011; Breeze *et, al*., 2011). Moreover, in most cases the TFs engross in regulation of stress hormone, and potentially act as molecular switches for the fine-tuning of hormonal responses.

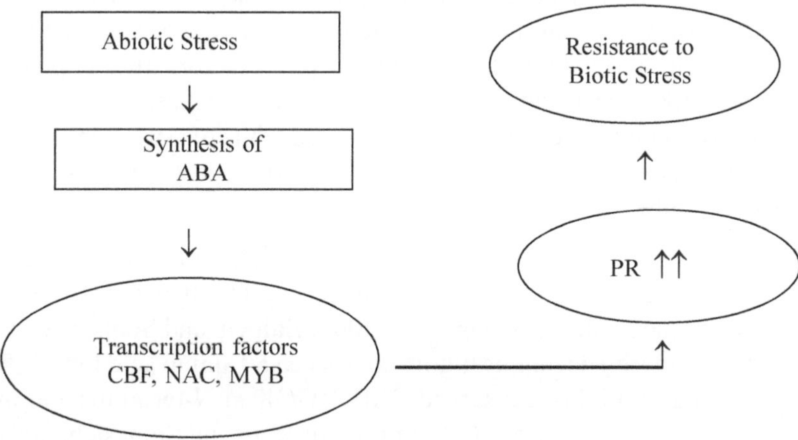

Fig. 3: Abiotic stress can enhance the expression of specific transcription factors (TFs) like C-repeat Binding Factors (*CBF*), No Apical meristem ATAF and Cup-Shaped Cotyledon (*NAC*), *MYB* mediated by abscisic acid (ABA). Although the exact role of ABA in plant pathogen interactions is still a matter of debate, in some specific cases it has been shown to promote resistance against biotic stress following abiotic stress. This is attributed to the over-expression of TFs inducing the up-regulation of *PR* genes (Modified from Rejeb *et al*., 2014).

The basic helix-loop-helix (bHLH) domain-containing transcription factor AtMYC2 is a regulator of ABA signaling and the genetic lession of *AtMYC2* results in elevated levels of basal and activated transcription from JA-ethylene responsive defense genes (Anderson *et al*., 2004). *BOTRYTIS SUSCEPTIBLE1* (*BOS1*) gene is mediated by both ABA and JA and induces resistance against osmotic stress and necrotrophic pathogens, whereas *bos1* mutant plants are more susceptible to both stresses (Mengiste *et al*., 2003). In Arabidopsis, the transcription factor *MYB96* plays an important role in plant protection under pathogen infection by mediating the molecular linkage between ABA induced by drought stress and SA expressed following pathogen infection (Seo, *et al*., 2010). Transcription factor *SlAIM1* in tomato responds positively to the combination of abiotic stress and infection with *Botrytis cinerea* (Abuqamar *et al*., 2009) and *OsMAPK5,* which has kinase activity, is a positive regulator of the rice response to drought, salt, and cold tolerance and disease resistance (Xiong and Yang, 2003). Four members of the *NAC* family of genes that encodes plant-specific transcription factors involved in various bioprocesses. For example, *OsNAC6, Arabidopsis transcription activation factor 1* (*ATAF1*), *ATAF2* and *dehydration 26* (*RD26*) are potentially involved in regulation of responses to abiotic and biotic stresses (Wu *et al*., 2009). It has been evident that, many *PR* genes induced in plants upon exposure to abiotic stress and also PR proteins are crucial for plant resistance against pathogens, and their expression is strongly up-regulated when plants are attacked (Seo *et al*., 2010). The up-regulation of some transcription factors after exposure to abiotic stress leads to an accumulation of PR proteins whereas, over-expression of certain transcription factors in plants confronted with cold stress and infection activates cold-responsive *PR* genes, thereby conferring protection against both stressors (Seo and Park, 2010).

WRKY proteins are a recently identified class of DNA-binding proteins that recognize the TTGAC(C/T) W-box elements and TFs that contain WRKY domains are unique to plants (Dong *et al*., 2003; Eulgem and Somssich, 2007). WRKY family is derived from its highly conserved 60 amino acid long WRKY domain, comprising highly conserved WRKYGQK at N-terminus and metal chelating zinc finger signature at C-terminus (Eulgem and Somssich, 2007). It has been suggested that WRKY factors play a key role in regulating the pathogen induced defense program. WRKY transcription factors plays vital role in regulating several different plant processes and it is anticipated that WRKY transcription factor benefitted to both abiotic and biotic stress pathways with different signal transduction pathways. For example rice *WRKY45 (OsWRKY45)* gene expression is markedly induced in response to ABA and various abiotic stresses such as NaCl, dehydration and by pathogens such as *Pyricularia oryzae Cav*. and *Xanthomonas oryzae* pv. oryzae. Whereas, *OsWRKY45*-over-expressing plants exhibited several changes: (I) expression of ABA-induced responses and abiotic-related stress factors, (II) enhanced drought resistance and (III) increased expression of *PR* genes and resistance to the bacterial pathogen *Pseudomonas syringae*. Here, OsWRKY45 acting as a regulator and also as a protective molecule upon water deficit and pathogen attack (Qiu and Yu, 2009). *VvWRKY11* gene from *Vitis vinnifera* is involved in the response to dehydration and biotic stress (Liu *et al*., 2011) and transgenic *Arabidopsis* seedlings over-expressing *VvWRKY11* have higher tolerance to water stress induced by mannitol than wild-type plants. Some other well known WRKY transcription factor family, such as expression of *VvWRKY11*, *AtWRKY39* and *AtWRKY53* indicate that these genes are co -regulator of the plant response against pathogens and hydric and heat stress and some WRKY transcription factors (OsWRKY24 and OsWRKY45) antagonize ABA function, repressing an ABA-inducible promoter, indicated that these molecules works with their own capabilities.

Conclusions

Crop plants encounters number of abiotic and biotic stresses resulting negative impact on growth, development and yields. The environmental changes occurs from day of seed sowing to harvesting of crop, thus plants must respond to stress over the course of each day and often must respond to quite a lot of stresses at same time. There is a need to understand physiological, biochemical and molecular responses to face challenges throughout the abiotic and biotic stress. We can understand stress as a stimulus or influence that is outside the normal range of homeostatic control in a given organism: when stress tolerance is improved, certain mechanisms are activated at molecular, biochemical, physiological and morphological levels and once stress is controlled, a new physiological state is established, and homeostasis is re-established. Under post stress condition, crop plants may return to the original state or a new physiological state. As we discussed in this chapter, multiple pathways of cellular signaling and their interactions are activated after recognition of foreign stimulus from the environment. Specific factors including transcription factors such as *WRKYs, ATAF1 and 2, MYC2, RD2, BOS1, OsNAC6* and OsMPK5 kinase are molecular player, common to multiple networks of stress signaling pathways regulated by abscisic acid, salicylic acid, jasmonic acid and ethylene as well as ROS signaling. Untangling this network of interconnected signal pathways and understanding the principal mechanisms remains a challenge for the future that possibly will help in developing more resilient plants.

References

Abuqamar, S., Luo, H. L., Laluk, K., Mickelbart, M. V. and Mengiste, T. 2009. Crosstalk between biotic and abiotic stress responses in tomato is mediated by the AIM1 transcription factor. *Plant Journal*, 58: 347–360.

Achard, P., Renou, J. P., Berthome, R., Harberd, N. P. and Genschik, P. 2008. Plant DELLA sre strain growth and promote survival of adversity by reducing the levels of reactive oxygen species. *Current Biology*, 18: 656–660.

Achuo, E. A., Prinsen, E. and Hofle, M. 2006. Influence of drought; salt stress and abscisic acid on the resistance of tomato to *Botrytis cinerea* and *Oidium neolycopersici. Plant Pathol.* 55: 178–186.

Ali, M. A., Abbas, A., Kreil, D. P. and Bohlmann, H. 2013. Over expression of the transcription factor RAP2.6 leads to enhanced callose deposition in syncytia and enhanced resistance against the beet cyst nematode *Heterodera schachtii* in Arabidopsis roots. *BMC Plant Biology*, 13: 47.

Al-Karaki, G. 2000. Growth of mycorrhizal tomato and mineral acquisition under salt stress. *Mycoorhiza*, 10: 51-54.

Allen, G. J., Chu, S. P., Harrington, C. L., Schumaker, K., Hoffmann, T., Tang, Y. Y., Grill, E. and Schroeder, J. I. 2001. A defined range of guard cell calcium oscillation parameters encodes stomatal movements. *Nature*, 411: 1053-1057.

Anderson, J. P., Badruzsaufari, E., Schenk, P. M., Manners, J. M., Desmond, O. J., Ehlert, C., Maclean, D. J., Ebert, P. R. and Kazan, K. 2004. Antagonistic interaction between abscisic acid and jasmonate– ethylene signaling pathways modulates defense gene expression and disease resistance in Arabidopsis. *The Plant Cell*, 16: 3460–3479.

Arimura, G. I. and Maffei, M. E. 2010. Calcium and secondary CPK signaling in plants in response to herbivore attack. *Biochemical and Biophysical Research Communications*, 400: 455-460.

Atkinson, N. J. and Urwin, P. E. 2012. The interaction of plant biotic and abiotic stresses: from genes to the field. *Journal of Experimental Botany*, 63: 3523–43.

Avarro, L., Bari, R., Achard, P., Lison, P., Nemri, A., Harberd, N. P. 2008. DELLAs control plant immune responses by modulating the balance of jasmonic acid and salicylic acid signaling. *Current Biology*, 18: 650–655.

Bari, R. and Jones, J. D. 2009. .Role of plant hormones in plant defence responses. *Plant Molecular Biology*, 69: 473–488.

Blodgett, J. T., Kruger, E. L. and Stanosz, G. R. 1997. Effects of moderate water stress on disease development by *Sphaeropsis sapinea* on red pine. *Phytopathology*, 87: 422-428.

Bowler, C. and Fluhr, R. 2000. The role of calcium and activated oxygens as signals for controlling cross-tolerance. *Trends in Plant Science*, 5: 241–246.

Boyer, J. S. 1995. Biochemical and biophysical aspects of water deficits and the predisposition to disease. *Annual Review of Plant Phytopathology*, 33: 251-274.

Brader, G., Djamei, A., Teige, M., Palva, E. T. and Hirt, H. 2007. The MAP kinase kinase MKK2 affects disease resistance in *Arabidopsis*. *Mol. Plant Microbe Interact*, 20: 589–596.

Breeze, E., Harrison, E., Mchattie, S., Hughes, L., Hickman, R. and Hill, C. 2011. High-resolution temporal profiling of transcripts during *Arabidopsis* leaf senescence reveals a distinct chronology of processes and regulation. *Plant Cell*, 23: 873–894.

Capiati, D. A., Pais, S. M. and Tellez-Iñon, M.T. 2006. Wounding increases salt tolerance in tomato plants: Evidence on the participation of calmodulin-like activities in cross-tolerance signaling. *J. Exp.Bot.* 57: 2391–2400.

Cheng, S. H., Willmann, M. R. and Chen, H. C. and Sheen, J. 2002. Calcium Signaling through Protein Kinases. The Arabidopsis Calcium-Dependent Protein Kinase Gene Family. *Plant Physiol*, 129: 469-485.

Cheong, Y. H., Kim, K. N., Pandey, G. K., Gupta, R., Grant, J. J. and Luan, S. 2003. Cbl1, a calcium sensor that differentially regulates salt, drought, and cold responses in *Arabidopsis*. *Plant Cell*, 15: 1833-45.

Chinchilla, D., Zipfel, C., Robatzek, S., Kemmerling, B., Nurnberger, T. and Jones, J. D. G., Felix, G. and Boller, T. 2007. A flagellin-induced complex of the receptor FLS2 and BAK1 initiates plant defence. *Nature*, 448: 497–500.

Chini, A., Grant, J. J., Seki, M., Shinozaki, K. and Loake, G. J. 2004. Drought tolerance established by enhanced expression of the *CC-NBS-LRR* gene, *ADR1*, requires salicylic acid, EDS1 and ABI1. *Plant Journal*, 38: 810-822.

Choudhury, S., Panda, P., Sahoo, L. And Panda, S. K. 2013. Reactive oxygen species signaling in plants under abiotic stress. *Plant Signal. Behav*, 8: e23681.

Colditz, F., Nyamsuren, O., Niehaus, K., Eubel, H., Braun, H. P. and Krajinski, F. 2004. Proteomic approach: Identification of *Medicago truncatula* proteins induced in roots after infection with pathogenic oomycete *Aphanomyces euteiches*. *Plant Molecular Biology*, 55:109-120.

Davletova, S., Schlauch, K., Coutu, J. and Mittler, R. 2005. The Zinc-Finger protein Zat12 plays a central role in reactive oxygen and abiotic stress signaling in Arabidopsis. *Plant Physiol,* 139: 847–856.

De Vos, M., Van Oosten, V. R. , Van Poecke, R. M. P., Van Pelt, J. A, Pozo, M. J., Mueller, M. J., Buchala, A. J., Métraux, J. P., Van Loon, L. C., Dicke, M. and Pieterse, C. M. J. 2005. Signal Signature and Transcriptome Changes of Arabidopsis During Pathogen and Insect Attack. *Molecular Plant-Microbe Interactions*, 18: 923-937.

Dey, S., Ghose, K. and Basu, D. 2010. Fusarium elicitor-dependent calcium influx and associated ros generation in tomato is independent of cell death. *Eur J Plant Pathol,* 126: 217-228.

Dong, J., Chen, C. and Chen, Z. 2003. Expression profiles of the Arabidopsis; WRKY gene super family during plant defense response. *Plant molecular biology*, 51: 21-37.

Edwards, H. and Allen, J. 1983. Effect of kinetin, abscisic acid, and cations on host-parasite realtions of barley inoculated with *Erysiphe graminis* f.sp. *hordei. Phytopathology,* 107: 22-30.

Ellis, C. and Turner, J. G. 2001. The Arabidopsis mutant cev1 has constitutively active jasmonate and ethylene signal pathways and enhanced resistance to pathogens. *Plant Cell*, 13: 1025–1033.

Errakhi, R., Dauphin, A., Meimoun, P., Lehner, A., Reboutier, D., Vatsa, P., Briand, J., Madiona, K., Rona, J., Barakate, M., Wendehenne, D., Beaulieu, C. and Bouteau, F. 2008. An early Ca2+ influx is a prerequisite to thaxtomin A-induced cell death in Arabidopsis thaliana cells. *J Exp Botany*, 59: 4259-4270.

Eulgem, T. and Somssich, I. E. 2007. Networks of WRKY transcription factors in defense signaling. *Current Opinion in Plant Biology*, 10: 366-371.

Finkelstein, R. 2013. Abscisic acid synthesis and response. *Arabidopsis Book,* 11: e0166.

Fujita, M., Fujita, Y., Noutoshi, Y., Takahashi, F., Narusaka, Y. and Yamaguchi-Shinozaki, K. 2006. Crosstalk between abiotic and biotic stress responses: a current view from the points of convergence in the stress signaling networks. *Current Opinion in Plant Biology*, 9: 436–42.

Gechev, T.S., van Breusegem, F., Stone, J.M., Denev, I. and Laloi, C. 2006. Reactive oxygen species as signals that modulate plant stress responses and programmed cell death. *Bioessays,* 28: 1091–1101.

Goel, A. K., Lundberg, D., Torres, M. A., Matthews, R., Akimoto-Tomiyama, C., Farmer, L., Dangl, J. L. and Grant, S. R. 2008. The *Pseudomonas syringae* type III effector HopAM1 enhances virulence on water-stressed plants. *Mol. Plant Microbe Interact.* 21: 361–370.

Gonzalez, A., Vera, J., Castro, J., Dennett, G., Mellado, M., Morales, B., Correa, J. A. and Moenne, A. 2011.Co-occurring increases of calcium and organellar reactive oxygen species determine differential activation of antioxidant and defense enzymes in Ulva compressa (Chlorophyta) exposed to copper excess. *Plant Cell Environ,* 33: 1627- 1640.

Gudesblat, G. E., Iusem, N. D. and Morris, P. C. 2007. Guard cell-specific inhibition of *Arabidopsis* MPK3 expression causes abnormal stomatal responses to abscisic acid and hydrogen peroxide. *NewPhytol1*, 73: 713–721.

Holley, S. R., Yalamanchili, R. D., Moura, D. S., Ryan, C. A. and Stratmann, J. W. 2003. Convergence of signaling pathways induced by system in, oligosaccharide elicitors, and ultraviolet-B radiation at the level of mitogen-activated protein kinases in Lycopersicon peruvianum suspension-cultured cells. *Plant Physiology*, 132: 1728–1738.

Hong, J. K. and Hwang, B. K. 2005. Induction of enhanced disease resistance and oxidative stress tolerance by overexpression of pepper basic PR-1 gene in *Arabidopsis*. *Physiologia Plantarum*, 124: 267-277.

Hou, X., Lee, L. Y., Xia, K., Yan, Y. and Yu, H. 2010. DELLAs modulate jasmonate signaling via competitive binding to JAZs. *Dev. Cell*, 19: 884–894.

Howe, G. A. and Jander, G. 2008. Plant Immunity to Insect Herbivores. *Annual Review of Plant Biology*, 59: 41-66.

Hussain, S. S., Kayani, M. A. and Amjad. M. 2011. Transcription factors as tools to engineer enhanced drought stress tolerance in plants. *Biotechnology Progress*, 27: 297-306.

Idänheimo, N., Gauthier, A., Salojärvi, J., Siligato, R., Brosché, M. and Kollist, H. 2014. The *Arabidopsis thaliana* cysteine-rich receptor-like kinases CRK6 and CRK7protect against apoplastic oxidative stress. *Biochemisty Biophysics Research Communication*, 445: 457–62.

Jabs, T., Tschope, M., Colling, C., Hahlbrock, K. and Scheel, D. 1997. Elicitor-stimulated ion fluxes and O2- from the oxidative burst are essential components in triggering defense gene activation and phytoalexin synthesis in parsley. *Proceedings of the National Academy of Sciences USA* 94: 4800-4805.

Jakab, G., Cottier, V., Toquin, V., Rigoli, G., Zimmerli, L., Metraux, J. P. and Mauch-Mani B. 2001. â-aminobutyric acid-induced resistance in plants. *European Journal of Plant Pathology.* 107: 29-37.

Jonak, C., Kiegerl, S., Ligterink, W., Barker, P. J., Huskisson, N. S. and Hirt, H. 1996. Stress signaling in plants: a mitogen-activated protein kinase pathway is activated by cold and drought. *Proceedings of the National Academy of Sciences USA*, 93: 11274-11279.

Jonak, C., Okresz, L., Bogre, L. and Hirt, H. 2002. Complexity, cross talk and integration of plant map kinase signalling. *Current Opinion in Plant Biology*, 5: 415-24.

Kaliff, M., Staal, J., Myrenas, M. and Dixelius, C. 2007. ABA Is Required for *Leptosphaeria maculans* Resistance via ABI1- and ABI4-Dependent Signaling. Molecular *Plant-Microbe Interactions*, 20: 335-345.

Kissoudis, C., van de Wiel, C., Visser, R. G. F. and Van Der Linden, G. 2014. Enhancing crop resilience to combined abiotic and biotic stress through the dissection of physiological and molecular crosstalk. *Front Plant Sci*, 5: e207

Kluesner, B., Young, J. J., Murata, Y., Allen, G. J., Mori, I. C., Hugouvieux, V. and Schroeder, J. I. 2002. Convergence of calcium signaling pathways of pathogenic elicitirs and abscisic acid in *Arabidopsis* guard cells. *Plant Physiology*, 130: 2152-63.

Larkindale, J. and Knight, M. 2002. Protection against heat stress-induced oxidative damage in *Arabidopsis* involves calcium, abscisic acid, ethylene, and salicylic acid. *Plant Physiology*, 128: 682-695.

Levitt, J. 1980. Water, radiation, salt and other stresses. In: Kozlowski TT (ed) *Responses of Plants to Environmental Stresses* (Vol 2), Academic Press NY, pp 365-488.

Li, Z., Zhang, L., Yu, Y., Quan, R., Zhang, Z. and Zhang, H. 2011. The ethylene response fac-tor AtERF11 that is transcriptionally modulated by the bZIP transcription factorHY5 is a crucial repressor for ethylene biosynthesis in Arabidopsis. *Plant journal*, 68: 88–99.

Liu, H., Yang, W., Liu, D., Han, Y., Zhang, A. and Li, S. 2011. Ectopic expression of a grapevine transcription factor <i>VvWRKY11</i> contributes to osmotic stress tolerance in Arabidopsis. *Molecular Biology Reports*, 38: 417-427.

Ma, S. and Bohnert, H. J. 2007. Integration of *Arabidopsis thaliana* stress-related transcript profiles, promoter structures and cell-specific expression. *Genome Biol.* 8: R49.doi:10.1186/gb-2007-8-4-r49

Mantyla, E., Lång, V. and Palva, E. 1995. Role of abscisic acid in drought-induced freezing tolerance, cold acclimation, and accumulation of LT178 and RAB18 proteins in *Arabidopsis thaliana*. *Plant Physiology*, 107: 141-148.

Mauch-Mani, B. and Mauch, F. 2005. The role of abscisic acid in plant–pathogen interactions.*Curr Opin Plant Biol*, 8:409–14.

Mengiste, T., Chen, X., Salmeron, J. and Dietrich, R. 2003. The *BOTRYTIS SUSCEPTABLE1* gene encodes an R2R3MYB transcription factor protein that is required for biotic and abiotic stress responses in *Arabidopsis*. *Plant Cell*, 15: 2551-2565.

Mittler, R. 2002. Oxidative stress, antioxidants and stress tolerance. *Trends in Plant Science*, 7: 405-410.

Mittler, R., Feng, X. and Cohen, M. 1998. Post-transcriptional suppression of cytosolic ascorbate peroxidase expression during pathogen-induced programmed cell death in tobacco. *Plant Cell,* 10: 461-474.

Mizoguchi, T., Irie, K., Hirayama, T., Hayashida, N., Yamaguchi-Shinozaki, K., Matsumoto, K. and Shinozaki, K. 1996. A gene encoding a mitogen-activated protein kinase kinase kinase is induced simultaneously with genes for a mitogen-activated protein kinase and an S6 ribosomal protein kinase by touch, cold, and water stress in *Arabidopsis thaliana*. *Proceedings of the National Academy of Sciences USA*, 93: 765-769.

Mohr, P. and Cahill, D. 2007. Suppression by ABA of salicylic acid and lignin accumulation and the expression of multiple genes, in <i>Arabidopsis</i> infected with <i>Pseudomonas syringae</i> pv. <i>tomato</i>. *Functional & Integrative Genomics*,7: 181-191.

Mohr, P. G. and Cahill, D. M. 2003. Abscisic acid influences the susceptibility of Arabidopsis thalianato Pseudomonas syringae pv. tomato and *Peronospora parasitica. Funct Plant Biol*, 30:461–9.

Nayyar, H., Bains, T. and Kumar, S. 2005. Low temperature induced floral abortion in chickpea: relationship to abscisic acid and cryoprotectants in reproductive organs. *Environmental and Experimantal Botany,* 1: 39-47.

Parker, J. E. 2009. The Quest for Long-Distance Signals in Plant Systemic Immunity. *Sci. Signal*, 2: 31.

Paschold, A., Halitschke, R. and Baldwin, I. T. 2007. Co(i)-ordinating defenses: NaCOI1 mediates herbivore- induced resistance in Nicotiana attenuata and reveals the role of herbivore movement in avoiding defenses. *The Plant Journal*, 51: 79-91.

Peleg, Z. and Blumwald, E. 2011. Hormone balance and abiotic stress tolerance in crop plants. *Curr.Opin.Plant Biol.* 14: 290–295.

Pieterse,C. M., Leon-Reyes, A.,VanDerEnt, S.andVanWees, S. C. 2009. Networking by small-molecule hormones in plant immunity. *Nat.Chem.Biol.* 5: 308–316.

Pitzschke, A., Djamel, A., Teige, M. and Hirt, H. 2009. VIP1 response elements mediate mitogen-activated protein kinase 3-induced stress gene expression. *Proc. Natl. Acad. Sci. USA*, 106: 18414–18419.

Qiu, Y and Yu. D. 2009. Over-expression of the stress-induced OsWRKY45 enhances disease resistance and drought tolerance in Arabidopsis. *Environmental and Experimental Botany*, 65: 35-47.

Rejeb, I. B., Pastor, V. and Mani, B. M. 2014. Plant responses to simultaneous biotic and abiotic stress: molecular mechanisms. *Plants,* 3: 458-475.

Rentel, M. C. and Knight, M. R. 2004. Oxidative stress induced calcium signalling in *Arabidopsis. Plant Physiology*, 135: 1471-1479.

Reusche, M., Thole, K., Janz, D., Truskina, J., Rindfleisch, S., Drübert, C., Polle, A., Lipka, V. and Teichmann, T. 2012. Verticillium infection triggers VASCULAR-RELATED NAC DOMAIN7-dependent *de novo* xylem formation and enhances drought tolerance in Arabidopsis. *Plant Cell*, 24:3823–3837.

Reymond, P., Bodenhausen, N., Van Poecke, R. M. P., Krishnamurthy, V., Dicke, M. and Farmer, E. E. 2004. A Conserved Transcript Pattern in Response to a Specialist and a Generalist Herbivore. *The Plant Cell Online*, 16: 3132-3147.

Reymond, P., Weber, H., Damond, M. and Farmer, E. E. 2000. Differential gene expression in response to mechanical wounding and insect feeding in Arabidopsis. *Plant Cell*, 12: 707–719.

Ruggiero, B., Koiwa, H., Manabe, Y., Quist, T. M., Inan, G., Saccardo, F., Joly, R. J., Hasegawa, P. M., Bressan, R. A. and Maggio, A. 2004. Uncoupling the effects of abscisic acid on plant growth and water relations. Analysis of *sto1/nced3*, an abscisic acid-deficient but salt stress-tolerant mutant in *Arabidopsis. Plant Physiology,* 136: 3134-3147.

Šamajová, O., Plíhal, O., Al-Yousif, M., Hirt, H. and Šamaj, J. 2013. Improvement of stress tolerance in plants by genetic manipulation of mitogen-activated protein kinases. *Biotechnol. Adv.*, 31: 118–128.

Scarpeci, T. E., Zanor, M. I., Mueller-Roeber, B. and Valle, E. M. 2013. Over expression of AtWRKY30 enhances abiotic stress tolerance during early growth stages in *Arabidopsis thaliana. Plant Molecular Biology*, 83: 265–77.

Schon, M., Toller, A., Diezel, C., Roth, C., Westphal, L. and Wiermer, M. 2013. Analyses of wrky18wrky40 plants reveal critical roles of SA/EDS1 signaling and indole-glucosinolate biosynthesis for *Golovinomyces orontii* resistance and a loss-of resistance towards *Pseudomonas syringae* pv. tomato AvrRPS4. *Mol Plant Microbe Interact,* 26:758–67.

Schön, M., Töller, A., Diezel, C., Roth, C., Westphal, L. and Wiermer, M. 2013. Analyses of wrky18wrky40 plants reveal critical roles of SA/EDS1 signaling and indole-glucosinolate biosynthesis for *Golovinomyces orontii* resistance and a loss-of resistance towards *Pseudomonas syringae* pv. tomato AvrRPS4. *Molecular Plant Microbe Interaction*. 26: 758–67.

Seo, P. J. and Park, C. M. 2010. *MYB96*-mediated abscisic acid signals induce pathogen resistance response by promoting salicylic acid biosynthesis in *Arabidopsis. New Phythol*, 186: 471–483.

Seo, P. J., Kim, M. J., Park, J. Y., Kim, S. Y., Jeon, J., Lee, Y. H., Kim, J. and Park, C. M. 2010. Cold activation of a plasma membrane-tethered NAC transcription factor induces a pathogen resistance response in Arabidopsis. *Plant journal,* 61: 661–671.

Shaik, R. and Ramakrishna, W. 2013. Genes and co-expression modules common to drought and bacterial stress responses in *Arabidopsis* and rice. *PLoSONE* 8 : e77261.

Shaik, R. and Ramakrishna, W. 2014. Machine learning approaches distinguish multiple stress conditions using stress-responsive genes and identify candi- date genes for broad resistance in rice. *PlantPhysiol*, 164: 481–495.

Sharma, Y., Leon, J., Raskin, I. and Davis, K. R. 1996. Ozone-induced responses in *Arabidopsis thaliana*: The role of salicylic acid in the accumulation of defense-related transcripts and induced resistance. *Plant Biology*, 93: 5099–5104.

Stanosz, G. R., Blodgett, J. T., Smith, D. R. and Kruger, E. L. 2001. Integrative biology: dissecting cross-talk between signalling pathways. *New Phytologist*, 149: 531-538.

Stout, M. J., Fidantsef, A. L., Duffey, S. S. and Bostock, R. M. 1999. Signal interactions in pathogen and insect attack: Systemic plant-mediated interactions between pathogens and herbivores of the tomato. *Lycopersicum esculentum. Physiol. Mol. Plant Pathol.* 54: 115–130.

Suzuki, N., Koussevitzky, S., Mittler, R. and Miller, G. 2012. ROS and redox signalling in the response of plants to abiotic stress. *Plant Cell Environ.* 35: 259–270.

Syvertsen, J. and Levy, Y. 2005. Salinity interactions with other abiotic and biotic stresses in citrus. *HortTechnology*, 15: 100-103.

Takahashi, F., Mizoguchi, T., Yoshida, R., Ichimura, K. and Shinozaki, K. 2011. Calmodulin-Dependent Activation of MAP Kinase for ROS Homeostasis in Arabidopsis, *Mol Cell,* 41: 649-660.

Teige, M., Scheikl, E., Euglem, T., Doczi, F., Ichimura, K., Shinozaki, K., Dangl, J. L. and Hirt, H. 2004. The MKK2 pathway mediates cold and salt stress signaling in *Arabidopsis*. *Molecular Cell*, 15: 141-152.

Tena, G., Asai, T., Chiu, W. L. and Sheen, J. 2001. Plant mitogen-activated protein kinase signaling cascades. *Current Opinion in Plant Biology*, 4: 392-400.

Thaler, J. S. and Bostock, R. M. 2004. Interactions between abscisic-acid-mediated responses and plant resistance to pathogens and insects. *Ecology,*1: 48-58.

Timmusk, S. and Wagner, E. 1999. The plant growth-promoting rhizobacterium *Paenibacillus polymyxa* induces changes in *Arabidopsis thaliana* gene expression: a possible connection between biotic and abiotic stress responses. *Molecular Plant Microbe Interaction*, 12: 951-959.

Tippmann, H. F., Schliiter, U. and Collinge, D. B. 2006. Common themes in biotic and abiotic Stress signalling in plants. *Floriculture, Ornamental and Plant Biotechnology.* 03: 52-67.

Ton, J. and Mauch-Mani, B. 2004. Beta-amino-butyric acid-induced resistance against necrotrophic pathogens is based on ABA-dependent priming for callose. *Plant Journal,* 38: 119-130.

Uno, Y., Rodriguez Milla, M. A., Maher, E. and Cushman, J. C. 2009. Identification of proteins that interact with catalytically active calcium-dependent protein kinases from Arabidopsis. *Mol. Genet. Genomics*, 281: 375-390.

Vlot, A. C., Dempsey, D. M. A. and Klessig, D. F. 2009. Salicylic Acid, a Multifaceted Hormone to Combat Disease. *Annual Review of Phytopathology*, 47: 177-206.

Wani, S. H. And Kumar, V. 2015. Plant stress tolerance: engineering ABA: a Potent phytohormone. *Transcriptomics:AnOpenAccess,* 3:1000113.

Wasilewska, A., Vlad, F., Sirichandra, C., Redko, Y., Jammes, F., Valon, C., Frey, N. F. d. and Leung, J. 2008. An Update on Abscisic Acid Signaling in Plants and More, *Molecular Plant*, 1: 198-217.

Wiese, J., Kranz, T. and Schubert, S. 2004. Induction of pathogen resistance in barley by abiotic stress. *Plant Biology*, 5: 529-536.

Williams, M. E. 2010. Introduction to phytohormones. *The Plant Cell*, 22: 1-9.

Williams, M. E. 2011. Jasmonates: Defense and More. *The Plant Cell*, 23: 1-11.

Wu, Y., Deng, Z., Lai, J., Zhang, Y., Yang, C., Yin, B., Zhao, Q., Zhang, L., Li, Y., Yang, C. and Xie, Q. 2009. Dual function of Arabidopsis ATAF1 in abiotic and biotic stress responses. *Cell Res*, 19: 1279-1290.

Xiong, L. and Yang, Y. 2003. Disease resistance and abiotic stress tolerance in rice are inversely modulated by an abscisic acid-inducible mitogen-activated protein kinase. *Plant Cell*, 15: 745–759.

Xiong, L., Schumarker, K. S. and Zhu, J. 2002. Cell signaling during cold, drought and salt stress. *Plant Cell*. 14: 165-183.

Xu, P., Chen, F., Mannas, J. P., Feldman, T., Sumner, L.W. and Roossinck, M. J. 2008. Virus infection improves drought tolerance. *New Phytol*, 180: 911–921.

Yalpani, N., Enyedi, A. J., León, J. and Raskin, I. 1994. Ultraviolet light and ozone stimulate accumulation of salicylic acid, pathogenesis-related proteins and virus resistance in tobacco. *Planta,* 193: 372-376.

Yalpani, N., Leon, J., Lawton, M. A. and Raskin, I. 1993. Pathway of salicylic acid biosynthesis in healthy and virus-inoculated tobacco. *Plant Physiology*, 103: 315-321.

Yano-Melo, A. M., Saggin, O. J. and Maia, L. C. 2003. Tolerance of mycorrhized banana (*Musa sp*. cv. *Pacovan*) plantlets to saline stress. *Agriculture, Ecosystems and Environment*, 95: 343- 348.

Yasuda, M., Ishikawa, A., Jikumaru, Y., Seki, M., Umezawa, T., Asami, T., Maruyama- Nakashita, A., Kudo, T., Shinozaki, K. and Yoshida, S. 2008. Antagonistic interaction between systemic acquired resistance and the abscisic acid-mediated abiotic stress response in Arabidopsis. *Plant Cell*, 20: 1678–1692.

Yu, X. M., Griffith, M. and Wiseman, S. B. 2001. Ethylene induces antifreeze activity in winter rye leaves. *Plant Physiology*, 126: 1232-1240.

Zhou, M. L., Ma, J. T., Pang, J. F., Zhang, Z. L., Tang, Y. X. and Wu, Y. M. 2010. Regulation of plant stress response by dehydration responsive element binding (DREB) transcription factors. *African Journal of Biotechnology*, 9: 9255-9279.

Zhu, Q., Zhang, J., Gao, X., Tong, J., Xiao, L. and Li, W. 2010. The Arabidopsis AP2/ERF transcription factor RAP2.6 participates in ABA, salt and osmotic stress responses. *Gene*, 457: 1–12.

2

Salinity Stress Induced Metabolic Changes and its Management

Mallikarjun Awati and S. M. Prasanna

Globally, agriculture productivity is challenged by abiotic and biotic stresses, but abiotic stresses in particular (Gong *et al*., 2013) affect spreading of plant species across different environmental zones (Chaves *et al*., 2003). The changing climate is expected to worsen abiotic factors globally and adaptation strategies need to be established for target crops to specific environments (Beebe *et al*., 2011). Connect between different stress factors will likely surge harm to crop yields (Beebe, 2012). Soil salinity is a major environmental constraint to agricultural productivity (Greenway and Munns, 1980; Rhoades and Loveday, 1990). High concentrations of different types of salts, including chlorides, carbonates, and sulfates of magnesium, calcium, potassium, and sodium, characterize different saline soil areas around the world. Moreover, for each type of naturally occurring salinity, constantly changing environmental conditions, such as temperature and precipitation, as well as agricultural practices cause rapid modifications in levels of salinity and salt distribution patterns. Sodium and chloride are the predominant ions in the vast majority of saline areas.

During evolution, various species of plants, known as halophytes, readapted to life in high-salinity environments, but a large majority of plant species grown in non-saline areas are salt-sensitive (referred to as glycophytes). These glycophytic plants, including the majority of crop species, differ greatly in their tolerance to salt stress (Greenway and Munns, 1980; Flowers and Colmer, 2008).

Soil salinity is a major environmental constraint to crop production, affecting an estimated 45 million hectares of irrigated land, and is expected to increase due to global climate changes and as a consequence of many irrigation practices (Rangasamy, 2010; Munns and Tester, 2008). The deleterious effects of salt stress on agricultural yield are significant, mainly because crops exhibit slower growth rates, reduced tillering and, over months, reproductive development is affected (Munns and Tester, 2008). The ultimate aim of salinity tolerance research is to increase the ability of plants to maintain growth and productivity in saline soils relative to their growth in non-saline soils — that is, to reduce effects of salinity on growth and yield. A range of biotechnologies can facilitate this by speeding gene (and allele) discovery and speeding the delivery of crops with improved salt tolerance (using both marker-assisted selection and genetic modification).

The most important aspect of plant responses leading to salt stress tolerance is the regulation of uptake and distribution of Na+ ions (Tester and Davenport, 2003). Along osmotic homeostasis, maintenance of ionic homeostasis is an important strategy for achieving enhanced tolerance to environmental stresses (Sun *et al.,* 2009). Salinity is one of the major severe abiotic factors affecting crop growth and productivity (Munns and Tester, 2008). Salt's negative effects on plant growth have initially been associated with the osmotic stress component caused by decreases in soil water potential and, consequently, restriction of water uptake by roots (Evelin *et al.,* 2009).

Zhang *et al.* (2012) have summarized 2171 salt- responsive proteins from proteomic analysis of 34 different plant species (19 crops and other plants) and functionally categorized these proteins as follows; photosynthesis, carbohydrate and energy metabolism, metabolism, stress and defence, transcription, protein synthesis, protein folding and transport, protein degradation, signalling, membrane and transport, cell structure, cell division/differentiation and fate, miscellaneous, and unknown function. Transgenic development is another straight forward technology to improve crop yield in abiotic stress affected land (Roy and Basu, 2009).

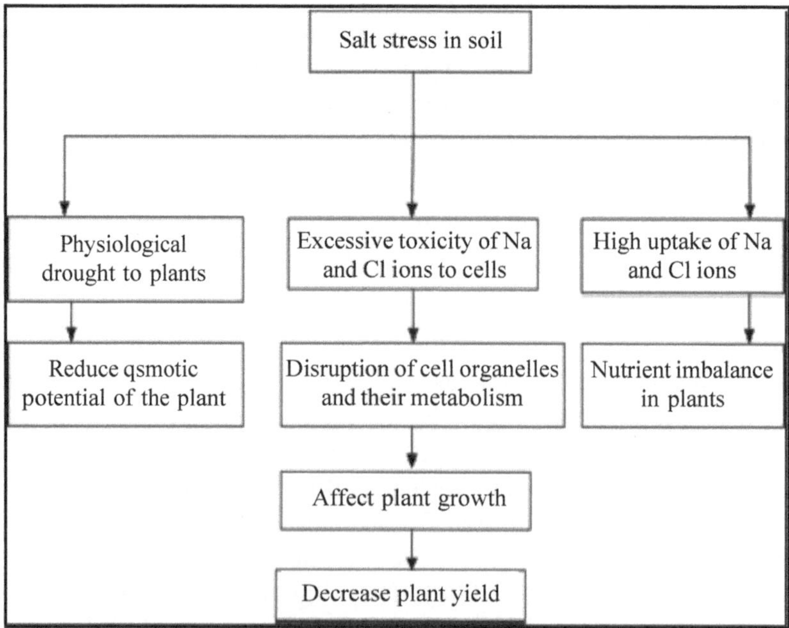

Causes of soil salinization

A significant decline in soil quality has occurred throughout the entire world as a result of adverse changes in its physical, chemical and biological properties. Salinization consists of an accumulation of water soluble salts in the soil. These salts include the ions potassium (K^+), magnesium (Mg^{2+}), calcium (Ca^{2+}), chloride (Cl^-), sulphate (SO_4^{2-}), carbonate (CO_3^{2-}), bicarbonate (HCO_3^-) and sodium (Na^+). Sodium accumulation is also called sodification. High sodium contents result in destruction of the soil structure which, due to a lack of oxygen, becomes incapable of assuring plant growth (Freire, 2009).

Salinity stress in a tree can be caused by a number of factors (Fig. 1). Most commonly stress is caused by the toxicity of ions, usually sodium, chloride and boron, which can affect the uptake of positively charged ions i.e. potassium and calcium. In addition, there is also an osmotic effect whereby if the concentration of toxic ions is high, uptake occurs more readily resulting in leaf tissue damage. The symptoms are usually noticed on the margins of the leaf and include necrotic spots, leaf bronzing and possibly defoliation. The most common observation of salt damage in the orchard is the 'typical leaf-tip burn' (Fig. 1). Both sodium and chloride toxicity result in necrosis (death) of the terminal (bottom) end of the leaf with sodium toxicity showing striations that are perpendicular to the midrib. The best method to determine whether sodium or chloride is the toxic element is through a standard leaf analysis test.

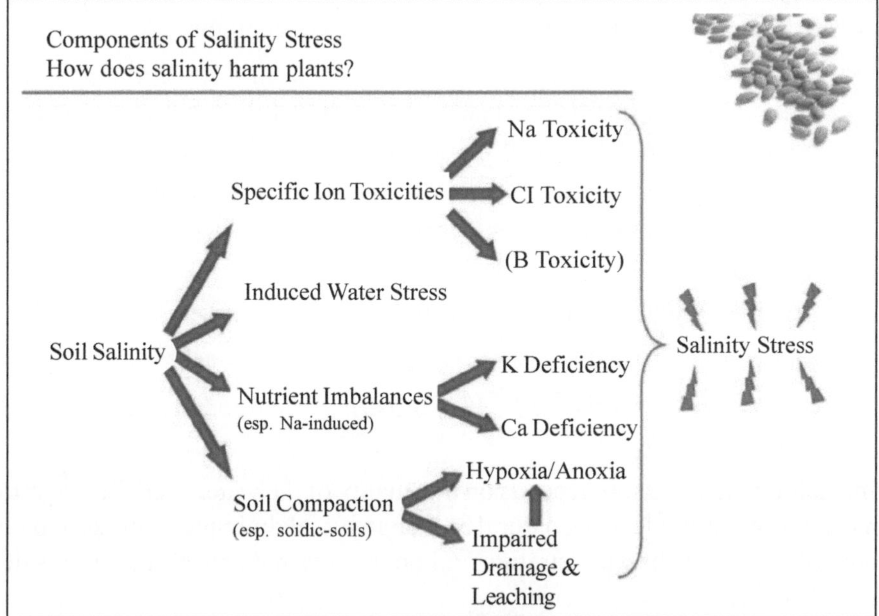

Fig. 1: Components of Salinity Stress

Natural soil salinization and sodification factors

The natural factors influencing soil salinity are

- Geological phenomena which increase the salts concentration in groundwater and consequently in the soil;
- Natural factors capable of bringing groundwater containing elevated salt contents to the surface;
- Infiltration of groundwater in below sea-level zones (micro-depressions with reduced or absent drainage);
- Drainage of waters from zones with geological substrates capable of liberating large amounts of salts;
- Action of winds, which, in coastal zones, can transport moderate amounts of salts to the interior.
- The salinization process occurs in soils situated in regions of low rainfall and which have a water-bearing stratum near the surface. In coastal zones, salinization could be associated with the over-exploitation of ground waters due to the demand induced by increased urbanization, or by industry and agriculture.

Secondary factors leading to soil salinization and sodification

The most influential anthropogenic factors are:

- Irrigation with water containing elevated salt contents;
- rise in phreatic water level due to human activities (infiltration of water from unlined channels and reservoirs, irregular distribution of irrigation water, deficient irrigation practices, inadequate drainage);
- use of fertilizers and other production factors, namely for intensive agriculture in land with low permeability and reduced possibilities for leaching;
- Irrigation with residual waters with high salt contents;
- Elimination of residual waters with high salt contents by way of the soil;
- Contamination of the soil with industrial water and sub-products with high salt contents.

Thus salinization of a soil depends on the quality of the water used for irrigation, on the existence and level of natural and/or artificial drainage of the soil, on the depth of the water- bearing stratum and on the original concentration of salt in the soil profile.

Salinity and its effect on crop plants

Salinity limits the output of food crops and growth reduction is the main morphological effect on many biochemical mechanisms of the plant. Plants under high saline unable take up adequate water for metabolic processes or maintain turgidity due to low osmotic potential. Naturally, salt-alkalinized soils are complex that include various ions creating soil-salt-alkalization complex (Läuchli and Lüttge, 2002). Alkaline salts ($NaHCO_3$ and Na_2CO_3) were shown more damaging to plants than neutral salts (NaCl and Na_2SO_4) (Yang *et al.*, 2007). Salt stress generally involves osmotic stress and ion injury (Ge and Li, 1990). Differential response of plants to salt and alkali stresses are largely due to high-pH associated stress (Munns, 2002; Li *et al.*, 2012). Increased uptake of cations, such as Na, Mg, Ca, cause different kinds of nutritional imbalances leads to different ranges of toxicity. Under general NaCl toxic conditions, plants absorb a higher amount of Na, which thus decreases the K:Na ratio (Ahmad and Umar, 2011; Ahmad and Prasad, 2012).

High salinity affects plants in two main ways: high concentrations of salts in the soil disturb the capacity of roots to extract water, and high concentrations of salts within the plant itself can be toxic, resulting in an inhibition of many physiological and biochemical processes such as nutrient uptake and assimilation (Hasegawa, Bressan, Zhu, & Bohnert, 2000, R. Munns, 2002, R Munns, Schachtman, & Condon, 1995, R Munns & Tester, 2008). Together, these effects reduce plant growth, development and survival. A two-phase model describing the osmotic and ionic effects of salt stress was proposed by Munns (1995) (Fig. 2).

Plants sensitive or tolerant to salinity differ in the rate at which salt reaches toxic levels in leaves. Timescale is days or weeks or months, depending on the species and the salinity level. During Phase 1, growth of both type of plants is reduced because of the osmotic effect of the saline solution outside the roots. During Phase 2, old leaves in the sensitive plant die and reduce the photosynthetic capacity of the plant. This exerts an additional effect on growth (Fig. 2).

In the first, osmotic phase starts immediately after the salt concentration around the roots increases to a threshold level making it harder for the roots to extract water, the rate of shoot growth falls significantly. An immediate response to this effect, which also mitigates ion flux to the shoot, is stomatal closure. However, because of the water potential difference between the atmosphere and leaf cells and the need for carbon fixation, this is an untenable long-term strategy of tolerance (Hasegawa *et al.*, 2000). Shoot growth is more sensitive than root growth to salt- induced osmotic stress probably because a reduction in the leaf area development relative to root growth would decrease the water use by the plant, thus allowing it to conserve soil moisture and prevent salt concentration in the soil (Munns and Tester, 2008). Reduction in shoot growth due to salinity is commonly expressed by a reduced leaf area and stunted shoots (Läuchli and Epstein, 1990). The growth inhibition of leaves sensitive to salt stress appears to be also a consequence of inhibition by salt of symplastic xylem loading of Ca^{2+} in the root (Läuchli and Grattan, 2007). Final leaf size depends on both cell division and cell elongation. Leaf initiation, which is governed by cell division, was shown to be unaffected by salt stress in sugar beet, but leaf extension was found to be a salt-sensitive process (Papp *et al.,* 1983), depending on Ca^{2+} status. Moreover the salt-induced inhibition of the uptake of important mineral nutrients, such as K^+ and Ca^{2+}, further reduces root cell growth (Larcher, 1980) and, in particular, compromises root tips expansion (Fig. 2).

The second phase, ion specific, corresponds to the accumulation of ions, in particular Na^+, in the leaf blade, where Na^+ accumulates after being deposited in the transpiration stream, rather than in the roots (Munns, 2002). Na^+ accumulation turns out to be toxic especially in old leaves, which are no longer expanding and so no longer diluting the salt arriving in them as young growing leaves do. If the rate at which they die is greater than the rate at which new leaves are produced, the photosynthetic capacity of the plant will no longer be able to supply the carbohydrate requirement of the young leaves, which

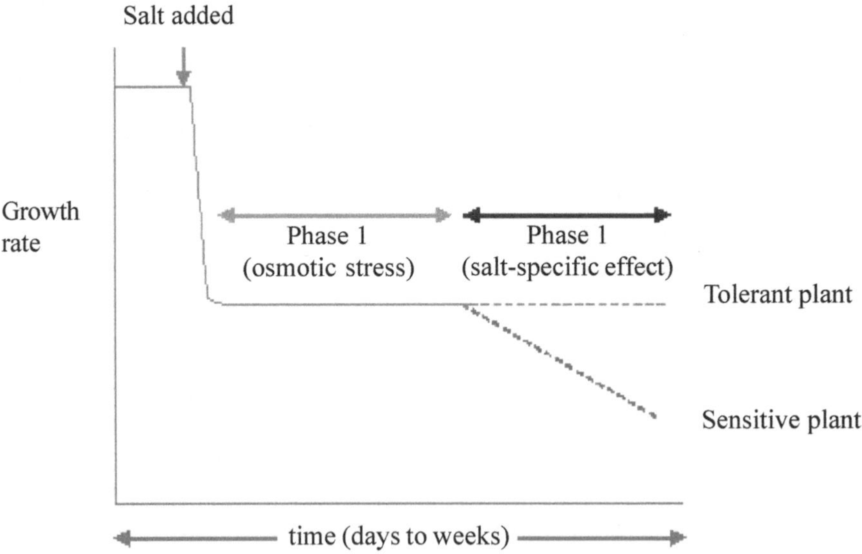

Fig. 2: Scheme of the two-phase growth response to salinity. Adapted from Munns (1995)

further reduces their growth rate (Munns and Tester, 2008). In photosynthetic tissues, in fact, Na^+ accumulation affects photosynthetic components such as enzymes, chlorophylls, and carotenoids (Davenport *et al.,* 2005). The derived reduction in photosynthetic rate in the salt sensitive plants can increase also the production of reactive oxygen species (ROS). Normally, ROS are rapidly removed by antioxidative mechanisms, but this removal can be impaired by salt stress (Allan and Fluhr, 1997; Foyer and Noctor, 2003).

Effect of Salinity on Plants

Physiological and Biochemical Mechanisms

Salinity stress involves changes in various physiological and metabolic processes, varies with stress severity and its duration and ultimately inhibits crop production (Munns, 2005; Rozema and Flowers, 2008). Initially, soil salinity represses plant growth through osmotic stress, which is then followed by ion toxicity (Rahnama *et al.,* 2010; James *et al.,* 2011). During initial phases, the water absorption capacity of the root system decreases and water loss from leaves is accelerated due to osmotic stress, and therefore salinity stress is also considered hyper osmotic stress (Munns, 2005). Osmotic stress at the initial stage causes various physiological changes, such as interruption of membranes, nutrient imbalance, impaired ability to detoxify reactive oxygen species (ROS), differences in antioxidant enzymes, and decreased photosynthetic activity (Munns and Tester, 2008; Pang *et al*., 2010). One of the most damaging effects is accumulation of Na^+ and Cl^- ions in tissues of plants exposed to soils with high NaCl concentrations. Higher Na^+ blocks the K+ uptake, results in lower productivity and may even lead to cell death (Ahmad and Umar, 2011; James *et al.,* 2011).

Plant adaptation or tolerance to salinity stress involves complex physiological traits, metabolic pathways and molecular or gene networks. Comprehensive knowledge of how plants respond to salinity stress at different levels and an integrated approach combining molecular, physiological, and biochemical techniques (Palao *et al*., 2014) are imperative for developing salt- tolerant varieties in salt-affected areas (Ashraf, 2014). Recent research revealed various adaptive responses to salinity stress at cellular, metabolic and physiological levels, although mechanisms underlying salinity tolerance are yet to be clearly understood (Gupta and Huang, 2014). Plants develop various physiological and biochemical mechanisms to survive in soils with high salt concentration. Principal mechanisms include, but are not limited to, ion transport, uptake and compartmentalization biosynthesis of osmoprotectants and compatible solutes, activation and synthesis of antioxidant enzymes/ compounds, polyamines and hormonal modulation (Reddy *et al*., 1992; Roy *et al*., 2014) (Fig. 3).

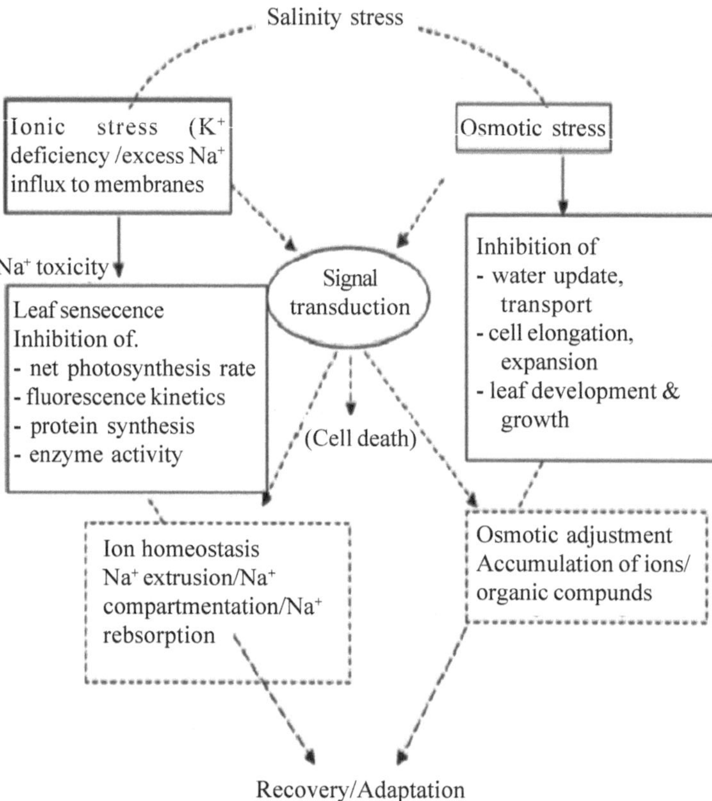

Fig. 3: Schematic summary of salinity stress in plants and corresponding intrinsic physiological responses (Partly adapted from Horie *et al.*, 2012)

Salinity and Hormonal Regulation

Among the plant hormones, ABA plays a key role with ameliorating stress effects. As a first line of defense, ABA has long been recognized to synthesize in roots during soil water deficit (Popova *et al.*, 1995; He and Cramer, 1996). ABA can mitigate inhibitory effects of salinity on photosynthesis (Yong *et al.*, 2014), growth, and translocation of assimilates (Cabot *et al.*, 2009). The positive link of ABA with salinity tolerance partially credited to K+&Ca2+ levels (Hussain *et al.*, 2016) and compatible solutes in cytocol, which offset Na+ and Cl^- uptake (Chen *et al.*, 2001; Gurmani *et al.*, 2011). Other compounds such as, salicylic acid (SA) and brassinosteroids (BR) also share abiotic stress responses (Fragnire *et al.*, 2011; Khan *et al.*, 2015). Ashraf *et al.* (2010) reviewed a possible role of BRs and SA in mitigating harmful effects of salt stress and discussed their exogenous applications in regulating various biochemical and physiological processes. Reddy *et al.* (2015) suggest over-production of proline eases salt stress and protects photosynthetic and antioxidant enzyme activities in transgenic sorghum [*Sorghum bicolor* (L.)]. The effect of salt stress on intrinsic physiological traits, malondialdehyde (MDA) levels and antioxidant enzyme activities were evaluated in 40-day-old transgenic lines and compared with wild type plants. The photosynthetic rate was reduced in wild type plants almost completely. Salinity induced cent percent stomatal closure in wild type, while it did only 64–81% in transgenic plants (after 4 days), indicating transgenic lines were better in coping up with salt stress than wild type.

Mechanisms of salinity tolerance

Salinity have many different effects on a plant, so there are also many mechanisms for plants to tolerate this stress. These mechanisms can be classified into three main categories: firstly, osmotic tolerance, which is regulated by long distance signals that reduce shoot growth and is triggered before shoot Na^+ accumulation; secondly, ion exclusion, where Na^+ and Cl^- transport processes in roots reduce the accumulation of toxic concentrations of Na^+ and Cl^- within leaves; and finally, tissue tolerance, where high salt concentrations are found in leaves but are compartmentalized at the cellular and intracellular level (especially in the vacuole) (Fig. 4).

Very little, if anything, is known about tolerance to the 'osmotic phase'. This process must involve rapid, long-distance signalling, perhaps via processes such as ROS waves (Mittler *et al.*, 2011), Ca^{2+} waves or even long distance electrical signalling (Maischak et al., 2010). Differences in osmotic tolerance may be due to differences in this long-distance signalling, or they may involve differences in the initial perception of the salt or differences in the response to the signals. This is still an area of salinity research with many unknowns, and further research is required to obtain a better understanding of osmotic tolerance (Fig. 4).

Much more is known about the 'ionic phase', which is due to the accumulation of Na^+ and Cl^- in the leaf blade (a trait that is relatively easy to phenotype). Plants can reduce toxicity during the ionic phase by reducing accumulation of toxic ions in the leaf blades (Na^+ and Cl^- exclusion), and/or by increasing their ability to tolerate the salts that they have failed to exclude from the shoot, such as by compartmentation into vacuoles (tissue tolerance) (Fig. 4). Both of these processes involve a range of transporters and their controllers at both plasma membrane and tonoplast (Plett and Moller, 2010; Tester and Davenport, 2003). Tissue tolerance, involving the removal of Na^+ from the cytosol and compartmentalizing it in the vacuole before the ion has a detrimental effect on cellular processes, is also likely to require the synthesis of compatible solutes and higher level controls to coordinate transport and biochemical processes, thus having a role in both osmoprotection and osmotic adjustment (Munns and Tester, 2008; Flowers and Colmer, 2008) (Fig. 4).

It is clear from this brief overview that there are many mechanisms of salinity tolerance, and many of these can be present in a particular plant. To date, there is neither evidence that these mechanisms are mutually exclusive (*e.g.*, ion exclusion prevents tolerance to the 'osmotic phase' of salt toxicity), nor that a particular plant is committed to only one strategy (*e.g.*, a plant may have ion exclusion as its primary tolerance mechanism at moderate salinity, but then has tissue tolerance as its main tolerance mechanism when the exclusion processes are 'swamped' at high salinity). It is possible that some tolerance mechanisms are more effective in particular circumstances. For example, Na^+ exclusion may be more effective in conditions of higher salinity (Munns, 2012), whereas 'osmotic tolerance' may be more important in moderately saline conditions. Interactions with other abiotic stresses, such as low water availability, are also likely to be important.

As such, it is clear that salinity tolerance can be complex and involve many genes, as has been pointed out for several decades — for example, programs designed to specifically introgress salinity tolerance using traditional breeding methods appear to have frequently failed (as measured by the apparent absence of commercial products), which has usually been attributed to the multigenic nature of salinity tolerance (Dewey, 1962; Flowers and Yeo, 1995). It is therefore necessary not to study the molecular genetic basis of salinity tolerance as a particular trait in itself, but to study the mechanisms of traits that are hypothesized to contribute to salinity tolerance. The most intensively studied of these traits is exclusion of Na^+ from leaf blades, mainly because it is relatively straightforward to phenotype. Focusing on this has led to significant increases in salinity tolerance, as measured by yield in the field, at least in durum wheat (which has poor Na^+ exclusion) when grown in highly saline sites where yield has already been greatly reduced (Munns *et al.,* 2012; James *et al.,* 2012). Differences in traits must then be correlated with differences in tolerance, as measured by performance in the field - yield in a saline site relative to yield in a less saline site. Traits can be measured in any system that enables the trait to be quantified, using necessary experimental manipulations, such as described earlier to access the 'osmotic tolerance' trait - key is to test the relevance of the trait being measured with salinity tolerance in field conditions.

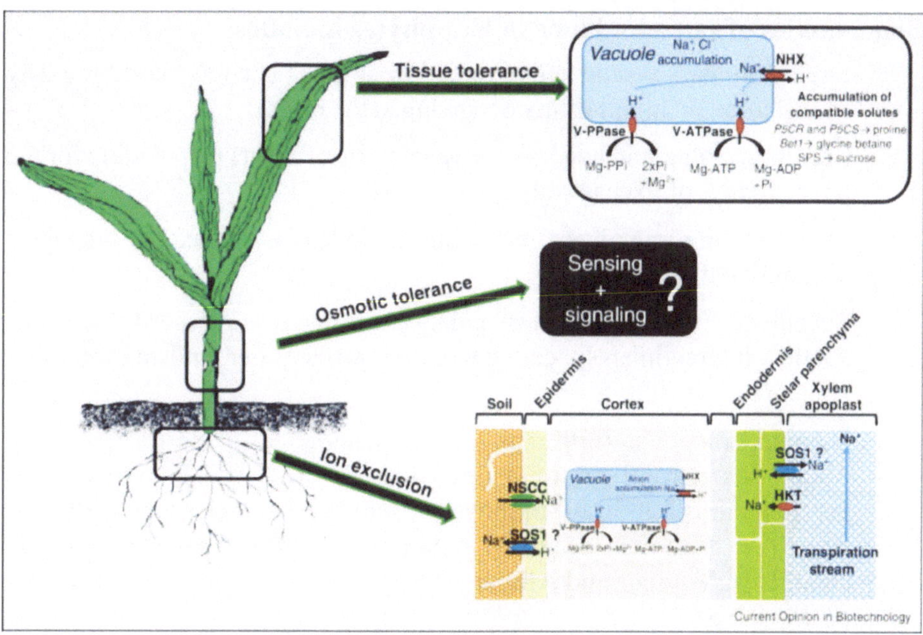

Figure 4. The three main mechanisms of salinity tolerance in a crop plant. Tissue tolerance, where high salt concentrations are found in leaves but are compartmentalized at the cellular and intracellular level (especially in the vacuole), a process involving ion transporters, proton pumps and synthesis of compatible solutes. Osmotic tolerance, which is related to minimizing the effects on the reduction of shoot growth, and may be related to as yet unknown sensing and signalling mechanisms. Ion exclusion, where Na^+ and Cl^- transport processes, predominantly in roots, prevent the accumulation of toxic concentrations of Na^+ and Cl^- within leaves. Mechanisms may include retrieval of Na^+ from the xylem, compartmentation of ions in vacuoles of cortical cells and/or efflux of ions back to the soil.

Plant adaption to salinity (avoidance and tolerance)

Plants with the power of living in saline environments may have avoidance (plant morphological, anatomical and physiological changes) or/and tolerance (cellular and molecular approaches) mechanisms (Fu *et al.*, 2011; Ashrafijou *et al.*, 2010; Vinocur and Altman, 2005). Salt resistance plants known as Halophytes, have an important mechanism of avoidance (plant tries to keep away the salt ions) and better at regulation Na transport than glygophytes. The avoidance mechanisms mostly result of morphological and physiological changes at the whole-plant level (Kumar Parida and Jha, 2010; Tester and Davenport, 2003).

Some tactics of salt avoidance in halophytes includes

1. Exclusion: passive and active rejection of ions (The most common way of sustaining in high salinity occurring at the roots)
2. Secretion (salt gland and hair on aerial parts of plants responsible for the concealment of excess salt)
3. Shedding (in some plants, removing of old leaves that excess salts have accumulated in them)
4. Succulence (thick leaves of succulent plants have larger mesophyll cells, smaller intercellular spaces, decreased surface area and higher water content)
5. Stomatal response (either guard cell uses K^+ in place of sodium to obtain it's normal turgor or may uses K^+ to limit Na^+ intake, this mechanism is important in those halophytes that lack glands) (Roy *et al.*, 2014; Batool *et al.*, 2014; Aslam *et al.*, 2011; Kumar Parida and Jha, 2010; Vinocur and Altman, 2005; Volkmar *et al.*, 1998; Flowers *et al.*, 1986).

Avoidance mechanisms chiefly are not very accountable to practical manipulations. Tolerance mechanism is plants ability to grow and develop on the saline condition through cellular and molecular biochemical modifications that is amenable to practical manipulations. Most of avoidance mechanisms are reliant upon some other mechanism at the cellular level (Vinocur and Altman, 2005). In recent decades, there are a numerous articles about successful development of salinity tolerance in plants by classical breeding, including Landrace assessment and screening (Roy and Sengupta, 2014; Sadat Noori, 2005; Sadat Noori and McNeilly, 1999), Transgressive Segregation (Shahbaz and Ashraf, 2013; Sadat Noori and Harati, 2005), Triple Test Cross (Sadat Noori and Sokhansanj, 2004) and even artificial neural network (Sadat Noori *et al.*, 2011a), although these methods are full of difficulties. As the first step for screening program for salinity tolerance or any kind of desirable traits it is important to investigate the heritability of the good traits (Izadi Darbandi *et al.*, 2013). Plant breeders need to deploy the new techniques such as Laser and biotechnological tools. Up to now there are several reports about improving salt tolerance in the plant by Laser (Sadat Noori *et al.*, 2011b; Ashrafijou *et al.*, 2010), but because of lack of enough information on its heritability and consequent effects, biotechnological tools for search in crucial problems of crop improvement for sustainable agriculture are more popular. Genetic engineering for developing tolerant plants provides a faster way to improving the crops (Hossain *et al.*, 2007). Genetic engineering as a solution to abiotic stress tolerance is based on introgression and expression of genes involving in signaling and controlling pathways or genes encoding proteins producing stress tolerance or enzymes related to pathways resulting in the functional and structural metabolite synthesis (Roy *et al.*, 2014; Vinocur and Altman, 2005).

In the way of introduction of Salinity tolerance by genetic engineering candidate genes are encoding

(1) antioxidants and detoxifying enzymes,

(2) ion transport,

(3) compatible solutes,

(4) transcription factors and

(5) late embryogenesis abundant proteins (Roy and Sengupta, 2014; Roy *et al.*, 2014; Shang *et al.*, 2012; Jenks *et al.*, 2007).

Because salinity mechanisms doesn't have 100% efficiency, hence plant breeders to improve salinity tolerance of plants should deal with that by using several mechanisms instead of using just one (Batool *et al.*, 2014; Yamaguchi and Blumwald, 2005; Chinnusamy *et al.*, 2005; Murkute *et al.*, 2005).

Physiological and biochemical mechanisms of salt tolerance

Plants develop various physiological and biochemical mechanisms in order to survive in soils with high salt concentration.

Principle mechanisms include, but are not limited to,

- Ion homeostasis and compartmentalization,
- Ion transport and uptake,
- Biosynthesis of osmoprotectants and compatible solutes,
- Activation of antioxidant enzyme and synthesis of antioxidant compounds,
- Synthesis of polyamines,
- Generation of nitric oxide (NO), and
- Hormone modulation.

Research advances elucidating these mechanisms are discussed below.

Ion homeostasis and salt tolerance

Maintaining ion homeostasis by ion uptake and compartmentalization is not only crucial for normal plant growth but is also an essential process for growth during salt stress (Niu Xiaomu *et al.,* 1995; Hasegawa, 2013). Irrespective of their nature, both glycophytes and halophytes cannot tolerate high salt concentration in their cytoplasm. Hence, the excess salt is either transported to the vacuole or sequestered in older tissues which eventually are sacrificed, thereby protecting the plant from salinity stress (Reddy *et al.*, 1992; Zhu, 2003).

Major form of salt present in the soil is NaCl, so the main focus of research is the study about the transport mechanism of Na^+ ion and its compartmentalization. The Na^+ ion that enters the cytoplasm is then transported to the vacuole via Na^+/H^+ antiporter. Two types of H^+ pumps are present in the Vacuolar membrane: vacuolar type H^+-ATPase (V-ATPase) and the vacuolar pyrophosphatase (V-PPase) (Dietz *et al.,* 2001; Wang *et al.,* 2001). Of these, V-ATPase is the most dominant H^+ pump present within the plant cell. During non-stress conditions it plays an important role in maintaining solute homeostasis, energizing secondary transport and facilitating vesicle fusion. Under stressed condition the survivability of the plant depends upon the activity of V-ATPase (Dietz *et al.*, 2001).

Increasing evidence demonstrates the roles of a Salt Overly Sensitive (SOS) stress signalling pathway in ion homeostasis and salt tolerance (Hasegawa *et al.*, 2000; Sanders, 2000). The SOS signalling pathway (Fig. 1) consists of three major proteins, SOS1, SOS2, and SOS3. *SOS1,* which encodes a plasma membrane Na^+/H^+ antiporter, is essential in regulating Na^+ efflux at cellular level. It also facilitates long distance transport of Na^+ from root to shoot. Over expression of this protein confers salt tolerance in plants Shi *et al.,* 2000; Shi *et al.*, 2002). *SOS2* gene, which encodes a serine/threonine kinase, is activated by salt stress elicited Ca^+ signals. This protein consists of a well-developed N-terminal catalytic domain and a C-terminal regulatory domain (Liu *et al.*, 2000). The third type of protein involved in the SOS stress signalling pathway is the SOS3 protein which is a myristoylated Ca^+ binding protein and contains a myristoylation site at its N-terminus.

This site plays an essential role in conferring salt tolerance (Ishitani *et al.*, 2000). C-terminal regulatory domain of SOS2 protein contains a FISL motif (also known as NAF domain), which is about 21 amino acid long sequence, and serves as a site of interaction for Ca^{2+} binding SOS3 protein.

Compatible solute accumulation and osmotic protection

Compatible solutes, also known as compatible osmolytes, are a group of chemically diverse organic compounds that are uncharged, polar, and soluble in nature and do not interfere with the cellular metabolism even at high concentration.

They mainly include proline (Hoque *et al.,* 2007; Tahir *et al.*, 2012), glycine betaine Khan *et al.*, 2000; Wang and Nii, 2000, sugar (Kerepesi and Galiba, 2000; Bohnert *et al.*, 1995), and polyols (Ford, 1984; Saxena *et al.,* 2013). Organic osmolytes are synthesised and accumulated in varying amounts amongst different plant species. For example, quaternary ammonium compound beta alanine betaine's accumulation is restricted among few members of Plumbaginaceae (Hanson *et al.*, 1994), whereas accumulation of amino acid proline occurs in taxonomically diverse sets of plants (Saxena *et al.*, 2013). The concentration of compatible solutes within the cell is maintained either by irreversible synthesis of the compounds or by a combination of synthesis and degradation. The biochemical pathways and genes involved in these processes have been thoroughly studied. As their accumulation is proportional to the external osmolarity, the major functions of these osmolytes are to protect the structure and to maintain osmotic balance within the cell via continuous water influx (Hasegawa *et al.*, 2000).

Amino acids such as cysteine, arginine, and methionine, which constitute about 55% of total free amino acids, decrease when exposed to salinity stress, whereas proline concentration rises in response to salinity stress (Shintinawy and Shourbagy, 2001). Proline accumulation is a well-known measure adopted for alleviation of salinity stress (Saxena *et al.*, 2013; Matysik *et al.*, 2002; Ben Ahmed *et al.*, 2010). Intracellular proline which is accumulated during salinity stress not only provides tolerance towards stress but also serves as an organic nitrogen reserve during stress recovery. Proline is synthesised either from glutamate or ornithine. In osmotically stressed cell glutamate functions as the primary precursor. The biosynthetic pathway comprises two major enzymes, pyrroline carboxylic acid synthetase and pyrroline carboxylic acid reductase. Both these regulatory steps are used to overproduce proline in plants (Sairam and Tyagi, 2004). It functions as an O_2 quencher thereby revealing its antioxidant capability. This was observed in a study carried out by Matysik *et al.* (2002). Ben Ahmed *et al.* (2010) observed that proline supplements enhanced salt tolerance in olive (*Olea europaea*) by amelioration of some antioxidative enzyme activities, photosynthetic activity, and plant growth and the preservation of a suitable plant water status under salinity conditions.

Antioxidant regulation of salinity tolerance

Abiotic and biotic stress in living organisms, including plants, can cause overflow, deregulation, or even disruption of electron transport chains (ETC) in chloroplasts and mitochondria. Under these conditions molecular oxygen (O_2) acts as an electron acceptor, giving rise to the accumulation of ROS. Singlet oxygen ($1O_2$), the hydroxyl radical (OH·), the superoxide radical ($O2^{·-}$), and hydrogen peroxide (H_2O_2) are all strongly oxidizing compounds and therefore potentially harmful for cell integrity (Grob *et al.,* 2013). Antioxidant metabolism, including antioxidant enzymes and non-enzymatic compounds, play critical parts in detoxifying ROS induced by salinity stress.

Salinity tolerance is positively correlated with the activity of antioxidant enzymes, such as superoxide dismutase (SOD), catalase (CAT), glutathione peroxidise (GPX), ascorbate peroxidise (APX), and glutathione reductase (GR) and with the accumulation of nonenzymatic antioxidant compounds (Asada, 1999; Gupta *et al*., 2005). Gill *et al*. (2013) and Tuteja *et al*. (2013) have recently reported a couple of helicase proteins (*e.g*., DESD-box helicase and OsSUV3 dual helicase) functioning in plant salinity tolerance by improving/maintaining photosynthesis and antioxidant machinery.

Roles of polyamines in salinity tolerance

Polyamines (PA) are small, low molecular weight, ubiquitous, polycationic aliphatic molecules widely distributed throughout the plant kingdom. Polyamines play a variety of roles in normal growth and development such as regulation of cell proliferation, somatic embryogenesis, differentiation and morphogenesis, dormancy breaking of tubers and seed germination, development of flowers and fruit, and senescence (Knott *et al.,* 2007; Galston *et al.,* 1997). It also plays a crucial role in abiotic stress tolerance including salinity and increases in the level of polyamines are correlated with stress tolerance in plants (Gupta *et al.,* 2013; Kovacs *et al*., 2010). The most common polyamines that are found within the plant system are diamine putrescine (PUT), triamine spermidine (SPD) and tetra-amine spermine (SPM) (Alcazar *et al*., 2011; Shu *et al*., 2012). The PA biosynthetic pathway has been thoroughly investigated in many organisms including plants and has been reviewed in details (Alet *et al*., 2012; Rambla *et al*., 2010). PUT is the smallest polyamine and is synthesised from either ornithine or arginine by the action of enzyme ornithine decarboxylase (ODC) and arginine decarboxylase (ADC), respectively (Gupta *et al*., 2013; Hasnuzzaman *et al*., 2014). N-carbamoyl-putrescine is converted to PUT by the enzyme N-carbamoyl-putrescine aminohydrolase (Alcazar *et al.,* 2010; Bouchereau *et al*., 1999). The PUT thus formed functions as a primary substrate for higher polyamines such as SPD and SPM biosynthesis. The triamine SPD and tetramine SPM are synthesized by successive addition of aminopropyl group to PUT and SPD, respectively, by the enzymes spermidine synthase (SPDS) and spermine synthase (SPMS) (Alcazar *et al*., 2006; Fluhr and Mattoo, 1996). ODC pathway is the most common pathway for synthesis of polyamine found in plants. Most of the genes involved in the ODC pathway have been identified and cloned.

However there are some plants where ODC pathway is absent; for instance in *Arabidopsis* polyamines are synthesized via ADC pathway (Kusano *et al.*, 2007; Hanfrey *et al.*, 2001).

All the genes involved in polyamine biosynthesis pathways have been identified from different plant species including *Arabidopsis* Urano *et al.*, 2003; Janowitz *et al.*, 2003). Polyamine biosynthesis pathway in *Arabidopsis* involves six major enzymes: ADC encoding genes (*ADC1* and *ADC2*); SPDS (*SPDS1* and *SPDS2*) and SAMDC (*SAMDC1, SAMDC2, SAMDC3, SAMDC4*) (Janowitz *et al.*, 2003; Hashimoto *et al.*, 1998). On the contrary, SPM synthase, thermospermine synthase, agmatine iminohydrolase and N-carbamoylputrescine amidohydrolase are represented by single genes only (Urano *et al.*, 2004; Hanzawa *et al.*, 2002).

Roles of nitric oxide in salinity tolerance

Nitric oxide (NO) is a small volatile gaseous molecule, which is involved in the regulation of various plant growth and developmental processes, such as root growth, respiration, stomata closure, flowering, cell death, seed germination and stress responses, as well as a stress signalling molecule (Delledonne *et al.*, 1998; Crawford, 2006). NO directly or indirectly triggers expression of many redox-regulated genes. NO reacts with lipid radicals thus preventing lipid oxidation, exerting a protective effect by scavenging superoxide radical and formation of peroxynitrite that can be neutralised by other cellular processes. It also helps in the activation of antioxidant enzymes (SOD, CAT, GPX, APX, and GR) (Bajgu, 2014).

Hormonal regulation of salinity tolerance

ABA is an important phytohormone whose application to plant ameliorates the effect of stress condition(s). It has long been recognized as a hormone which is upregulated due to soil water deficit around the root. Salinity stress causes osmotic stress and water deficit, increasing the production of ABA in shoots and roots (He and Cramer, 1996; Popova *et al.*, 1995). The accumulation of ABA can mitigate the inhibitory effect of salinity on photosynthesis, growth, and translocation of assimilates (Popova *et al.*, 1995; Jeschke *et al.*, 1997). The positive relationship between ABA accumulation and salinity tolerance has been at least partially attributed to the accumulation of K^+, Ca^{2+} and compatible solutes, such as proline and sugars, in vacuoles of roots, which counteract with the uptake of Na^+ and Cl^- (Chen *et al.*, 2001; Gurmani *et al.*, 2011). ABA is a vital cellular signal that modulates the expression of a number of salt and water deficit-responsive genes.

Amelioration of salinity

Salinization can be restricted by leaching of salt from root zone, changed farm management practices and use of salt tolerant plants. Irrigated agriculture can be sustained by better irrigation practices such as adoption of partial root zone drying methodology, and drip or micro-jet irrigation to optimize use of water. The spread of dry land salinity can be contained by reducing the amount of water passing beyond the roots. This can be done by re-introducing deep rooted perennial plants that continue to grow and use water during the seasons that do not support annual crop plants. This may restore the balance between rainfall and water use, thus preventing rising water tables and the movement of salt to the soil surface (Manchanda and Garg, 2008).

Farming systems can change to incorporate perennials in rotation with annual crops (phase farming), in mixed plantings (alley farming, intercropping), or in site-specific plantings (precision farming) (Munns *et al.,* 2002). Although the use of these approaches to sustainable management can ameliorate yield reduction under salinity stress, implementation is often limited because of cost and availability of good water quality or water resource. Evolving efficient, low cost, easily adaptable methods for the abiotic stress management is a major challenge. Worldwide, extensive research is being carried out, to develop strategies to cope with abiotic stresses, through development of salt and drought tolerant varieties, shifting the crop calendars, resource management practices etc. (Venkateswarlu and Shanker, 2009).

Combination of organic and chemical amendments

Bjugstad *et al.* (1981) found that mulching, fertilization, gypsum application, and physical amendment application (sawdust, straw, perlite, and vermiculite) all helped in promoting plant establishment on a saline-sodic bentonite spoil. The researchers found that the organic amendments provided rapid improvements in soil physical conditions often unattainable with chemical amendment only. The addition of organic matter in conjunction with gypsum has been successful in reducing adverse soil properties associated with sodic soils. Vance *et al.* (1998) found that the addition of organic matter and gypsum to the surface soil decreased dispersion and reduced EC in surface soils more effectively than gypsum alone. Wong *et al.* (2009) determined that the addition of organic material increased soil microbial biomass while added gypsum decreased pH. Chorum and Rengasamy (1997) found a larger decrease in pH in highly alkaline soil with the combined addition of gypsum and green manure, compared to the addition of green manure or gypsum alone. Wong *et al.* (2009) conclude that there is often a dormant population of salt- tolerant microorganisms in degraded soils and can multiply rapidly when substrate is made available. The large increase in EC with the addition of gypsum does not appear to affect microbial respiration rates. Decomposition processes in their study were apparently limited by microbial substrate (food source) rather than poor, salt-affected soil conditions.

Correcting salt-affected soils

Salt-affected soils can be corrected by

Improving drainage

In soils with poor drainage, deep tillage can be used to break up the soil surface as well as clay pans and hardpans, which are layers of clay or other hard soils that restrict the downward flow of water.

Leaching

Leaching can be used to reduce the salts in soils. You must add enough low-salt water to the soil surface to dissolve the salts and move them below the root zone. The water must be relatively free of salts (1,500 - 2,000 ppm total salts), particularly sodium salts.

Reducing evaporation

Applying residue or mulch to the soil can help lower evaporation rates.

Applying chemical treatments

Before leaching saline-sodic and sodic soils, you must first treat them with chemicals, to reduce the exchangeable sodium content. To remove or exchange with the sodium, add calcium in a soluble form such as gypsum.

Conclusions

Salinity is a significant problem affecting agriculture worldwide and is predicted to become a larger problem in the coming decades. The detrimental effects of high salinity on plants can be observed at the whole-plant level in terms of plant death and/or decrease in productivity. Some plant species are clearly more flexible than others in their requirements for survival in salty environments. An understanding of how single cell responses to salt are coordinated with organismal and whole-plant responses to maintain an optimal balance between salt uptake and compartmentation is fundamental to our knowledge of how plants successfully adapt to salt stress.

Use of both genetic manipulation and traditional breeding approaches will be required to unravel the mechanisms involved in salinity tolerance and to develop salt-tolerant cultivars better able to cope with the increasing soil salinity constraints. Salinity tolerance involves a complex of responses at molecular, cellular, metabolic, physiological, and whole-plant levels. Extensive research through cellular, metabolic, and physiological analysis has elucidated that among various salinity responses, mechanisms or strategies controlling ion uptake, transport and balance, osmotic regulation, hormone metabolism, antioxidant metabolism, and stress signalling play critical roles in plant adaptation to salinity stress. Taking advantage of the latest advancements in the field of genomic, transcriptomic, proteomic, and metabolomic techniques, plant biologists are focusing on the development of a complete profile of genes, proteins, and metabolites responsible for different mechanisms of salinity tolerance in different plant species. However, there is lack of the integration of information from genomic, transcriptomic, proteomic, and metabolomics studies, and the combined approach is essential for the determination of the key pathways or processes controlling salinity tolerance.

References

Ahmad, P. and Prasad, M. N. V. 2012. Abiotic Stress Responses in Plants *In:* Metabolism, Productivity and Sustainability. Springer, New York, NY.

Ahmad, P. and Umar, S. 2011. Oxidative Stress *In*: Role of Antioxidants in Plants. Studium Press, New Delhi.

Alc′azar, R., Cuevas, J. C. and Planas, J. 2011. Integration of polyamines in the cold acclimation response. *Pl. Sci.* 180(1): 31–38.

Alc′azar, R., Marco, F. and Cuevas J. C. 2006. Involvement of polyamines in plant response to abiotic stress. *Biotech. Letters* 28(23): 1867–1876.

Alc′azar, R., Planas, J. and Saxena, T. 2010. Putrescine accumulation confers drought tolerance in transgenic *Arabidopsis* plants over expressing the homologous Arginine decarboxylase 2 gene. *Pl. Physiol. and Biochem.* 48(7): 547–552.

Alet, A. I., S′anchez, D. H. and Cuevas, J. C. 2012. New insights into the role of spermine in *Arabidopsis thaliana* under long-term salt stress. *Pl. Sci.* 182(1): 94–100.

Allan, A. C. and Fluhr, R. 1997. Two distinct sources of elicited reactive oxygen species in tobacco epidermal cells. *Plant Cell*, 9: 1559- 1572.

Asada, K. 1999. The water-water cycle in chloroplasts: scavenging of active oxygens and dissipation of excess photons. *Ann. Rev. of Pl. Biol.* 50: 601–639.

Ashraf, M. 2014. Some important physiological selection criteria for salt tolerance in plants. *Flora* 199: 361–376.

Ashraf, M., Akram, N. A., Arteca, R. N. and Foolad, M. R. 2010. The physiological, biochemical and molecular roles of brassinosteroids and salicylic acid in plant processes and salt tolerance. *Crit. Rev. Plant Sci.* 29: 162–190.

Ashrafijou, M., Sadat Noori, S. A., Izadi Darbandi, A. and Saghafi, S. 2010. Effect of salinity and radiation on proline accumulation in seeds of canola (*Brassica napus* L.). *Plant Soil Environ.* 56 (7): 312-317.

Aslam, R., Bostan, N., Amen, N., Maria, M. and Safdar, W. 2011. A critical review on halophytes: Salt tolerant plants. *J. Med. Plants Res.* 5(33): 7108-7118.

Batool, N., Shahzad, A., Ilyas, N. and Noor, T. 2014. Plants and Salt stress. *Int. J. Agri. Crop Sci.* 7(9): 582-589.

Beebe, S. 2012. Common bean breeding in the tropics. *In:* Plant Breeding Reviews, Vol. 36. Janick J (ed). Hoboken, John Wiley & Sons, Inc., NJ. pp. 357–426.

Beebe, S., Ramirez, J., Jarvis, A., Rao, I., Mosquera, G. and Bueno, G. 2011. Genetic improvement of common beans and the challenges of climate change. *In*: Crop Adaptation of Climate Change. Yadav SS, Redden R.J, Hatfield JL, Lotze-Campen H, Hall AE (ed). Wiley, New York, NY. pp. 356–369.

Ben Ahmed, C., Ben Rouina, B., Sensoy, S., Boukhriss, M. and Ben Abdullah, F. 2010. Exogenous proline effects on photosynthetic performance and antioxidant defense system of young olive tree. *J. Agric. and Food Chem.* 58(7): 4216–4222.

Bjugstad, A. J., Yamamoto, T. and Uresk, D. W. 1981. Shrub establishment on coal and bentonite clay mines. *In*: Shrub Establishment on disturbed arid and semiarid lands. Wyoming Game and Fish Department, Laramie. pp 104-122.

Bohnert, H. J., Nelson, D. E. and Jensen, R. G. 1995. Adaptations to environmental stresses. *Plant Cell* 7(7): 1099–1111.

Bouchereau, A., Aziz, A., Larher, F. and Martin-Tanguy, J. 1999. Polyamines and environmental challenges: recent development. *Pl. Sci.* 140(2): 103–125.

Cabot, C., Sibole, J. V., Barceló, J. and Poschenrieder, C. 2009. Abscisic acid decreases leaf Na+ exclusion in salt-treated *Phaseolus vulgaris* L. *J. Plant Growth Regul.* 28: 187–192.

Chaves, M. M., Maroco, J. P. and Pereira, J. S. 2003. Understanding plant responses to drought– from genes to the whole plant. *Funct. Plant Biol.* 30: 239–264.

Chen, S., Li, J., Wang, S., Hüttermann, A. and Altman, A. 2001. Salt, nutrient uptake and transport, and ABA of *Populus euphratica*: a hybrid in response to increasing soil NaCl. *Trees Struct. Funct.* 15: 186–194.

Chinnusamy, V., Jagendorf, A. and Zhu, J. K. 2005. Understanding and improving salt tolerance in plants. *Crop Sci.* 45: 437–448.

Chorum, M. and Rengasamy, P. 1997. Carbonate chemistry, pH and physical properties of alkaline sodic soil as affected by various amendments. *Australian J. of Soil Res.* 35: 149-161.

Davenport, R. James, R. Zakrisson-Plogander, A., Tester, M. and Munns, R. 2005. Control of sodium transport in durum wheat. Pl. Physiol., 137: 807- 818.

Dewey, D. R. 1962. Breeding crested wheatgrass Agropyron desertorum for salt tolerance. *Crop Sci.* 2:403-407.

Dietz, K. J., Tavakoli, N. and Kluge, C. 2001. Significance of the Vtype ATPase for the adaptation to stressful growth conditions and its regulation on the molecular and biochemical level. *J. Exptl. Botany* 52(363): 1969–1980.

El-Shintinawy, F. and El-Shourbagy, M. N. 2001. Alleviation of changes in protein metabolism in NaCl-stressed wheat seedlings by thiamine. *Biologia Plantarum* 44(4): 541–545.

Flowers, T. J. and Colmer, T. D. 2008. Salinity tolerance in halophytes. *New Phytol.* 179: 945–963.

Flowers, T. J. and Yeo, A. R. 1995. Breeding for salinity resistance in crop plants: where next? *Aust. J. Plant Physiol.* 22:875-884.

Flowers, T. J., Hajibagheri, M. A. and Clipson, N. J. W. 1986. Halophytes. *Q. Rev. Biol.* 61: 313–336.

Fluhr, R. and. Mattoo, A. K. 1996. Ethylene-biosynthesis and perception. *Critical Rev. in Pl. Sci.* 15: 479–523.

Ford, C. W. 1984. Accumulation of low molecular weight solutes in water-stressed tropical legumes. *Phytochemistry* 23(5): 1007–1015.

Foyer, C and Noctor, G. 2003. Redox sensing and signaling associated with reactive oxygen in chloroplast, peroxisome and mitochondria. Physiologia Plantarum, 119: 355-364.

Fragnire, C., Serrano, M., Abou-Mansour, E., Métraux, J.-P. and L'Haridon, F. 2011. Salicylic acid and its location in response to biotic and abiotic stress. *FEBS Lett.* 585: 1847–1852.

Fu, X. Z., Ullah Khan, E., Hu, S. S., Fan, Q. J. and Liu, J. H. 2011. Over expression of the betaine aldehyde dehydrogenase gene from *Atriplex hortensis* enhances salt tolerance in the transgenic trifoliate orange (*Poncirus trifoliata* L. Raf.). *Environ. Exp. Bot.* 74: 106–113.

Galston, A. W., Kaur-Sawhney, R., Altabella, T. and Tiburcio, A. F. 1997. Plant polyamines in reproductive activity and response to abiotic stress. *Botanica Acta* 110(3): 197–207.

Ge, Y. and Li, J. D. 1990. A preliminary study on the effects of halophytes on salt accumulation and desalination in the soil of Songnen Plain, northeast China. *Acta. Prat. Sin.* 1: 70–76.

Gill, S. S., Tajrishi, M., Madan, M. and Tuteja, N. 2013. A DESD box helicase functions in salinity stress tolerance by improving photosynthesis and antioxidant machinery in rice (*Oryza sativa* L. cv. PB1). *Pl. Mol. Biol.* 82(1-2): 1–22.

Gong, Y., Rao, L. and Yu, D. 2013. Abiotic stress in plants. *In*: Agricultural Chemistry. Stoytcheva M (ed). InTech, Rijeka, Croatia.

Greenway, H. and Munns, R. 1980. Mechanisms of salt tolerance in non halophytes. *Annu. Rev. Plant Physiol.* 31: 149–190.

Groß, F., Durner, J. and Gaupels, F. 2013. Nitric oxide, antioxidants and prooxidants in plant defence responses. *Frontiers in Pl. Sci.* 4: 419.

Gupta, B. and Huang, B. 2014. Mechanism of salinity tolerance in plants: physiological, biochemical, and molecular characterization. *Int. J. Genomics* 2014: 701596.

Gupta, K. Dey, A. and Gupta, B. 2013. Plant polyamines in abiotic stress responses. *Acta Physiol. Plantarum* 35(7): 2015–2036.

Gupta, K. J., Stoimenova, M. and Kaiser, W. M. 2005. In higher plants, only root mitochondria, but not leaf mitochondria reduce nitrite to NO, in vitro and in situ. *J. Exptl. Bot.* 56(420): 2601–2609.

Gupta, K., Dey, A. and Gupta, B. 2013. Polyamines and their role in plant osmotic stress tolerance. *In*: Climate Change and Plant Abiotic Stress Tolerance. Tuteja N and Gill SS (ed). Wiley-VCH, Weinheim, Germany. pp. 1053–1072,

Gurmani, A. R., Bano, A., Khan, S. U., Din, J. and Zhang, J. L. 2011. Alleviation of salt stress by seed treatment with abscisic acid (ABA), 6-benzylamino purine (BA) and chlormequat chloride(CCC) optimizes ion and organic matter accumulation and increases yield of rice(*Oryza sativa* L.). *Aust. J. Crop Sci.* 5: 1278–1285.

Hanfrey, C., Sommer, S., Mayer, M. J., Burtin, D. and Michael, A. J. 2001. *Arabidopsis* polyamine biosynthesis: absence of ornithine decarboxylase and the mechanism of arginine decarboxylase activity. *Pl. Journal* 27(6): 551–560.

Hanson, A. D., Rathinasabapathi, B., Rivoal, J., Burnet, M., Dillon, M. O. and Gage, D. A. 1994. Osmoprotective compounds in the Plumbaginaceae: a natural experiment in metabolic engineering of stress tolerance. *In*: Proceedings of the National Academy of Sciences of the United States of America, 91(1). pp. 306–310.

Hanzawa, Y., Imai, A., Michael, A. J., Komeda, Y. and Takahashi, T. 2002. Characterization of the spermidine synthase-related gene family in *Arabidopsis thaliana. FEBS Letters* 527(1–3): 176–180.

Hasanuzzaman, M., Nahar, K. and. Fujita, M. 2014. Regulatory role of polyamines in growth, development and abiotic stress tolerance in plants. *In*: Plant Adaptation to Environmental Change: Significance of Amino Acids and Their Derivatives. pp. 157–193.

Hasegawa, P. M. 2013. Sodium (Na+) homeostasis and salt tolerance of plants. *Environ. and Exptl. Bot.* 92: 19–31.

Hasegawa, P. M., Bressan, R. A., Zhu, J. K. and Bohnert, H. J. 2000. Plant cellular and molecular responses to high salinity. *Ann. Rev. Pl. Biology* 51: 463–499.

Hashimoto, T., Tamaki, K., Suzuki, K. I. and Yamada, Y. 1998. Molecular cloning of plant spermidine synthases. *Pl. and Cell Physiol.* 39(1): 73–79.

He, T. and Cramer, G. R. 1996. Abscisic acid concentrations are correlated with leaf area reductions in two salt-stressed rapid cycling *Brassica* species. *Plant Soil* 179: 25–33.

Hoque, M. A., Banu, M. N. A. and Okuma, E. 2007. Exogenous proline and glycinebetaine increase NaCl-induced ascorbate glutathione cycle enzyme activities, and proline improves salt tolerance more than glycine betaine in tobacco Bright Yellow-2 suspension-cultured cells. *J. Pl. Physiol.* 164(11): 1457–1468.

Horie, T., Karahara, I. and Katsuhara, M. 2012. Salinity tolerance mechanisms in glycophytes: an overview with the central focus on rice plants. *Rice J.* 5: 1–18.

Hossain, Z., Mandal, A. K. A., Datta, S. K. and Biswas, A. K. 2007. Development of NaCl tolerant line in *Chrysanthemum morifolium* Ramat. through shoot organogenesis of selected callus line. *J. Biotechnol.* 129: 658–667.

Hussain, M. I., Lyra, A. A., Farooq, M., Nikolaos, N. and Khalid, N. 2016. Salt and drought stresses in safflower: a review. *Agron. Sustain. Dev.* 36(4): 344-348.

Izadi Darbandi, A., Bahmani, K., Ramshini, H. A. and Moradi, N. 2013. Heritability Estimates of Agronomic Traits and Essential Oil Content in Iranian Fennels. *J. Agri. Sci. Technol.* 15: 1275-1283.

James, R. A., Blake, C., Byrt, C. S. and Munns, R. 2011. Major genes for Na+ exclusion, Nax1 and Nax2 wheat HKT1;4 and HKT1;5, decrease Na+ accumulation in bread wheat leaves under saline and water logged conditions. *J. Exp. Bot.* 62: 2939–2947.

James, R. A., Blake, C., Zwart, A. B., Hare, R. A., Rathjen, A. J. and Munns, R. 2012. Impact of ancestral wheat sodium exclusion genes Nax1 and Nax2 on grain yield of durum wheat on saline soils. *Funct. Plant Biol.* 39:609-618.

Janowitz, T., Kneifel, H. and Piotrowski, M. 2003. Identification and characterization of plant agmatine iminohydrolase, the last missing link in polyamine biosynthesis of plants. *FEBS Letters* 544(1-2): 258–261.

Jenks, M. A., Hasegawa, P. M. and Jain, S. M. 2007. Advances in molecular breeding toward drought and salt tolerant crops: plant growth and development under salinity stress. *Springer.* 1–32.

Kerepesi, I. and Galiba, G. 2000. Osmotic and salt stress-induced alteration in soluble carbohydrate content in wheat seedlings. *Crop Sci.* 40(2): 482–487.

Khan, M. A., Ungar, I. A. and Showalter, A. M. 2000. Effects of sodium chloride treatments on growth and ion accumulation of the halophyte *Haloxylon recurvum. Commun. Soil Sci. Pl. Analysis* 31(17-18): 2763–2774.

Khan, M., Izbal, R., Mehar, F., Tasir, S. P., Naser, A. A. and Nafees, A. K. 2015. Salicilic acid inducted abiotic stress tolerance and underlying mechanisms in plants. *Front. Plant Sci.* 6:462.

Knott, J. M., R¨omer, P. and Sumper, M. 2007. Putative spermine synthases from *Thalassiosira pseudonana* and *Arabidopsis thaliana* synthesize thermospermine rather than spermine. *FEBS Letters*, 581(16): 3081–3086.

Kov´acs, Z., Simon-Sarkadi, L., Szucs, A. and Kocsy, G. 2010. Differential effects of cold, osmotic stress and abscisic acid on polyamine accumulation in wheat. *Amino Acids* 38(2): 623–631.

Kumar Parida, A. and Jha, B. 2010. Salt tolerance mechanisms in mangroves: a review. *Trees.* 24: 199–217.

Kusano, T., Yamaguchi, K., Berberich, T. and Takahashi, Y. 2007. Advances in polyamine research in 2007. *J. of Plant Res.* 120(3): 345–350.

Larcher, W.1980. Physiological Plant ecology. *In.* 2nd totally rev. edition(ed). Springer-Verlag Berlin and New York, pp. 303.

Lauchi, A. and Epstein, E. 1990. Plant responses to saline and sodic conditions. *In*: Agricultural salinity assessment and management. Tanji KK (ed). American Society of Civil Engineers, New York. pp. 113-137.

Lauchi, A. and Grattan, S. R.. 2007. Plant growth and development under saline stress. In: Advances in Molecular Breeding Toward Drought and Salt Tolerant Crops. Jenks MA, Hasegawa, P. M.., Jain, S. M. (ed). Springer, The Netherlands. Pp. 1-32.

Läuchli, A. and Lüttge, U. 2002. Salinity in the soil environment. *In*: Salinity: Environment-Plants-Molecules. KK Tanji (ed). Boston Kluwer Academic Publishers Boston, MA. pp. 21–23.

Li, C., Wang, X., Wang, H., Ni, F. and Shi, D. 2012. Comparative investigation of single salts stresses and their mixtures on Eragrostioid (*Chlorisvirgata*) to demonstrate the relaxation effect of mixed anions. *Aust. J. Crop Sci.* 6: 839–845.

Liu, J., Ishitani, M., Halfter, U., Kim, C. S. and Zhu, J. K. 2000. The *Arabidopsis thaliana* SOS2 gene encodes a protein kinase that is required for salt tolerance. *In*: Proceedings of the National Academy of Sciences of the United States of America, 97(7). pp. 3730–3734.

Maischak, H, Zimmermann, M. R., Felle, H. H., Boland, W. and Mithofer, A. 2010. Alamethicin-induced electrical long distance signaling in plants. *Plant Signal Behav*. 5:988-990.

Manchanda, G. and Garg, N. 2008. Salinity and its effects on the functional biology of legumes. *Acta Physiol. Plant.* 30: 595–618.

Matysik, J., Alia, A., Bhalu, B. and Mohanty, P. 2002. Molecular mechanisms of quenching of reactive oxygen species by proline under stress in plants. *Curr. Science* 82(5): 525–532.

Mittler, R., Vanderauwera, S., Suzuki, N., Miller, G., Tognetti, V. B., Vandepoele, K., Gollery, M., Shulaev, V. and Van Breusegem, F. 2011. ROS signaling: the new wave? *Trends Plant Sci.* 16:300-309.

Munns, R. 2002.Comparative physiology of salt and watrer stress, *Plant, Cell and Environment,* 25(2): 239-250.

Munns, R. 2005. Genes and salt tolerance: bringing them together. *New Phytol.* 167: 645–663.

Munns, R. and Tester, M. 2008. Mechanisms of salinity tolerance. *Annu. Rev. Plant Biol.* 59: 651–681.

Munns, R., Husain, S., Rivelli, A. R., James, R. A., Condon, A. G., Lindsay, M. P., Lagudah, E. S., Schachtman, D. P. and Hare, R. A. 2002. Avenues for increasing salt tolerance of crops, and the role of physiologically based selection traits. *Plant Soil* 247(1): 93–105.

Munns, R., James, R. A., Xu, B., Athman, A., Conn, S.J., Jordans, C., Byrt, C. S., Hare, R. A., Tyerman, S. D. and Tester, M. 2012. Wheat grain yield on saline soils is improved by an ancestral Na+ transporter gene. *Nat. Biotechnol.* 30:360-364.

Munns, R., Schachtman, D. and Condon, A. 1995. The significance of a two-phase growth response to salinity in Wheat and Barley. *Funct. Plant Biol*., 22(4): 561-569.

Murkute, A. A., Sharma, S. and Singh, S. K. 2005. Citrus in term of soil and water salinity: a review. *J. Sci. Ind. Res.* 64: 393-402.

Niu Xiaomu, N. X., Bressan, R. A., Hasegawa, P. M. and Pardo, J. M. 1995. Ion homeostasis in NaCl stress environments. *Plant Physiology* 109(3): 735–742.

Palao, D. C., De La Viña, C. B., Aiza Vispo, N. and Singh, R. K. 2014. New phenotyping technique for salinity tolerance at reproductive stage in rice. *In*: Proceedings of the 3rd International Plant Phenotyping Symposium, Chennai, India.

Pang, Q., Chen, S., Dai, S., Chen, Y., Wang, Y. and Yan, X. 2010. Comparative proteomics of salt tolerance in *Arabidopsis thaliana* and *Thellungiella halophila*. *J. Proteome Res.* 9: 2584–2599.

Papp, J.C., Ball, M. C. and Terry, N. 1983. A comparive study of the effects of NaCl salinity on respiration, photosynthesis and leaf extension growth in *Beta vulgaris* L.(sugar beet). *Plant, Cell and Environment* 6(8): 675-677.

Plett, D. C. and Moller, I. S. 2010. Na+ transport in glycophytic plants: what we know and would like to know. *Plant Cell Environ.* 33: 612-626.

Popova, L. P., Stoinova, Z. G. and Maslenkova, L. T. 1995. Involvement of abscisic acid in photosynthetic process in *Hordeum vulgare* L. during salinity stress. *J. Plant Growth Regul.* 14: 211–218.

Rahnama, A., James, R. A., Poustini, K. and Munns, R. 2010. Stomatal conductance as a screen for osmotic stress tolerance in durum wheat growing in saline soil. *Funct. Plant Biol.* 37: 255–263.

Rambla, J. L., Vera-Sirera, F., Bl´azquez, M. A., Carbonell, J. and Granell, A. 2010. Quantitation of biogenic tetraamines in *Arabidopsis thaliana. Analytical Biochem.* 397(2): 208–211.

Reddy, M. P., Sanish, S. and Iyengar, E. R. R. 1992. Photosynthetic studies and compartmentation of ions in different tissues of *Salicornia brachiata* Roxb under saline conditions. *Photosynthetica* 26: 173–179.

Reddy, P. S., Jogeswar, G., Rasineni,G. K., Maheswari, M., Reddy, A. R. and Varshney, R. K. 2015. Proline over-accumulation alleviates salt stress and protects photosynthetic and antioxidant enzyme activities in transgenic sorghum (*Sorghum bicolour* (L.) Moench. *Plant Physiol. Biochem.* 94: 104–113.

Rengasamy, P. 2010. Soil processes affecting crop production in salt-affected soils. *Funct. Plant Biol*. 37: 613-620.

Rhoades, J. D. and Loveday, J. 1990. Salinity in irrigated agriculture. *In:* Irrigation of agricultural crops. Stewart BA, Nielsen DR (ed). Madison, Wisconsin, ASA. pp. 1089–1142.

Roy, C. and Sengupta, D. N. 2014. Effect of Short Term NaCl Stress on Cultivars of *Solanum lycopersicum*: A Comparative Biochemical Approach. *J. Stress Physiol. Biochem.* 10(1): 59-81.

Roy, S. J., Negrão, S. and Tester, M. 2014. Salt resistant crop plants. *Curr. Opin. Biotechnol.* 26: 115–124.

Rozema, J. and Flowers, T. 2008. Ecology: crops for a salinized world. *Science* 322: 1478–1480.

Sadat Noori, S. A. 2005. Assessment for salinity tolerance through intergeneric hybridisation: *Triticum durum* × *Aegilops speltoides*. *Euphytica* 146: 149–155.

Sadat Noori, S. A. and Harati, M. 2005. Breeding for Salt-Resistance Using Transgressive Segregation in Spring Wheat. *J. Sci.* 16(3): 217-222.

Sadat Noori, S. A. and McNeilly, T. 1999. Assessment of variability in salt tolerance in diploid *Aegilops* spp. *J. Genet. Breed.* 183-188.Sadat Noori, S. A. and Sokhansanj, A. 2004. Triple test cross analysis for genetic components of salinity tolerance in spring wheat. *J. Sci.* 15(1): 13-19.

Sadat Noori, S. A., Ebrahimi, M., Khazaei, J. and Khalaj, H. 2011a. Predicting yield of wheat genotypes at different salinity by artificial neural network. *Afr. J. Agric. Res.* 6(12): 2660-2675.

Sadat Noori, S. A., Ferdosizadeh, L., Izadi Darbandi, A., Mortazavian, S. M. M. and Saghafi, S. 2011b. Effects of Salinity and Laser Radiation on Proline Accumulation in Seeds of Spring Wheat. *J. Plant Physiol. Breed.* 1(2): 11-20.

Sairam, R. K. and Tyagi, A. 2004. Physiology and molecular biology of salinity stress tolerance in plants. *Curr. Sci.* 86(3): 407–421.

Sanders, D. 2000. Plant biology: the salty tale of *Arabidopsis. Curr. Biology* 10(13): R486–R488.

Saxena, S. C., Kaur, H. and Verma, P. 2013. Osmoprotectants: potential for crop improvement under adverse conditions. *In:* Plant Acclimation to Environmental Stress. Springer, New York, NY, USA. pp. 197–232.

Shahbaz, M. and Ashraf, M. 2013. Improving Salinity Tolerance in Cereals. *Crit. Rev. Plant Sci.* 32: 237–249.

Shang, G., Li, Y., Hong, Z., Cheng long, L., Shao, Z. and Qing chang, L. 2012. Over expression of SOS genes enhanced salt tolerance in sweet potato. J. *Integr. Agri.* 11(3): 378-386.

Shi, H., Ishitani, M., Kim, C. and Zhu, J. K. 2000. The *Arabidopsis thaliana* salt tolerance gene SOS1 encodes a putative Na+/H+ antiporter. *In*: Proceedings of the National Academy of the Sciences of the United States of America, 97(12). pp. 6896–6901.

Shi, H., Quintero, F. J., Pardo, J. M. and. Zhu, J. K. 2002. The putative plasma membrane Na+/H+ antiporter SOS1 controls long distance Na+ transport in plants. *Plant Cell* 14(2): 465–477.

Shu, S., Guo, S. R. and Yuan, L. Y. 2012. A review: polyamines and photosynthesis. *In:* Advances in Photosynthesis—Fundamental Aspects. Najafpour MM (ed). InTech, Rijeka, Croatia. pp. 439–464.

Tahir, M. A., Aziz, T., Farooq, M. and Sarwar, G. 2012. Silicon induced changes in growth, ionic composition, water relations, chlorophyll contents and membrane permeability in two salt stressed wheat genotypes. *Arch. Agron. and Soil Sci.* 58(3): 247–256.

Tester, M. and Davenport, R. 2003. Na+ tolerance and Na+ transport in higher plants. *Ann. Bot.* 91:503-527.

Tuteja, N., Sahoo, R. K., Garg, B. and Tuteja, R. 2013. OsSUV3 dual helicase functions in salinity stress tolerance by maintaining photosynthesis and antioxidant machinery in rice (*Oryza sativa* L. cv. IR64). *The Pl. Journal* 76(1): 115–127.

Urano, K., Yoshiba, Y. Nanjo, T. 2003. Characterization of *Arabidopsis* genes involved in biosynthesis of polyamines in abiotic stress responses and developmental stages. *Pl. Cell and Environ.* 26(11): 1917–1926.

Urano, K., Yoshiba, Y., Nanjo, T., Ito, T., Yamaguchi-Shinozaki, K. and Shinozaki, K. 2004. *Arabidopsis* stress-inducible gene for arginine decarboxylase AtADC2 is required for accumulation of putrescine in salt tolerance. *Biochem. and Biophysical Res. Communications* 313(2): 369–375.

Vance, W. H., Tisdeel, J. M. and McKenzie, B. M., 1998. Residual effects of surface application of organic matter and calcium salts on the sub-soil of a red-brown earth. *Australian J. of Exptl. Agric.* 38: 595-600.

Venkateswarlu, B. and Shanker, A. K. 2009. Climate change and agriculture: adaptation and mitigation strategies. *Indian J. Agron.* 54: 226–230.

Vinocur, B. and Altman, A. 2005. Recent advances in engineering plant tolerance to abiotic stress: achievements and limitations. *Curr. Opin. Biotech.* 16: 123–132.

Volkmar, K. M., Hu, Y. and Steppuhn, H. 1998. Physiological responses of plants to salinity: A review. *Can. J. Plant Sci.* 78: 19-27.

Wang, B., L¨uttge, U. and Ratajczak, R. 2001. Effects of salt treatment and osmotic stress on V-ATPase and V-PPase in leaves of the halophyte *Suaeda salsa*. *J. Exptl. Botany* 52(365): 2355–2365.

Wang, Y. and Nii, N. 2000. Changes in chlorophyll, ribulose bisphosphate carboxylase-oxygenase, glycine betaine content, photosynthesis and transpiration in *Amaranthus tricolor* leaves during salt stress. *J. Hort. Sci. and Biotech.*, 75(6): 623–627.

Wong, N. L., Dalal, R. C. and Greene, R. S. B. 2009. Carbon Dynamics of sodic and saline soils following gypsum and organic material additions: A laboratory incubation. *Applied Soil Ecology,* 40: 29-40.

Yamaguchi, T. and Blumwald, E. 2005. Developing salt tolerant crop plants: challenges and opportunities. *Trends Plant Sci.*, 10: 615–620.

Yang, C., Chong, J., Kim, C., Li, C., Shi, D. and Wang, D. 2007. Osmotic adjustment and ion balance traits of an alkali resistant halophyte *Kochia sieversiana* during adaptation to salt and alkali conditions. *Plant Soil*, 294: 263–276.

Yong, H., Shuya, Y. and Hyang, L. 2014. Towards plant salinity tolerance implications from ion transporters and biochemical regulation. *Plant Growth Regul.* 35: 133–143.

Zhu, J. K. 2003. Regulation of ion homeostasis under salt stress. *Curr. Opinion in Pl. Biology* 6(5): 441–445.

3

High Temperature Stress: Responses Mechanism and Management

Gurdev Chand, Bhav Kumar Sinha, Magdeshwar Sharma and Sapalika Dogra

Growth and development of plants are dependent upon the temperature surrounding the plant and each species has a specific temperature range represented by a minimum, maximum, and optimum. These values were summarized by Hatfield *et al.* (2008, 2011) for a number of different species typical of grain and fruit production. The expected changes in temperature over the next 30 to 50 years are predicted to be in the range of 2 to 3°C Intergovernmental Panel Climate Change (IPCC) (2007). Heat waves or extreme temperature events are projected to become more intense, more frequent, and last longer than what is being currently been observed in recent years (Meehl *et al.*, 2007). Extreme temperature events may have short-term durations of a few days with temperature increases of over 5°C above the normal temperatures. A recent review by Barlow *et al.* (2015) on the effect of temperature extremes, frost and heat, in wheat (*Triticum aestivum* L.) revealed that frost caused sterility and abortion of formed grains while excessive heat caused reduction in grain number and reduced duration of the grain- filling period. Analysis by Meehl *et al.* (2007) revealed that daily minimum temperatures will increase more rapidly than daily maximum temperatures leading to the increase in the daily mean temperatures and a greater likelihood of extreme events and these changes could have detrimental effects on grain yield. If these changes in temperature are expected to occur over the next 30 years then understanding the potential impacts on plant growth and development will help develop adaptation strategies to offset these impacts.

At very high temperatures, severe cellular injury and even cell death may occur within minutes, which could be attributed to a catastrophic collapse of cellular organization. At moderately high temperatures, injuries or death may occur only after long-term exposure. Direct injuries due to high temperatures include protein denaturation and aggregation, and increased fluidity of membrane lipids.

Indirect or slower heat injuries include inactivation of enzymes in chloroplast and mitochondria, inhibition of protein synthesis, protein degradation and loss of membrane integrity. Heat stress also affects the organization of microtubules by splitting and/or elongation of spindles, formation of microtubule asters in mitotic cells, and elongation of phragmoplast microtubules. These injuries eventually lead to starvation, inhibition of growth, reduced ion flux, production of toxic compounds and reactive oxygen species (ROS) (Howarth, 2005). Immediately after exposure to high temperatures and perception of signals, changes occur at the molecular level altering the expression of genes and accumulation of transcripts, thereby leading to the synthesis of stress-related proteins as a stress tolerance strategy. Expression of heat shock proteins (HSPs) is known to be an important adaptive strategy in this regard. The HSPs, ranging in molecular mass from about 10 to 200 kDa, have chaperone like functions and are involved in signal transduction during heat stress The tolerance conferred by HSPs results in improved physiological phenomena such as photosynthesis, assimilate partitioning, water and nutrient use efficiency, and membrane stability. Such improvements make plant growth and development possible under heat stress. However, not all plant species or genotypes within species have similar capabilities in coping with the heat stress. There exists tremendous variation within and between species, providing opportunities to improve crop heat-stress tolerance through genetic means. Some attempts to develop heat-tolerant genotypes via conventional plant breeding protocols have been successful (Camejo *et al*., 2005). Recently, however, advanced techniques of molecular breeding and genetic engineering have provided additional tools, which could be employed to develop crops with improved heat tolerance and to combat this universal environmental adversary. However, to assure achievement of success in this strategy, concerted efforts of plant physiologist, molecular biologists and crop breeders are imperative. In this chapter accentuates on plant responses and adaptations to heat stress at the whole plant, cellular and sub-cellular levels, tolerance mechanisms and strategies for genetic improvement of crops with heat-stress tolerance.

Table 1: Threshold high temperatures for some crop plants

Crop plants	Threshold tempertare (°C)	Growth stage	References
Wheat	26	Post anthesis	Stone and nicolas (1994)
Corn	38	Grain filling	Thompson(1986)
Cotton	45	Reproductive	Rehman *et al.*,(2004)
Pearlmillet	35	Seedling	Ashraf and Hafeez(2004)
Tomato	30	Emergence	Camejo *et al* .(2005),
Brassica	29	Flowering	Morrison and Stewart(2002)
Cool season pulses	25	Flowering	Siddiqui (1999)
Groundnut	34	Pollen production	Vara Parshad *et al.*,(2000)
Cowpea	41	Flowering	Patel and Hall(1990)
Rice	34	Grain yield	Morita *et al.*,(2004)

Heat-stress threshold

A threshold temperature refers to a value of daily mean temperature at which a detectable reduction in growth begins. Upper and lower developmental threshold temperatures have been determined for many plant species through controlled laboratory and field experiments. A lower developmental threshold or a base temperature is one below which plant growth and development stop. Similarly, an upper developmental threshold is the temperature above which growth and development cease. Knowledge of lower threshold temperatures is important in physiological research as well as for crop production. Base threshold temperatures vary with plant species, but for cool season crops 0°C is often the best-predicted base temperature (Miller *et al.*, 2001). Cool season and temperate crops often have lower threshold temperature values compared to tropical crops. Upper threshold temperatures also differ for different plant species and genotypes within species. However, determining a consistent upper threshold temperature is difficult because the plant behavior may differ depending on other environmental conditions (Miller *et al.*, 2001). In tomato, for example, when the ambient temperature exceeds 35°C, its seed germination, seedling and vegetative growth, flowering and fruit set, and fruit ripening are adversely affected. For other plant species, the higher threshold temperature may be lower or higher than 35°C. Upper threshold temperatures for some major crop species are displayed in (Table 1.) High temperature sensitivity is particularly important in tropical and subtropical climates as heat stress may become a major limiting factor for field crop production. Brief exposure of plants to high temperatures during seed filling can accelerate senescence, diminish seed set and seed weight, and reduce yield. This is because under such conditions plants tend to divert resources to cope with the heat stress and thus limited photosynthates would be available for reproductive development. Another effect of heat stress in many plant species is induced sterility when heat is imposed immediately before or during anthesis. Pulse legumes are particularly sensitive to heat stress at the bloom stage; only a few days of exposure to high temperatures (30–35°C) can cause heavy yield losses through flower drop or pod abortion. In general, base and upper threshold temperatures vary in plant species belonging to different habitats. Thus, it is highly desirable to appraise threshold temperatures of new cultivars to prevent damages by unfavorable temperatures during the plant ontogeny.

Temperature responses in plants

Responses to temperature differ among crop species throughout their life cycle and are primarily the phenological responses, i.e., stages of plant development. For each species, a defined range of maximum and minimum temperatures form the boundaries of observable growth. Vegetative development (node and leaf appearance rate) increases as temperatures rise to the species optimum level. For most plant species, vegetative development usually has a higher optimum temperature than for reproductive development. Cardinal temperature values for selected annual (non-perennial) crops are given in (Hatfield *et al.* 2008, 2011) for different species. If we depict the range of temperatures in the following diagram (Fig. 1) then the definition of extreme temperatures affecting plant response will be species dependent.

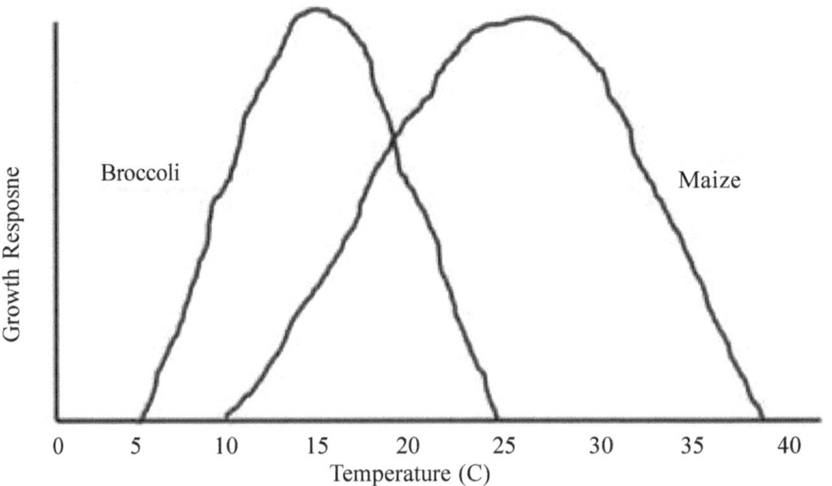

Fig. 1: Temperature response for maize and broccoli plants showing the lower, upper and optimum temperature limits for the vegetative growth phase

For example, an extreme event for maize (*Zea mays* L.) will be warmer than for a cool season vegetable (*broccoli, Brassica oleracea* L.) where the maximum temperature for growth is 25°C compared to 38°C. In understanding extreme events and their impact on plants we will have to consider the plant temperature response relative to the meteorological temperature. Faster development of non-perennial crops results in a shorter life cycle resulting in smaller plants, shorter reproductive duration, and lower yield potential. Temperatures which would be considered extreme and fall below or above specific thresholds at critical times during development can significantly impact productivity. Photoperiod sensitive crops, e.g., soybean, would also interact with temperature causing a disruption in phenological development. In general, extreme high temperatures during the reproductive stage will affect pollen viability, fertilization, and grain or fruit formation (Hatfield *et al.*, 2008, 2011). Chronic exposures to extreme temperatures during the pollination stage of initial grain or fruit set will reduce yield potential. However, acute exposure to extreme events may be most detrimental during the reproductive stages of development. The impacts of climate change are most evident in crop productivity because this parameter represents the component of greatest concern to producers, as well as consumers. Changes in the length of the growth cycle are of little consequence as long as the crop yield remains relatively consistent. Yield responses to temperature vary among species based on the crop's cardinal temperature requirements. Warming temperatures associated with climate change will affect plant growth and development along with crop yield.

Temperature extremes in climate

One of the more susceptible phenological stages to high temperatures is the pollination stage. Maize pollen viability decreases with exposure to temperatures above 35°C. The effect of temperature is enhanced under high vapor pressure deficits because pollen viability (prior to silk reception) is a function of pollen moisture content which is strongly dependent on vapor pressure deficit. During the endosperm division phase, as temperatures increased to 35°C from 30°C the potential kernel growth rate was reduced along with final kernel size, even after the plants were returned to 30°C. Exposure to temperatures above 30°C damaged cell division and amyloplast replication in maize kernels which reduced the size of the grain sink and ultimately yield. Rice (*Orzya sativa* L.) shows a similar temperature response to maize because pollen viability and production declines as daytime maximum temperature (T_{max}) exceeds 33°C and ceases when Tmax exceeds 40°C. Current cultivars of rice flower near mid-day which makes T max a good indicator of heat-stress on spikelet sterility. These exposure times occur quickly after anthesis and exposure to temperatures above 33°C within 13 h after anthesis (dehiscence of the anther, shedding of pollen, germination of pollen grains on stigma, and elongation of pollen tubes) cause negative impacts on reproduction. Current observations in rice reveal that anthesis occurs between about 9 to 11 am in rice and exposure to high temperatures may already be occurring and will increase in the future. There is emerging evidence that differences exist among rice cultivars for flowering times during the day. Given the negative impacts of high temperatures on pollen viability, recent observations from Shah *et al.* (2011) suggest flowering at cooler times of the day would be beneficial to rice grown in warm environments. They proposed that variation in flowering times during the day would be a valuable phenotypic marker for high-temperature tolerance. As daytime temperatures increased from 30 to 35°C, seed set on male-sterile, female fertile soybean (*Glycine max* (L.) Merr.) plants decreased. This confirms earlier observations on partially male-sterile soybean in which complete sterility was observed when the daytime temperatures exceeded 35°C regardless of the night temperatures and concluded that daytime temperatures were the primary factor affecting pod set. Crop sensitivity to temperature extremes depends upon the length of anthesis. Maize, for example, has a highly compressed phase of anthesis for 3to 5 days, while rice, sorghum (*Sorghum bicolor* L. *Moench*.) and other small grains may extend anthesis over a period of a week or more. Inn soybean, peanut (*Arachis hypogaea* L.), and cotton (*Gossypium hirsutum* L.) anthesis occurs over several weeks and avoid a single occurrence of an extreme event affecting all of the pollening flowers. For peanut (and potentially other legumes) the sensitivity to elevated temperature for a given flower, extends from 6 days prior to opening (pollen cell division and formation) up through the day of anthesis. Therefore, several days of elevated temperature may affect fertility of flowers in their formative 6-day phase or anthesis. Singh *et al.* (2015) found differences in the threshold temperature for grain sorghum among genotypes and differences in the percentage of seed set in response to high temperatures. Pollination processes in other cereals, maize and sorghum, may have a similar sensitivity to elevated daytime temperature as rice. Rice and sorghum have exhibited similar sensitivities of grain yield, seed harvest index, pollen viability, and success in grain formation in which pollen viability and percent fertility is first reduced at instantaneous hourly air temperature above 33°C and reaches zero at 40°C. Diurnal max/min day/night temperatures of 40/30°C (35°C mean) cause zero yields for those two species with the same expected response for maize.

Plant responses to heat stress

Morphological changes

In tropical climates, excess of radiation and high temperatures are often the most limiting factors affecting plant growth and final crop yield. High temperatures can cause considerable pre- and post-harvest damages, including scorching of leaves and twigs, sunburns on leaves, branches and stems, leaf senescence and abscission, shoot and root growth inhibition, fruit discoloration and damage, and reduced yield. Similarly, in temperate regions, heat stress has been reported as one of the most important causes of reduction in yield and dry matter production in many crops, including maize.

High-temperature-induced modifications in plants may be direct as on existing physiological processes or indirect in altering the pattern of development. These responses may differ from one phenological stage to another. For example, long-term effects of heat stress on developing seeds may include delayed germination or loss of vigor, ultimately leading to reduced emergence and seedling establishment. Under diurnally varying temperatures, coleoptile growth in maize reduced at 40°C and ceased at 45°C. High temperatures caused significant declines in shoot dry mass, relative growth rate and net assimilation rate in maize, pearl millet and sugarcane, though leaf expansion was minimally affected (Wahid, 2007). Major impact of high temperatures on shoot growth is a severe reduction in the first internode length resulting in premature death of plants. For example, sugarcane plants grown under high temperatures exhibited smaller internodes, increased tillering, early senescence, and reduced total biomass.

Heat stress, singly or in combination with drought, is a common constraint during anthesis and grain filling stages in many cereal crops of temperate regions. For example, heat stress lengthened the duration of grain filling with reduction in kernel growth leading to losses in kernel density and weight by up to 7% in spring wheat. Similar reductions occurred in starch, protein and oil contents of the maize kernel and grain quality in other cereals under heat stress (Maestri *et al*., 2002). In wheat, both grain weight and grain number appeared to be sensitive to heat stress, as the number of grains per ear at maturity declined with increasing temperature In temperate and tropical lowlands, heat susceptibility is a cause of yield loss in common bean, *Phaseolus vulgaris* and groundnut, *Arachis hypogea*. In tomato, reproductive processes were adversely affected by high temperature, which included meiosis in both male and female organs, pollen germination and pollen tube growth, ovule viability, stigmatic and style positions, number of pollen grains retained by the stigma, fertilization and post-fertilization processes, growth of the endosperm, proembryo and fertilized embryo. Also, the most noticeable effect of high temperatures on reproductive processes in tomato is the production of an exserted style (i.e., stigma is elongated beyond the anther cone), which may prevent self-pollination. Poor fruit set at high temperature has also been associated with low levels of carbohydrates and growth regulators released in plant sink tissues. Growth chamber and greenhouse studies suggest that high temperature is most deleterious when flowers are first visible and sensitivity continues for 10–15 days. Reproductive phases most sensitive to high temperature are gametogenesis (8–9 days before anthesis) and fertilization (1–3 days after anthesis) in various plants. Both male and female gametophytes are sensitive to high temperature and response varies with genotype; however, ovules are generally less heat sensitive than pollen. Overall, based on the available studies, it seems that plant responses to high temperature vary with plant species and phenological stages. Reproductive processes are markedly affected by high temperatures in most plants, which ultimately affect fertilization and post-fertilization processes leading to reduced crop yield.

Anatomical changes

Although limited details are available, anatomical changes under high ambient temperatures are generally similar to those under drought stress. At the whole plant level, there is a general tendency of reduced cell size, closure of stomata and curtailed water loss, increased stomatal and trichomatous densities, and greater xylem vessels of both root and shoot. In grapes (*Vitis vinifera*), heat stress severely damaged the mesophyll cells and increased permeability of plasma membrane. With the onset of high temperature regime, *Zygophyllum qatarense* produced polymorphic leaves and tended to reduce transpirational water loss by showing bimodal stomatal behavior. At the sub-cellular level, major modifications occur in chloroplasts, leading to significant changes in photosynthesis. For example, high temperatures reduced photosynthesis by changing the structural organization of thylakoids. Studies have revealed that specific effects of high temperatures on photosynthetic membranes result in the loss of grana stacking or its swelling. In response to heat stress, chloroplasts in the mesophyll cells of grape plants became round in shape, the stroma lamellae became swollen, and the contents of vacuoles formed clumps, whilst the cristae were disrupted and mitochondria became empty. Such changes result in the formation of antenna-depleted photosystem-II (PSII) and hence reduced photosynthetic and respiratory activities (Zhang *et al.*, 2005). In general, it is evident that high temperature considerably affects anatomical structures not only at the tissue and cellular levels but also at the sub-cellular level. The cumulative effects of all these changes under high temperature stress may result in poor plant growth and productivity.

Phenological changes

Observation of changes in plant phenology in response to heat stress can reveal a better understanding of interactions between stress atmosphere and the plant. Different phonological stages differ in their sensitivity to high temperature; however, this depends on species and genotype as there are great inter and intra-specific variations. Heat stress is a major factor affecting the rate of plant development, which may be increasing to a certain limit and decreasing afterwards (Howarth, 2005).

The developmental stage at which the plant is exposed to the stress may determine the severity of possible damages experienced by the crop. It is, however, unknown whether damaging effects of heat episodes occurring at different developmental stages are cumulative. Vulnerability of species and cultivars to high temperatures may vary with the stage of plant development, but all vegetative and reproductive stages are affected by heat stress to some extent. During vegetative stage, for example, high day temperature can damage leaf gas exchange properties. During reproduction, a short period of heat stress can cause significant increases in floral buds and opened flowers abortion; however, there are great variations in sensitivity within and among plant species. The staple cereal crops can tolerate only narrow temperature ranges, which if exceeded during the flowering phase can damage fertilization and seed production, resulting in reduced yield (Porter, 2005). Thus, for crop production under high temperatures, it is important to know the developmental stages and plant processes that are most sensitive to heat stress, as well as whether high day or high night temperatures are more injurious. Such insights are important in determining heat-tolerance potential of crop plants.

Physiological responses

Water relations

Plant water status is the most important variable under changing ambient temperatures (Mazorra *et al*., 2002). In general, plants tend to maintain stable tissue water status regardless of temperature when moisture is ample; however, high temperatures severely impair this tendency when water is limiting. Under field conditions, high temperature stress is frequently associated with reduced water availability. In *Lotus creticus*, for example, elevated night temperatures caused a greater reduction in leaf water potential of water-stressed as compared to non-stressed plants (A˜non *et al*., 2004). In general, during daytime enhanced transpiration induces water deficiency in plants, causing a decrease in water potential and leading to perturbation of many physiological processes. High temperatures seem to cause water loss in plants more during daytime than nighttime.

Accumulation of compatible osmolytes

A key adaptive mechanism in many plants grown under abiotic stresses, including salinity, water deficit and extreme temperatures, is accumulation of certain organic compounds of low molecular mass, generally referred to as compatible osmolytes. Under stress, different plant species may accumulate a variety of osmolytes such as sugars and sugar alcohols (polyols), proline, tertiary and quaternary ammonium compounds, and tertiary sulphonium compounds (Sairam and Tyagi, 2004). Accumulation of such solutes may contribute to enhanced stress tolerance of plants, as briefly described in below.

Glycinebetaine (GB), an amphoteric quaternary amine, plays an important role as a compatible solute in plants under various stresses, such as salinity or high temperature. For example, high level of GB accumulation was reported in maize and sugarcane due to desiccating conditions of water deficit or high temperature. In contrast, plant species such as rice (*Oryza sativa*), mustard (*Brassica* spp.), Arabidopsis (*Arabidopsis thaliana*) and tobacco (*Nicotiana tabacum*) naturally do not produce GB under stress conditions. However, genetic engineering has allowed the introduction of GB-biosynthetic pathways into GB-deficit species (Quan *et al*., 2004).

Like GB, proline is also known to occur widely in higher plants and normally accumulates in large quantities in response to environmental stresses (Kavi Kishore *et al*., 2005). In assessing the functional significance of accumulation of compatible solutes, it is suggested that proline or GB synthesis may buffer cellular redox potential under heat and other environmental stresses. Similarly, accumulation of soluble sugars under heat stress has been reported in sugarcane, which entails great implications for heat tolerance. Under high temperatures, fruit set in tomato plants failed due to the disruption of sugar metabolism and proline transport during the narrow window of male reproductive development. Hexose sensing in transgenic plants engineered to produce trehalose, fructans or mannitol may be an important contributory factor to the stress-tolerant phenotypes.

Among other osmolytes, -4-aminobutyric acid (GABA), a non-protein amino acid, is widely distributed throughout the biological world to act as a compatible solute. GABA is synthesized from the glutamic acid by a single step reaction catalyzed by glutamate decarboxylase (GAD). An acidic pH activates GAD, a key enzyme in the biosynthesis of GABA. Episodes of high temperatures increase the cytosolic level of Ca^{+2}, which leads to calmodulin-mediated activation of GAD (Taiz and Zeiger, 2006). Several other studies show that various environmental stresses increase GABA accumulation through metabolic or mechanical disruptions, thus leading to cytosolic acidification. Kinetics of GABA in plants show a stress-specific pattern of accumulation, which is consistent with its physiological role in the mitigation of stress effects. Rapid accumulation of GABA in stressed tissues may provide a critical link in the chain of events stemming from perception of environmental stresses to timely physiological responses.

Photosynthesis

Alterations in various photosynthetic attributes under heat stress are good indicators of thermotolerance of the plant as they show correlations with growth. Any constraint in photosynthesis can limit plant growth at high temperatures. Photochemical reactions in thylakoid lamellae and carbon metabolism in the stroma of chloroplast have been suggested as the primary sites of injury at high temperatures. Chlorophyll fluorescence, the ratio of variable fluorescence to maximum fluorescence (Fv/Fm), and the base fluorescence (F_0) are physiological parameters that have been shown to correlate with heat tolerance. Increasing leaf temperatures and photosynthetic photon flux density influence thermotolerance adjustments of PSII, indicating their potential to optimize photosynthesis under varying environmental conditions as long as the upper thermal limits do not exceed. In tomato genotypes differing in their capacity for thermotolerance as well as in sugarcane, an increased chlorophyll *a*:*b* ratio and a decreased chlorophyll: carotenoids ratio were observed in the tolerant genotypes under high temperatures, indicating that these changes were related to thermotolerance of tomato. Furthermore, under high temperatures, degradation of chlorophyll *a* and *b* was more pronounced in developed compared to developing leaves Such effects on chlorophyll or photosynthetic apparatus were suggested to be associated with the production of active oxygen species (Camejo *et al.*, 2006).

PSII is highly thermolabile, and its activity is greatly reduced or even partially stopped under high temperatures, which may be due to the properties of thylakoid membranes where PSII is located. Heat stress may lead to the dissociation of oxygen evolving complex (OEC), resulting in an imbalance between the electron flow from OEC toward the acceptor side of PSII in the direction of PSI reaction center (Fig.2) (De Ronde *et al.*, 2004). Heat stress causes dissociation of manganese (Mn)-stabilizing 33-kDa protein at PSII reaction center complex followed by the release of Mn atoms. Heat stress may also impair other parts of the reaction center, e.g., the D_1 and/or the D_2 proteins. Heat induced inhibition of oxygen evolution and PSII activity. Heat stress leads to either (1) dissociation or (2) inhibition of the oxygen evolving complexes (OEC). This enables an alternative internal e^- donor such as proline instead of H2O to donate electrons to PSII. In barley, heat pulses abruptly damaged the PSII units and caused loss of their capacity of oxygen evolution leading to a restricted electron transport, which was totally abolished after four hours. This implied that the degradation of the impaired PSII units occurred in the light during this period of time. Following this, *de novo* synthesis of PSII units in the light gave a gradual rise to the observed PSII activities.

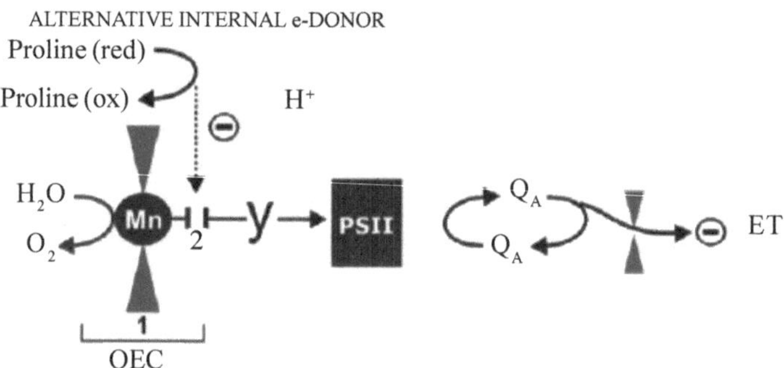

Fig. 2: Heat induced inhibition of oxygen evolution and PSII activity. Heat stress leads to either (1) dissociation or (2) inhibition of the oxygen evolving complexes (OEC). This enables an alternative internal e⁻-donor such as proline instead of H_2O to donate electrons to PSII. Reproduced with permission from De Ronde *et al.* (2004).

These effects can result from different events, including inhibition of electron transport activity and limited generation of reducing powers for metabolic functions (Allakhverdieva *et al.*, 2001). In field-grown Pima cotton under high temperatures, leaf photosynthesis was functionally limited by photosynthetic electron transport and ribulose-1,5-bisphosphate (RuBP) regeneration capacity, but not rubisco activity. On the other hand, under high temperatures, PSI stromal enzymes and chloroplast envelops are thermo stable and in fact PSI driven cyclic electron pathway, capable of contributing to thylakoid proton gradient, is activated.

High temperature influences the photosynthetic capacity of C_3 plants more strongly than in C_4 plants. It alters the energy distribution and changes the activities of carbon metabolism enzymes, particularly the rubisco, thereby altering the rate of RuBP regeneration by the disruption of electron transport and inactivation of the oxygen evolving enzymes of PSII. Heat shock reduces the amount of photosynthetic pigment, soluble proteins, rubisco binding proteins (RBP) and large- (LS) and small subunits (SS) of rubisco in darkness but increases them in light, indicating their roles as chaperones and HSPs (Kepova *et al.,* 2005). Moreover, under heat stress, starch or sucrose synthesis is greatly influenced as observed from reduced activities of sucrose phosphate synthase, ADP glucose pyrophosphorylase and invertase.

In any plant species, the ability to sustain leaf gas exchange under heat stress has a direct relationship with heat tolerance. During the vegetative stage, high day temperature can cause damage to compensated leaf photosynthesis, reducing CO_2 assimilation rates. Increased temperatures curtail photosynthesis and increase CO_2 transfer conductance between intercellular spaces and carboxylation sites. Stomatal conductance (gs) and net photosynthesis (Pn) are inhibited by moderate heat stress in many plant species due to decreases in the activation state of rubisco. Although with an increase in temperature rubisco catalytic activity increases, a low affinity of the enzyme for CO_2 and its dual nature as an oxygenase limits the possible increases in Pn. For example, in maize the Pn was inhibited at leaf temperatures above 38°C and inhibition was much more severe when temperature was increased abruptly rather than gradually. However, this inhibition was independent of stomatal response to high temperature (Crafts-Brander and Salvucci, 2002). Despite observed negative effects of high temperature, the optimum temperature for leaf photosynthesis is likely to increase with elevated levels of atmospheric CO_2. Several studies have concluded that CO_2-induced increases in crop yields are much more plausible in warm- than in cool-season crops. Thus, despite its other potential negative implications, global warming may not greatly affect the overall Pn.

A well-known consequence of elevated temperature in plants is the damage caused by heat-induced imbalance in photosynthesis and respiration; in general the rate of photosynthesis decreases while dark- and photo-respiration rates increase considerably under high temperatures. Also, rate of biochemical reactions decreases and enzyme inactivation and denaturation take place as the temperature increases leading to severely reduced photosynthesis. However, the magnitude of such alterations in response to heat stress differs with species and genotypes. Furthermore, it has been determined that the photosynthetic CO_2 assimilation rate is less affected by heat stress in developing leaves than in completely developed leaves. Heat stress normally decreases the duration of developmental phases leading to smaller organs, reduced light perception and carbon assimilation processes including transpiration, photosynthesis and respiration (Stone, 2001). Nonetheless, photosynthesis is considered as the physiological process most sensitive to high temperatures, and that rising atmospheric CO_2 content will drive temperature increases in many already stressful environments. This CO_2-induced increase in plant high-temperature tolerance may have a substantial impact on both the productivity and distribution of many crop species in future.

Assimilate partitioning

Under low to moderate heat stress, a reduction in source and sink activities may occur leading to severe reductions in growth, economic yield and harvest index. Assimilate partitioning, taking place via apoplastic and symplastic pathways under high temperatures has significant effects on transport and transfer processes in plants (Taiz and Zeiger, 2006). However, considerable genotypic variation exists in crop plants for assimilate partitioning, as for example among wheat genotypes. To elucidate causal agents of reduced grain filling in wheat under high temperatures, examined three main components of the plant system including source (flag leaf blade), sink (ear), and transport pathway (peduncle). It was determined that photosynthesis had a broad temperature optimum from 20 to 30°C; however it declined rapidly at temperatures above 30°C. The rate of 14°C assimilate movement out of the flag leaf (phloem loading), was optimum around 30°C, however, the rate of movement through the stem was independent of temperature from 1 to 50°C. It was concluded that, at least in wheat, temperature effects on translocation result indirectly from temperature effects on source and sink activities. From such results, increased mobilization efficiency of reserves from leaves, stem or other plant parts has been suggested as a potential strategy to improve grain filling and yield in wheat under heat stress. This suggestion, however, is based on present limited knowledge of physiological basis of assimilate partitioning under high temperature stress. Further investigation in this area may lead to improved crop production efficiency under high-temperature stress.

Cell membrane thermostability

Sustained function of cellular membranes under stress is pivotal for processes such as photosynthesis and respiration. Heat stress accelerates the kinetic energy and movement of molecules across membranes thereby loosening chemical bonds within molecules of biological membranes. This makes the lipid bilayer of biological membranes more fluid by either denaturation of proteins or an increase in unsaturated fatty acids. The integrity and functions of biological membranes are sensitive to high temperature, as heat stress alters the tertiary and quaternary structures of membrane proteins. Such alterations enhance the permeability of membranes, as evident from increased loss of electrolytes. The increased solute leakage, as an indication of decreased cell membrane thermostability (CMT), has long been used as an indirect measure of heat-stress tolerance in diverse plant species, including soybean , potato and tomato, wheat (Blum *et al*., 2001), cotton , sorghum, cowpea and barley (Wahid and Shabbir, 2005). Electrolyte leakage is influenced by plant/tissue age, sampling organ, developmental stage, growing season, degree of hardening and plant species. In maize, injuries to plasmalemma due to heat stress were much less severe in developing than in mature leaves. It was determined that an increase in saturated fatty acids in mature leaves elevated melting temperature of plasma membranes and thus reducing heat tolerance of the plant. In *Arabidopsis* plants grown under high temperature, total lipid content in membranes decreased to about one-half and the ratio of unsaturated to saturated fatty acids decreased to one-third of the levels at normal temperatures. It should be noted, however, that in some species heat tolerance does not correlate with the degree of lipid saturation, suggesting that factors other than membrane stability might be limiting the growth at high temperatures.

The relationship between CMT and crop yield under high temperatures may vary from plant to plant and invokes for study of individual crops before using it as an important physiological selection criterion. For example, whereas a significant relationship between CMT and yield was observed in a few plant species such as sorghum, no such relationship was observed in soybean or wheat. Thus, the major cause(s) of yield suppression under heat stress remain largely elusive and deserve further experimentation.

Hormonal changes

Plants have the ability to monitor and adapt to adverse environmental conditions, though the degree of adaptability or tolerance to specific stresses varies among species and genotypes. Hormones play an important role in this regard. Cross-talk in hormone signaling reflects an organism's ability to integrate different inputs and respond appropriately. Hormonal homeostasis, stability, content, biosynthesis and compartmentalization are altered under heat stress (Maestri *et al.*, 2002).

Abscisic acid (ABA) and ethylene (C_2H_4), as stress hormones, are involved in the regulation of many physiological properties by acting as signal molecules. Different environmental stresses, including high temperature, result in increased levels of ABA. For example, recently it was determined that in creeping bentgrass (*Agrostis palustris*), ABA level did not rise during heat stress, but it accumulated upon recovery from stress suggesting a role during the latter period (Larkindale and Huang, 2005). However, the action of ABA in response to stress involves modification of gene expression. Analysis of ABA-responsive promoters revealed several potential *cis*- and *trans*-acting regulatory elements. ABA mediates acclimation/adaptation of plants to desiccation by modulating the up- or down-regulation of numerous genes (Xiong *et al.*, 2002). Under field conditions, where heat and drought stresses usually coincide, ABA induction is an important component of thermotolerance, suggesting its involvement in biochemical pathways essential for survival under heat-induced desiccation stress. Other studies also suggest that induction of several HSPs (e.g., HSP70) by ABA may be one mechanism whereby it confers thermotolerance. More so, heat shock transcription factor 3 acts synergistically with chimeric genes with a small HSP promoter, which is ABA inducible. A gaseous hormone, ethylene regulates almost all growth and developmental processes in plants, ranging from seed germination to flowering and fruiting as well as tolerance to environmental stresses. Measurement of the rate of ethylene released per unit amount of tissue provides information on the relative changes in cellular concentration of C_2H_4. However, the levels of ethylene or the enzymes involved in ethylene biosynthesis vary at different time intervals during the day. For instance, the endogenous concentration of 1-amino-cyclopropane-1-carboxylic acid (ACC), a precursor of ethylene biosynthesis, measured at predawn and at maximum solar radiation during a summer drought in rosemary (*Rosmarinus officinalis*) showed a sharp distinction between the two times, which was positively correlated with the intensity of incident solar radiations (Munne-Bosch *et al.*, 2002).

Heat stress changes ethylene production differently in different plant species (Arshad and Frankenberger, 2002). For example, while ethylene production in wheat leaves was inhibited slightly at 35°C and severely at 40°C, in soybean ethylene production in hypocotyls increased by increasing temperature up to 40°C and it showed inhibition at 45°C. Despite the fact that ACC accumulated in both species at 40°C, its conversion into ethylene occurred only in soybean hypocotyls but not in wheat. Wheat leaves transferred to 18°C followed by a short exposure to 40°C showed an increase in ethylene production after 1 h lag period, possibly due to conversion of accumulated ACC to ethylene during that period. Similarly, creeping bentgrass showed ethylene production upon recovery, but not when under heat stress (Larkindale and Huang, 2005). Temperatures up to 35°C have been shown to increase ethylene production and ripening of propylene-treated kiwifruit, but temperature above 35°C inhibits ripening by inhibiting ethylene production, although respiration continues until the tissue disintegration. In pepper (*Piper nigrum*), increase in the level of ACC was positively correlated with high temperatures. Exposure of imbibed sunflower seed to 45°C for 48 h induced a state of thermo dormancy, which appeared to associate with the loss of seed's ability to convert ACC to ethylene. However, treatment with 2.5 ml Methephon or 55 ml L^{-1} ethylene improved germination of the seed at 25°C. Ethylene may overcome the inhibitory effect of high temperature on thermosensitive lettuce seed due to increased mannanase activity, which helps weakening of the endosperm and facilitates germination. High temperature-induced abscission of reproductive organs relates to an increased ACC level; this is accompanied with both reduced levels and transport capacity of auxins to reproductive organs. The effect of pre-harvest temperature on ripening characteristics of shaded and sun-exposed apple fruits indicated that the former treatment produced up to 90% less ethylene than the latter (Klein *et al.*, 2001). In maize, ethylene production was highest at the top ear and lowest at the middle ear, suggesting that ethylene plays an important role in assimilates partitioning to grain filling.

Among other hormones, salicylic acid (SA) has been suggested to be involved in heat-stress responses elicited by plants. SA is an important component of signaling pathways in response to systemic acquired resistance (SAR) and the hypersensitive response (HR). SA stabilizes the trimers of heat shock transcription factors and aids them bind heat shock elements to the promoter of heat shock related genes. Long term thermotolerance can be induced by SA, in which both Ca^{2+} homeostasis and antioxidant systems are thought to be involved (Wang and Li, 2006b). Sulphosalicylic acid (SSA), a derivative of SA, treatment can effectively remove H_2O_2 and increase heat tolerance. In this regard, catalase (CAT) plays a key role in removing H_2O_2 in cucumber (*Cucumus sativus*) seedlings treated with SSA under heat stress. In contrast, while glutathione peroxidase (GPX), ascorbate peroxidase (APX) and glutathione reductase (GR) showed higher activities in all SSA treatments under heat stress, they were not key enzymes in removing H_2O_2 (Shi *et al*., 2006). Thermotolerance of plants also can be enhanced by spraying leaves with acetyl-SA . Likewise, methyl salicylate (MeSA), a derivative of SA, has multiple functions. In addition to acting as signal molecule, it gives thermotolerance to holm oak (*Quercus ilex*) by enhanced xanthophylls de-epoxidation and content of ascorbate, antioxidants and tocopherol in leaves (Wang and Li, 2006a).

The effects of gibberellins and cytokinins on high temperature tolerance are opposite to that of ABA. An inherently heat-tolerant dwarf mutant of barley impaired in the synthesis of gibberellins was repaired by application of gibberellic acid, whereas application of triazole paclobutrazol, a gibberellins antagonist, conferred heat tolerance. In creeping bentgrass, the levels of various cytokinins, zeatin (Z), zeatin riboside (ZR), dihydrogen zeatin riboside (DHZR) and isopentinyl adenosine (IPA), showed dramatic decreases by the 5th day in root and 10th day in shoot, which were correlated with decreased dry matter production (Liu and Huang, 2005). In a dwarf wheat variety, high temperature-induced decrease in cytokinin content was found to be responsible for reduced kernel filling and its dry weight.

Another class of hormones, brassinosteroids have recently been shown to confer thermotolerance to tomato and oilseed rape (*Brassica napus*), but not to cereals. The potential roles of other phytohormones in plant thermo tolerance are yet unknown.

Secondary metabolites

Most of the secondary metabolites are synthesized from the intermediates of primary carbon metabolism via phenylpropanoid, shikimate, mevalonate or methyl erythritol phosphate (MEP) pathways. High-temperature stress induces production of phenolic compounds such as flavonoids and phenylpropanoids. Phenylalanine ammonia-lyase (PAL) is considered to be the principal enzyme of the phenylpropanoid pathway. Increased activity of PAL in response to thermal stress is considered as the main acclamatory response of cells to heat stress. Thermal stress induces the biosynthesis of phenolics and suppresses their oxidation, which is considered to trigger the acclimation to heat stress for example as in watermelon, *Citrulus vulgaris* (Rivero *et al*., 2001).

Carotenoids are widely known to protect cellular structures in various plant species irrespective of the stress type (Wahid, 2007). For example, the xanthophyll cycle (the reversible interconversion of two particular carotenoids, violaxanthin and zeaxanthin) has evolved to play this essential role in photoprotection. Since zeaxanthin is hydrophobic, it is found mostly at the periphery of the light harvesting complexes, where it functions to prevent peroxidative damage to the membrane lipids triggered by ROS (Horton, 2002). Recent studies have revealed that carotenoids of the xanthophylls family and some other terpenoids, such as isoprene or tocopherol, stabilize and photoprotect the lipid phase of the thylakoid membranes. When plants are exposed to potentially harmful environmental conditions, such as strong light and/or elevated temperatures, the xanthophylls including violaxanthin, antheraxanthin and zeaxanthin partition between the light-harvesting complexes and the lipid phase of the thylakoid membranes. The resulting interaction of the xanthophyll molecules and the membrane lipids brings about a decreased fluidity (thermostability) of membrane and a lowered susceptibility to lipid peroxidation under high temperatures.

Phenolics, including flavonoids, anthocyanins, lignins, etc., are the most important class of secondary metabolites in plants and play a variety of roles including tolerance to abiotic stresses (Wahid, 2007). Studies suggest that accumulation of soluble phenolics under heat stress was accompanied with increased phenyl ammonia lyase (PAL) and decreased peroxidase and polyphenol lyase activities (Rivero *et al*., 2001). Anthocyanins, a subclass of flavonoid compounds, are greatly modulated in plant tissues by prevailing high temperature; low temperature increases and elevated temperature decreases their concentration in buds and fruits. For example, high temperature decreases synthesis of anthocyanins in reproductive parts of red apples chrysanthemums and asters. One of the causes of low anthocyanin concentration in plants at high temperatures is a decreased rate of its synthesis and stability. On the other hand, vegetative tissues under high temperature stress show an accumulation of anthocyanins including rose and sugarcane leaves. It has been suggested that in addition to their role as UV screen, anthocyanins serve to decrease leaf osmotic potential, which is linked to increased uptake and reduced transpirational loss of water under environmental stresses including high temperature. These properties may enable the leaves to respond quickly to changing environmental conditions.

Isoprenoids, another class of plant secondary products, are synthesized via mevalonate pathway. Being of low molecular weight and volatile in nature, their emission from leaves has been reported to confer heat-stress tolerance to photosynthesis apparati in different plants. Studies have revealed that their biosynthesis is cost effective. While deriving considerable amount of photosynthates, they show compensatory benefits as to heat tolerance (Funk *et al.*, 2004). Plants capable of emitting greater amounts of isoprene generally display better photosynthesis under heat stress, thus there is a relationship between isoprene emission and heat-stress tolerance. Sharkey (2005) opined that isoprene production protects the PSII from the damage caused by ROS, including H_2O_2, produced during heat-induced oxygenase action of rubisco, even though the photosynthetic rate approaches zero. It is proposed that endogenous production of isoprene protects the biological membranes from damaging effects by directly reacting with oxygen singlets ($^1O_2$) by means of isoprene-conjugate double bond.

Molecular responses

Oxidative stress and antioxidants

In addition to tissue dehydration, heat stress may induce oxidative stress. For example, generation and reactions of activated oxygen species (AOS) including singlet oxygen ($^1O_2$), superoxide radical ($O2^{\cdot -}$), hydrogen peroxide (H_2O_2) and hydroxyl radical ($OH^{\cdot}$) are symptoms of cellular injury due to high temperature (Liu and Huang, 2000). AOS cause the autocatalytic peroxidation of membrane lipids and pigments thus leading to the loss of membrane semi-permeability and modifying its functions (Xu *et al.*, 2006). Superoxide radical is regularly synthesized in the chloroplast and mitochondrion and some quantities are also produced in microbodies (Fig. 2). The scavenging of $O2^{\cdot -}$ by superoxide dismutase (SOD) results in the production of H_2O_2, which is removed by APX or CAT. However, both $O2^{\cdot -}$ and H_2O_2 are not as toxic as the ($.OH^-$), which is formed by the combination of $O2^{\cdot -}$ and H_2O_2 in the presence of trace amounts of Fe^{2+} and Fe^{3+} by the Haber–Weiss reaction. The $OH^{\cdot}$ can damage chlorophyll, protein, DNA, lipids and other important macromolecules, thus fatally affecting plant metabolism and limiting growth and yield (Sairam and Tyagi, 2004). As depicted in Fig. 3, plants have developed a series of both enzymatic and non-enzymatic detoxification systems to counteract AOS, thereby protecting cells from oxidative damage (Sairam and Tyagi, 2004). For example, over expression of SOD in plants affect a number of physiological phenomena, which include the removal of H_2O_2, oxidation of toxic reductants, biosynthesis and degradation of lignin in cell walls, auxin catabolism, defensive responses to wounding, defense against pathogen or insect attack, and some respiratory processes. More specifically, expression and activation of APX is related to the appearance of physiological injuries caused in plants by thermal stress (Mazorra *et al.*, 2002).

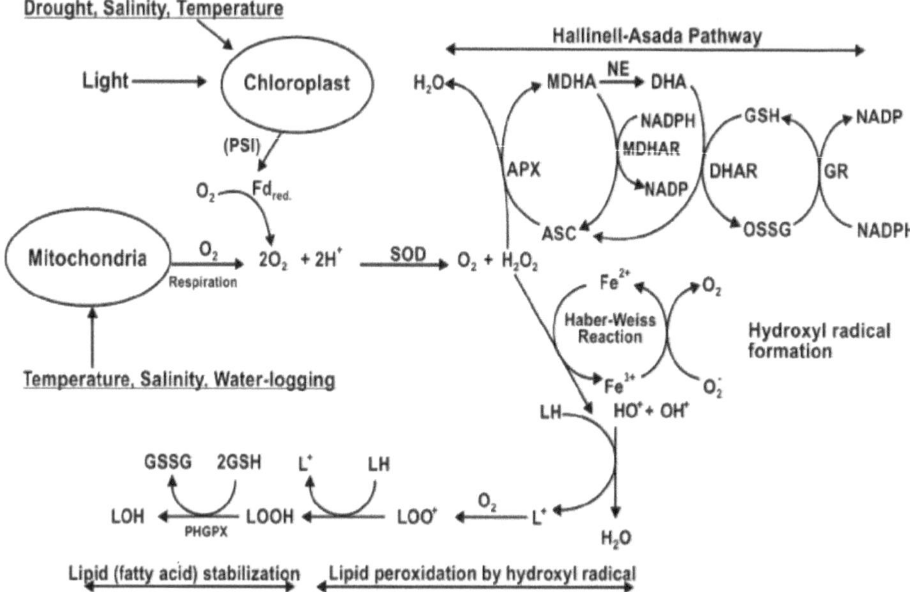

Fig.3: Schematic presentation for generation and scavenging of superoxide radical, hydrogen peroxide, hydroxyl radical-induced lipid peroxidation and glutathione peroxidase-mediated fatty acid stabilization under environmental stresses. APX, ascorbate peroxidase; ASC, ascorbate; DHA, dehydroascorbate; DHAR, dehydroascorbate reductase; Fd, ferredoxin; GR, glutathione reductase; GSH, red glutathione; GSSG, oxi-glutathione; HO, hydroxyl radical; LH, lipid; L, LOO; LOOH, unstable lipid radicals and hydroperoxides; LOH, stable lipid (fatty acid); MDHA, monodehydro-ascorbate; MDHAR, mono dehydro-ascorbate reductase; NE, non-enzymatic reaction; PHGPX, phospholipid-hydroperoxide glutathione peroxidase; SOD, superoxide dismutase. Reproduced with permission from Sairam and Tyagi (2004).

Decrease in antioxidant activity in stressed tissues results in higher levels of AOS that may contribute to injury. Protection against oxidative stress is an important component in determining the survival of a plant under heat stress. Studies on heat-acclimated versus non-acclimated cool season turfgrass species suggested that the former had lower production of ROS as a result of enhanced synthesis of ascorbate and glutathione (Xu *et al.*, 2006).Available data suggest that some signaling molecules may cause an increase in the antioxidant capacity of cells. Certainly further research is necessary to identify the signaling molecules, which enhanced production of antioxidants in cells exposed to heat stress.

Stress proteins

Expression of stress proteins is an important adaptation to cope with environmental stresses. Most of the stress proteins are soluble in water and therefore contribute to stress tolerance presumably via hydration of cellular structures. Although heat shock proteins (HSPs) are exclusively implicated in heat-stress response, certain other proteins are also involved.

Heat shock proteins

Synthesis and accumulation of specific proteins are ascertained during a rapid heat stress, and these proteins are designated as HSPs. Increased production of HSPs occurs when plants experience either abrupt or gradual increase in temperature. Induction of HSPs seems to be a universal response to temperature stress, being observed in all organisms ranging from bacteria to human. Plants of arid and semiarid regions may synthesize and accumulate substantial amounts of HSPs. Certain HSPs are also expressed in some cells under cyclic or developmental control. In higher plants, HSPs are usually induced under heat shock at any stage of development and major HSPs are highly homologous among distinct organisms. HSP-triggered thermotolerance is attributed to the observations that (a) their induction coincides with the organism under stress, (b) their biosynthesis is extremely fast and intensive, and (c) they are induced in a wide variety of cells and organisms.

Three classes of proteins, as distinguished by molecular weight, account for most HSPs, *viz.*, HSP90, HSP70 and low molecular weight proteins of 15–30 kDa. The special importance of small HSPs in plants is suggested by their unusual abundance and diversity. The proportions of these three classes differ among plant species. HSP70 and HSP90 mRNAs can increase ten-fold, while low molecular weight (LMW) HSPs can increase as much as 200-fold. Other proteins, such as 110 kDa polypeptides and ubiquitin, though less important, are also considered to be HSPs. All small-HSPs in plants are encoded by six nuclear gene families, each gene family corresponding to proteins found in distinct cellular compartments like cytosol, chloroplast, endoplasmic reticulum (ER), mitochondria and membranes. Some nuclear-encoded HSPs accumulate in the cytosol at low (27°C) and high (43°C) temperatures, but they accumulate in chloroplast at moderate (37°C) temperatures.

Immuno-localization studies have determined that HSPs normally associate with particular cellular structures, such as cell wall, chloroplasts, ribosomes and mitochondria (Yang *et al.*, 2006). When maize, wheat and rye seedlings were subjected to heat shocks (42°C), whereas five mitochondrial LMW-HSPs (28, 23, 22, 20 and 19 kDa) were expressed in maize, only one (20 kDa) was expressed in wheat and rye, suggesting the reason for higher heat tolerance in maize than in wheat and rye (Korotaeva *et al.*, 2001). In another study, a heat-tolerant maize line (ZPBL-304) exhibited increased amounts of chloroplast protein synthesis elongation factor under heat stress, which was related to the development of heat tolerance (Moriarty *et al*., 2002). Presence of HSPs can prevent denaturation of other proteins caused by high temperature. The ability of small HSPs to assemble into heat shock granules (HSGs) and their disintegration is a prerequisite for survival of plant cells under continuous stress conditions at sub lethal temperatures.

In response to high temperatures, specific HSPs have been identified in different plant species. For example, HSP68, which is localized in mitochondria and normally constitutively expressed, was determined to have increased expression under heat stress in cells of potato, maize, tomato, soybean and barley. Another HSP identified in maize is a nucleus-localized protein, HSP101, which belongs to the campylobacter invasion antigen B (CiaB) protein sub-family, whose members promote the renaturation of protein aggregates, and are essential for the induction of thermotolerance. Levels of HSP101 increased in response to heat shock, more abundantly in developing tassel, ear, silks, endosperm and embryo and less abundantly in vegetative and floral meristematic regions, mature pollen, roots and leaves (Young *et al*., 2001). In addition, heat treatment increases the level of other maize HSPs, which are associated with plant ability to withstand heat stress. For example, A 45-kDa HSP was found to play a major role in recovery from heat stress. Different studies have determined that cytosolic accumulations of nuclear encoded chloroplast proteins were reversible (within 3h) following return to normal growth temperature in many seed bearing species.

There are considerable variations in patterns of HSP production in different species and even among genotypes within species. Furthermore, the ability to synthesize characteristic proteins at 40°C and the intensity and duration of synthesis differ among various tissues examined within the same plant. Fast accumulation of HSPs in sensitive organs/tissues can play an important role in protection of metabolic apparati of the cell, thereby acting as a key factor for plant's adaptation to, and survival under, stress. In different plant species, elongating segments of primary roots exhibited a strong ability to synthesize nucleus-localized HSPs, which had roles in thermotolerance. Under heat-stress conditions, while synthesis of a typical set of HSPs was induced in male tissues of maize flowers undergoing pollen formation, the mature pollen showed no synthesis of HSPs and thus were sensitive to heat stress and responsible for the failure of fertilization at high temperatures. In another study, germinating maize pollen showed induction of 64 and 72 kDa peptides of HSPs under heat stress whilst in the whole plant expression of a 45 kDa HSP was responsible for the heat tolerance. Similar variation in the expression of HSPs can be found in other plant species. The mechanism by which HSPs contribute to heat tolerance is still enigmatic though several roles have been ascribed to them. Many studies assert that HSPs are molecular chaperones insuring the native configuration and functionality of cell proteins under heat stress. There is considerable evidence that acquisition of thermotolerance is directly related to the synthesis and accumulation of HSPs (Bowen *et al.*, 2002). For instance, HSPs provide for new or distorted proteins to fold into shapes essential for their normal functions. They also help shuttling proteins from one compartment to another and transporting old proteins to "garbage disposals" inside the cell. Among others, HSP70 has been extensively studied and is proposed to have a variety of functions such as protein translation and translocation, proteolysis, protein folding or chaperoning, suppressing aggregation, and reactivating denatured proteins (Zhang *et al.*, 2005). Recently, dual role of LMW HSP21 in tomato has been described as protecting PSII from oxidative damage and involvement in fruit color change during storage at low temperatures (Neta-Sharir *et al.*, 2005). In many plant species, thermotolerance of cells and tissues after a heat stress is pretty much dependent upon induction of HSP70, though HSP101 has also been shown to be essential. One hypothesis is that HSP70 participates in ATP-dependent protein unfolding or assembly/disassembly reactions and it prevents protein denaturation during heat stress. Evidence for the general protective roles of HSPs comes from the fact that mutants unable to synthesize them or the cells in which HSP70 synthesis is blocked or inactivated are more susceptible to heat injury (Burke, 2001). The level of HSP22, a member of the plant small HSP super-family, remained high under continuous heat stress. LMW-HSPs may play structural roles in maintaining cell membrane integrity. Localization of LMW-HSPs in chloroplast membranes further suggested that these proteins protect the PSII from adverse effects of heat stress and play a role in photosynthetic electron transport (Barua *et al.*, 2003).

Other heat induced proteins

Besides HSPs, there are a number of other plant proteins, including ubiquitin cytosolic Cu/Zn-SOD and Mn-POD, whose expressions are stimulated upon heat stress. For example, in *Prosopis chilensis* and soybean under heat stress, ubiquitin and conjugated-ubiquitin synthesis during the first 30 min of exposure emerged as an important mechanism of heat tolerance (Ortiz and Cardemil, 2001). Some studies have shown that heat shock induces Mn peroxidase, which plays a vital role in minimizing oxidative damages. In a study, a number of osmotin like proteins induced by heat and nitrogen stresses, collectively called Pir proteins, were found to be over expressed in the yeast cells under heat stress conferring them resistance to tobacco osmotin (an antifungal). Late embryogenesis abundant (LEA) proteins can prevent aggregation and protect the citrate synthase from desiccating conditions like heat- and drought-stress (Goyal *et al.*, 2005). Using proteomics tool, Majoul *et al.* (2003) determined enhanced expressions of25 LEA proteins in hexaploid wheat during grain filling. Geranium leaves exposed to drought and heat stress revealed expression of dehydrin proteins (25–60 kDa), which indicated a possible linkage between drought and heat-stress tolerance Recently, three low-molecular-weight dehydrins have been identified in sugarcane leaves with increased expression in response to heat stress. In essence, expression of stress proteins is an important adaptation toward heat-stress tolerance by plants. Of these, expression of low and high molecular weight HSPs, widely reported in a number of plant species, is the most important one. These proteins show organelle- and tissue-specific expression with deduced function like chaperones, folding and unfolding of cellular proteins, and protection of functional sites from the adverse effects of high temperature .Among other stress proteins, expression of ubiquitin, Pir proteins ,LEA and dehydrins has also been established under heat stress. A main function of these proteins appears to be protection of cellular and sub-cellular structures against oxidative damage and dehydrative forces.

Mechanism of heat tolerance

Plants manifest different mechanisms for surviving under elevated temperatures, including long-term evolutionary phonological and morphological adaptations and short-term avoidance or acclimation mechanisms such as changing leaf orientation, transpirational cooling, or alteration of membrane lipid compositions. In many crop plants, early maturation is closely correlated with smaller yield losses under high temperatures, which may be attributed to the engagement of an escape mechanism (Adams *et al.*, 2001). Plant's immobility limits the range of their behavioral responses to environmental cues and places a strong emphasis on cellular and physiological mechanisms of adaptation and protection. Also, plants may experience different types of stress at different developmental stages and their mechanisms of response to stress may vary in different tissues (Queitsch *et al.*, 2000). The initial stress signals (e.g., osmotic or ionic effects, or changes in temperature or membrane fluidity) would trigger downstream signaling processes and transcription controls, which activate stress-responsive mechanisms to reestablish homeostasis and protect and repair damaged proteins and membranes. Inadequate responses at one or more steps in the signaling and gene activation processes might ultimately result in irreversible damages in cellular homeostasis and destruction of functional and structural proteins and membranes, leading to cell death. Even plants growing in their natural distribution range may experience high temperatures that would be lethal in the absence of this rapid acclimation response. Furthermore, because plants can experience major diurnal temperature fluctuations, the acquisition of thermotolerance may reflect a more general mechanism that contributes to homeostasis of metabolism on a daily basis. Mild stress episodes, however, should be viewed as the acceleration of a program linked to the normal termination of phytomere production during the plant cycle, rather than as an abrupt event linked to stress.

Elucidating the various mechanisms of plant response to stress and their roles in acquired stress tolerance is of great practical and basic importance. Some major tolerance mechanisms, including ion transporters, osmoprotectants, free-radical scavengers, late embryogenesis abundant proteins and factors involved in signaling cascades and transcriptional control are essentially significant to counteract the stress effects (Wang *et al.*, 2004). Series of changes and mechanisms, beginning with the perception of heat and signaling and production of metabolites that enable plants to cope with adversaries of heat stress, have been proposed (Fig. 4). Heat stress effects are notable at various levels, including plasma membrane and biochemical pathways operative in the cytosol or cytoplasmic organelles (Sung *et al.*, 2003). Initial effects of heat stress, however, are on plasmalemma, which shows more fluidity of lipid bilayer under stress. This leads to the induction of Ca2+ influx and cytoskeletal reorganization, resulting in the upregulation of mitogen activated protein kinases (MAPK) and calcium dependent protein kinase (CDPK). Signaling of these cascades at nuclear level leads to the production of antioxidants and compatible osmolytes for cell water balance and osmotic adjustment. Production of ROS in the organelles (e.g., chloroplast and mitochondria) is of great significance for signaling as well as production of antioxidants (Bohnert *et al.*, 2006). The antioxidant defense mechanism is a part of heat-stress adaptation, and its strength is correlated with acquisition of thermotolerance (Maestri *et al.*, 2002). Accordingly, in a set of wheat genotypes, the capacity to acquire thermotolerance was correlated with activities of CAT and SOD, higher ascorbic acid content, and less oxidative damage (Sairam and Tyagi, 2004). One of the most closely studied mechanisms of thermotolerance is the induction of HSPs, which, as described in above, comprise several evolutionarily conserved protein families. However, each major HSP family has a unique mechanism of action with chaperonic activity. The protective effects of HSPs can be attributed to the network of the chaperone machinery, in which many chaperones act in concert. An increasing number of studies suggest that the HSPs/chaperones interact with other stress-response mechanisms (Wang *et al.*, 2004). The HSPs/ chaperones can play a role in stress signal transduction and gene activation (Nollen and Morimoto, 2002) as well as in regulating cellular redox state. They also interact with other stress-response mechanisms such as production of osmolytes (Diamant *et al.*, 2001) and antioxidants (Panchuk *et al.*, 2002). Membrane lipid saturation is considered as an important element in high temperature tolerance. In a mutant wheat line with increased heat resistance heat treatment increased relative quantities of linolenic acid (among galactolipids) and *trans*- 3-hexaldecanoic acid (among phospholipids), when compared with the wild type. The contribution of lipid and protein components to membrane function under heat stress needs further investigation. Localization of LMW-HSPs with chloroplastic membranes upon heat stress suggests that they play a role in protecting photosynthetic electron transport.

An important component of thermotolerance is changes in gene expression.

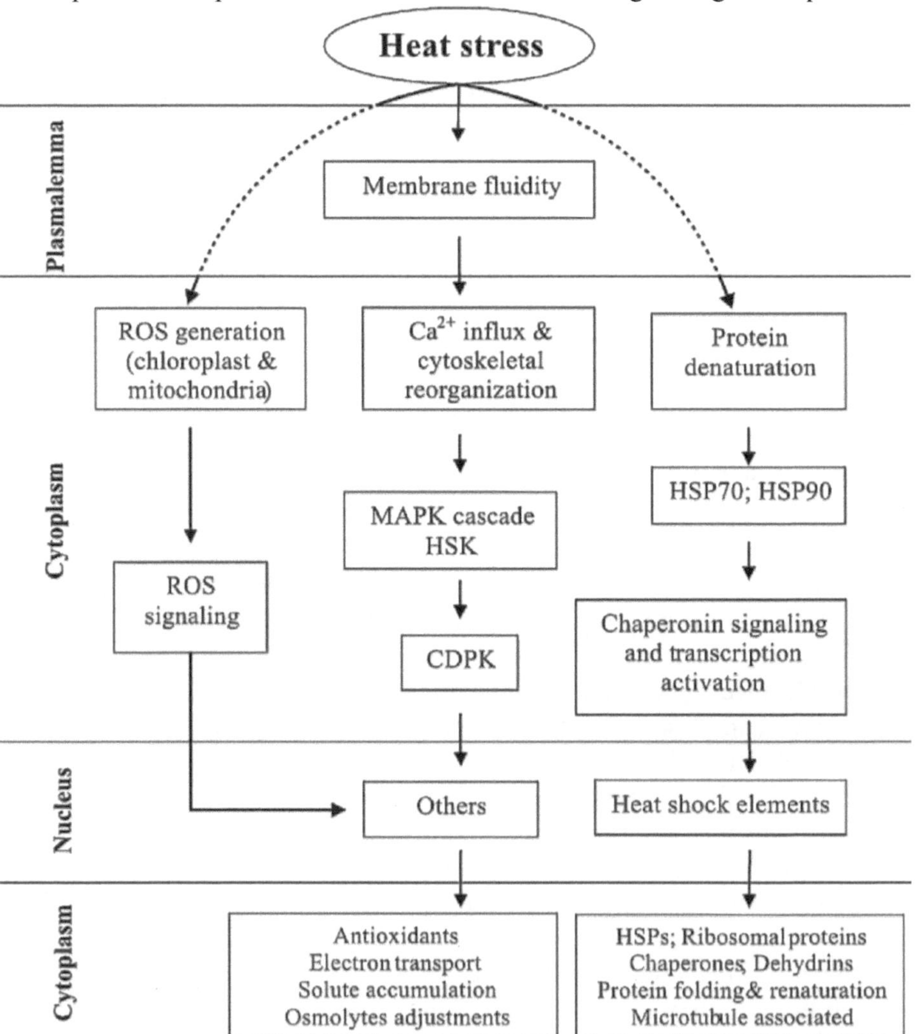

Fig. 4: Proposed heat-stress tolerance mechanisms in plants. MAPK, mitogen activated protein kinases; ROS, reactive oxygen species; HAMK, heat shock activated MAPK; HSE, heat shock element; HSPs, heat shock proteins; CDPK, calcium dependent protein kinase; HSK, histidine kinase. Partly adopted from Sung et al. (2003).

Heat stress is known to swiftly alter patterns of gene expression (Yang *et al*., 2006), inducing expression of the HSP complements and inhibiting expression of many other gene. The mRNAs encoding non heat- stress-induced proteins are destabilized during heat stress. Heat stress may also inhibit splicing of some mRNAs. Earlier it was hypothesized that HSP-encoding mRNAs could not be processed properly due to the absence of introns in the corresponding genes. Subsequently it was shown that some HSP-encoding genes have introns and, under heat-stress conditions, their mRNAs were correctly spliced. However, the mechanism of preferential post-transcription modification and translation of HSP-encoding mRNA under heat stress remains yet to be elucidated.

Acquired thermotolerance

Thermotolerance refers to the ability of an organism to cope with excessively high temperatures. It has long been known that plants, like other organisms, have the ability to acquire thermotolerance rather rapidly, may be within hours, so to survive under otherwise lethal high temperatures. The acquisition of thermotolerance is an autonomous cellular phenomenon and normally results from prior exposure to a conditioning pretreatment, which can be a short but sub lethal high temperature. The acquisition of high level of thermotolerance protects cells and organisms from a subsequent lethal heat stress. Thermotolerance can also be induced by a gradual increase in temperature to lethal highs, as would be experienced under natural conditions, and induction in this way involves a number of processes. Using Arabidopsis mutants, it was shown that, apart from heat shock proteins (HSP32 and HSP101), ABA, ROS and SA pathways are involved in the development and maintenance of acquired thermotolerance (Charng *et al*., 2006). Adaptive mechanisms that protect cells from protoxic effects of heat stress are key factors in acquisition of thermotolerance. Examination of the adverse effects caused by temperature extremes can reveal useful information, in particular as heat-stress responses in plants are similar to those of other stresses, including cold and drought (Rizhsky *et al*., 2002). The heat shock response (HSR), defined as a transient reprogramming of gene expression, is a conserved biological reaction of cells and organisms to elevated temperatures. HSR has been of great interest for studying molecular mechanisms of stress tolerance and regulation of gene expression in plants. The temperature for the induction of HSR coincides with optimum growth temperature for any given species, which is normally 5–10 °C above normal thermic conditions. The features of this response include induction of HSPs and subsequently acquisition of a higher level of thermotolerance. The transient synthesis of HSPs suggests that the signal triggering the response is lost, inactivated or no longer recognized under conditions of long-term heat treatment. The involvement of HSPs in heat-stress tolerance is a logical model but direct support for function of HSPs in promoting thermotolerance has been difficult to obtain (Sch¨offl, 2005).

In summary, thermotolerance acquired by plants via autonomous synthesis of pertinent compounds or induced via gradual exposure to heat episodes, though cost intensive, is an important and potentially vital strategy. This phenomenon is principally related to display of heat shock response and accomplished by reprogramming of gene expression, allowing plants to cope with the heat stress.

Temperature sensing and signaling

Perception of stress and relay of the signal for turning on adaptive response mechanisms are key steps towards plant stress tolerance. There are multiple stress perceptions and signaling pathways, some of which are specific while others may be involved in cross-talk at various steps (Chinnusamy *et al.*, 2004). General responses to stress involve signaling of the stress via the redox system. Chemical signals such as ROS, Ca^{2+} and plant hormones activate genomic re-programing via signal cascades (Suzuki and Mittler, 2006). Although the presence of a plant thermometer has not been established, it is suggested that changing membrane fluidity plays a central role in sensing and influencing gene expression both under high and low temperatures. This suggests that sensors are located in microdomains of membranes, which are capable of detecting physical phase transition, eventually leading to conformational changes and/or phosphorylation/ dephosphorylation cycles due to changes in temperature. In this regard, a model for temperature sensing and regulation of heat shock response integrates observed membrane alterations. Changes in the ratio of saturated to unsaturated fatty acid on the set point of temperature for the heat shock response (HSR) alters activities of HSFs. Rigidification of thylakoid membranes appears to invoke altered expression profiles of heat shock genes, suggesting that the temperature sensing mechanism may be located on the thylakoid membrane. The prospect of the thylakoid membrane acting as a heat sensor is physiologically crucial, because it is susceptible to temperature upshift, owing to its highly unsaturated character, and the presence of photosystems, which are fragile to temperature changes (Sung *et al.*, 2003).

Various signaling ions and molecules are involved in temperature sensing and signaling. As a signaling response to temperature stress, cytosolic Ca^{2+} sharply rises (Larkindale and Knight, 2002), which seems to be linked to the acquisition of tolerance possibly by transducing high temperature-induced signals to MAPK. MAPK cascades are important parts of signal transduction pathways in plants and thought to function ubiquitously in many responses to external signals (Kaur and Gupta, 2005). A heat-shock activated MAPK (HAMK) has been identified, the activation of which was triggered by apparent opposite changes in membrane fluidity coupled with cytoskeletal remodeling. Ca^{2+} influx and the action of Ca-dependent protein kinases (CDPK) have been closely correlated with the expression of HSPs (Sangwan and Dhindsa, 2002). However, another study suggested that Ca^{2+} is not required for production of HSPs in plants, despite the fact that heat stress induces uptake of Ca^{2+} and induction of some calmodulin (CaM) related genes. As a mediator of Ca^{2+} signal, CaM is activated by binding Ca^{2+}, inducing a cascade of regulatory events and regulation of many HSP genes (Liu *et al.*, 2003). Several studies have shown that Ca^{2+} is involved in the regulation of plant responses to various environmental stresses, including high temperature. Increasing cytosolic Ca^{2+} content under heat stress may alleviate heat injury, such as increased activity of antioxidants, turgor maintenance in the guard cells and enable plant cells to better survive.

Specific groups of potential signaling molecules like SA, ABA, $CaCl_2$, H_2O_2, and ACC may induce tolerance of plants to heat stress by reducing oxidative damage (Larkindale and Huang, 2004). Being molecules of somewhat novel interest in the last few years, H_2O_2 and NO have emerged to be central players in the world of plant cell signaling under stressful situations (Dat *et al.*, 2000). A protein phosphorylation cascade has been shown to be activated by H_2O_2 is a MAPK cascade.

Genetic improvement for heat-stress tolerance

Recent studies have suggested that plants experience oxidative stresses during the initial period of adjustment to any stress. Plant responses stress progress from general to specific. Specific responses require sustained expression of genes involved in processes specific to individual stresses (Yang *et al.*, 2006). These responses accommodate short-term reaction or tolerance to specific stresses. However, genome plasticity in plants, including genetic (e.g., directed mutation) and epigenetic (e.g., methylation, chromatin remodeling, histone acetylation) changes, allows long-term adaptation to environmental changes/ conditions. Such adaptations may be necessary for long-term survival or establishment of plant genotypes/species in particular environmental niches. Under agricultural systems, plants adaptation or their tolerance to environmental stresses can be manipulated by various approaches.

In general, the negative impacts of abiotic stresses on agricultural productivity can be reduced by a combination of genetic improvement and cultural practices. Genetic improvement entails development of cultivars which can tolerate environmental stresses and produce economic yield. Adjustment/ modifications in cultural practices, such as planting time, plant density, and soil and irrigation managements, however, can minimize stress effects, for example by synchronizing the stress sensitive stage of the plant with the most favorable time period of the season. In practice, to be successful in improving agricultural productivity in stress environments, both genetic improvement and adjustment in cultural practices must be employed simultaneously. Agriculturists have long been aware of desirable cultural practices to minimize adverse effects of environmental stresses on crop production. However, genetic improvement of crops for stress tolerance is relatively a new endeavor and has been considered only during the past 2–3 decades. Traditionally, most plant breeding programs have focused on development of cultivars with high yield potential in favorable (i.e., non stress) environments. Such efforts have been very successful in improving the efficiency of crop production per unit area and have resulted in significant increases in total agricultural production. However, genetic improvement of plants for stress tolerance can be an economically viable solution for crop production under stressful environments. The progress in breeding for stress tolerance depends upon an understanding of the physiological mechanisms and genetic bases of stress tolerance at the whole plant, cellular and molecular levels. Considerable information is presently available regarding the physiological and metabolic aspects of plant heat-stress tolerance, as discussed earlier. However, information regarding the genetic basis of heat tolerance is generally scarce, though the use of traditional plant breeding protocols and contemporary molecular biological techniques, including molecular marker technology and genetic transformation, have resulted in genetic characterization and/or development of plants with improved heat tolerance. In particular, the application of quantitative trait locus (QTL) mapping has contributed to a better understanding of the genetic relationship among tolerances to different stresses. In below, a summary of such efforts and progresses is presented and discussed.

Conventional breeding strategies

Physiological and genetic investigations indicate that most abiotic stress tolerance traits are complex, controlled by more than one gene, and highly influenced by environmental variation. The quantification of tolerance often poses serious difficulties. Direct selection under field conditions is generally difficult because uncontrollable environmental factors adversely affect the precision and repeatability of such trials. Often, no consistent high-temperature conditions can be guaranteed in field nurseries, as heat stress may or may not occur in the field. Furthermore, stress tolerance is a developmentally regulated, stage-specific phenomenon; tolerance at one stage of plant development may not be correlated with tolerance at other developmental stages. Individual stages throughout the ontogeny of the plant should be evaluated separately for the assessment of tolerance and for the identification, characterization and genetic manipulation of tolerance components. Moreover, species may show different sensitivity to heat stress at different developmental stages. For example, in tomato, though plants are sensitive to high temperatures throughout the plant ontogeny, flowering and fruit set are the most sensitive stages; fruit set is somewhat affected at day/night temperatures above 26/20°C and is severely affected above 35/26°C. Thus, partitioning of the tolerance into its developmental and physiological/genetic components may provide a better understanding of the plant's response to heat stress and facilitate development of plants with stress tolerance throughout the plant's life cycle.

A common method of selecting plants for heat-stress tolerance has been to grow breeding materials in a hot target production environment and identify individuals/lines with greater yield potential. Under such conditions, however, the presence of other stresses such as insect-pests has made the selection process very difficult, particularly during reproductive stage. A suggested approach has been to identify selection criteria during early stages of plant development, which may be correlated with heat tolerance during reproductive stages. Unfortunately, thus far no reliable criteria have been identified. A rather more effective approach has been development of glasshouses for heat tolerance screening. Theoretically, such nurseries can be utilized for screening throughout the plant life cycle, from seedling to reproductive stages. An advantage of glasshouse screening is that the required temperature conditions can be maintained consistently for the duration of the experiment. Also, because a key factor in screening for heat tolerance is maintaining high night temperatures, glasshouse nursery can provide such conditions more reliably than field nurseries. However, in many places in the world where high temperatures are a concern, such growth/glasshouse facilities are nonexistent or limited in size, precluding screening of large breeding populations. A major challenge in traditional breeding for heat tolerance is the identification of reliable screening methods and effective selection criteria to facilitate detection of heat-tolerant plants. Several screening methods and selection criteria have been developed/proposed by different researchers. For example, a heat tolerance index (HTI) for growth recovery after heat exposure was proposed for sorghum (Young *et al.*, 2001). The index is the ratio of the increase in coleoptile length after finite exposure to heat stress (e.g., 50°C) to the increase in coleoptiles length in the no-stress treatment. This approach allows a rapid and repeated recording of coleoptile length, which may be used to screen a large number of genotypes in a rather short period of time. Although this is a very cost effective and easy-to-assay technique of screening for heat tolerance, its correlation with performance under field conditions and its effectiveness in different crop species are yet unknown (Setimela *et al.*, 2005). In some crop species such as tomato, a strong positive correlation has been observed between fruit set and yield under high temperature. Thus, evaluation of germplasm to identify sources of heat tolerance has regularly been accomplished by screening for fruit set under high temperature. Furthermore, although poor fruit set at high temperature cannot be attributed to a single factor, decreases in pollen germination and/or pollen tube growth are among the most commonly reported factors. Therefore, pollen viability has been suggested as an additional indirect selection criterion for heat tolerance. Also, production of viable seed is often reduced under heat stress and thus high seed set has been arguably reported as an indication of heat tolerance.

Because pollen viability and fertility is adversely affected by high temperature, any type of fruit and seed production that may not need sexual hybridization and fertilization may provide heat tolerance. For example, apomixes, in particular the types that assure reproduction without the need for pollination may be very useful when developing cultivars for production under high temperatures. Through genetic engineering it may be possible to insert the cassette of genes needed to confer facultative apomixis. Currently, considerable research is underway to identify genes or enzymes that may be involved in production of apomixis (Albertini *et al*., 2005). Among many other traits which are affected by high temperature, the non-reproductive processes include photosynthetic efficiency, assimilate translocation, mesophyll resistance and disorganization of cellular membranes. Breeding to improve such traits under high temperatures may result in the development of cultivars with heat tolerance attributes.

Several other issues of concern when employing traditional breeding protocols to develop heat-tolerant crop plants are as follows:

1. Identification of genetic resources with heat tolerance attributes. In many plant species, for example soybeans and tomatoes, limited genetic variations exist within the cultivated species necessitating identification and use of wild accessions. However, often there are great difficulties in both the identification and successful use of wild accessions for stress tolerance breeding.
2. When screening different genotypes (in particular wild accessions) for growth under high temperatures, distinction must be made between heat tolerance and growth potential. Often plants with higher growth potential perform better regardless of the growing conditions.
3. When breeding for stress tolerance, often it is necessary that the derived lines/cultivars be able to perform well under both stress and non-stress conditions. Development of such genotypes is not without inherent difficulties. In some plant species, heat tolerance is often associated with some undesirable horticultural or agronomical characteristics. In tomato, for example, two undesirable characteristics commonly observed in heat-tolerant genotypes are small fruit and restricted foliar canopy. The

production of small fruit is most likely due to adverse effects of high temperature on the production of auxins in the fruit, and the poor canopy is due to the highly reproductive nature of the heat-tolerant genotypes.

In summary, breeding for heat tolerance is still in its infancy stage and warrants more attention than it has been given in the past. It is unfortunate that the literature contains relatively little information on breeding for heat tolerance in different crop species. However, despite all the complexity of heat tolerance and difficulties encountered during transfer of tolerance, some heat-tolerant inbred lines and hybrid cultivars with commercial acceptability have been developed and released, at least in a few crop species such as tomato. Nevertheless, to accelerate such progresses, major areas of emphasis in the future should be: (1) designing/development of accurate screening procedures; (2) identification and characterization of genetic resources with heat tolerance; (3) discerning the genetic basis of heat tolerance at each stage of plant development; (4) development and screening of large breeding populations to facilitate transfer of genes for heat tolerance to commercial cultivars. The use of advanced molecular biology techniques may facilitate development of plants with improved heat tolerance, as described in below.

Molecular and biotechnological strategies

Recent genetic studies and efforts to understand/improve high-temperature tolerance of crop plants using traditional protocols and transgenic approaches have largely determined that plant heat-stress tolerance is a multigenic trait. Different components of tolerance, controlled by different sets of genes, are critical for heat tolerance at different stages of plant development or in different tissues (Howarth, 2005). Thus, the use of genetic stocks with different degrees of heat tolerance, correlation and co-segregation analyses, molecular biology techniques and molecular markers to identify tolerance QTLs are promising approaches to dissect the genetic basis of thermotolerance (Maestri *et al.*, 2002). Most recently, biotechnology has contributed significantly to a better understanding of the genetic basis of heat tolerance. For example, several genes responsible for inducing the synthesis of HSPs have been identified and isolated in various plant species, including tomato and maize (Sun *et al.*, 2006;). The requirement of TATA box was earlier demonstrated by deletion analysis of soybean heat shock genes in sunflower. In addition, a number of other sequence motifs have been identified in plants that have quantitative effects on expression of certain heat shock genes. For example, there is evidence for the involvement of CCAAT box and AT-rich sequences. The HSEs are the binding sites for the transitive heat shock transcription factor (HSF), the activation of which in higher eukaryotes is a multi-step process. In response to heat stress, HSF is converted from a monomeric to trimeric form. The trimeric HSF is localized predominantly in the nucleus and has a high affinity of binding to HSEs. It is believed that interaction of HSF with HSP70 or other HSPs results in the activation of HSFvia conformational changes involving monomer to trimer transition and nuclear targeting. This expression of heat shock genes is modulated by the temperature within permissive range, thereby conferring plant thermotolerance (Yang *et al.*, 2006). Further research has demonstrated that thermotolerance of plants also can be modulated/ effected by changes in transcriptional and translational activities. Ongoing transcription is needed during stress to support a basal level of translational activity in the subsequent recovery from the stress, but it does not appear to be required for the heat-mediated increase in mRNA stability. In general, such activities in plants undergo rapid changes during developmental stages such as seed formation and germination, and also during abiotic stresses such as heat shock, hypoxia and wounding.

Two common biotechnological approaches to study and improve plant stress tolerance include marker-assisted selection (MAS) and genetic transformation. During the past two decades, the use of these approaches has contributed greatly to a better understanding of the genetic and biochemical bases of plant stress-tolerance and, in some cases, led to the development of plants with enhanced tolerance to abiotic stress. Because of the general complexity of abiotic stress tolerance and the difficulty in phenotypic selection for tolerance, MAS has been considered as an effective approach to improve plant stress tolerance. The use of this approach, however, requires identification of genetic markers that are associated with genes or QTLs affecting whole plant stress tolerance or individual components contributing to it. During the past two decades, substantial amounts of research has been conducted in different crop species to identify genetic markers associated with different environmental stresses, in particular drought, salinity and low temperatures. For example, molecular marker technology has allowed the identification and genetic characterization of QTLs with significant effects on stress tolerance during different stages of plant development and facilitated determination of genetic relationships among tolerance to different stresses. Comparatively, however, limited research has been conducted to identify genetic markers associated with heat tolerance in different plant species. In *Arabidopsis*, for example, four genomic loci (QTLs) determining its capacity to acquire thermotolerance were identified using a panel of heat-sensitive mutants. Use of restriction fragment length polymorphism (RFLP) revealed mapping of eleven QTLs for pollen germination and pollen tube growth under heat stress in maize.

Recent advances in genetic transformation and gene expression techniques have contributed greatly to a better understanding of the genetic and biochemical bases of plant stress-tolerance and, in some cases, led to the development of plants with enhanced tolerance to abiotic stresses. For example, a significant progress has been made in the identification of genes, enzymes or compounds with remarkable effects on plant stress tolerance at the cellular or organismal level (Bohnert *et al.*, 2006). Furthermore, manipulation of the expression or production of the identified genes, proteins, enzymes, or compounds through transgenic approaches has resulted in the development of plants with enhanced stress tolerance in different plant species (Rontein *et al.*, 2002). However, a major limitation in the use of such techniques for improving plant high-temperature tolerance is that the critical factors conferring the enhanced temperature tolerance in higher plants are still poorly understood.

Initial research on molecular manipulation to improve plant heat tolerance focused on production of enzymes that detoxify reactive oxygen species, including SOD. Reactive oxygen species are induced by most types of stresses (Sairam and Tyagi, 2004) and their production has been envisaged in stress cross-tolerance (Allan *et al.*, 2006). In addition to increased production of SOD, many other potential approaches can be utilized to detoxify ROS and produce plants tolerant of heat stress (Zidenga, 2005). If the critical components are determined, genetic engineering technology can be utilized to incorporate thermo-tolerance into adapted cultivars. Despite the very many limitations, some progress has been made. For example, transgenic tobacco plants with altered chloroplast membranes by silencing the gene encoding chloroplast omega- 3 fatty acid desaturase have been produced which produce less trienoic fatty acids and more dienoic fatty acids in their chloroplasts than the wild type. These plants exhibited greater photosynthesis and grew better than wild type plants under high temperatures (Murakami *et al.*, 2000). *Dnak1*, a gene responsible for high salt tolerance in the cyanobacterium *Aphanothece halophytica*, when transferred into tobacco was expressed and conferred high temperature resistance (Ono *et al.*, 2001). Development of plants capable of higher production of glycinebetaine through transformation with the *BADH* gene has been suggested as a potentially effective method to enhance heat tolerance in plants (Yang *et al.*, 2005). Thermal stability of rubisco activase, a molecular chaperone responsible for the activity of rubisco, is important in maintaining its activity (Salvucci and Crafts- Brandner, 2004). By transforming tobacco plants with rubisco activase gene, thermotolerance is achieved by reversible decarboxylation of rubisco a likely protective mechanism by which the plant protects its photosynthetic apparatus (Sharkey *et al.*, 2001).

Genetic engineering of heat shock factors (HSF) and antisense strategies are instrumental to the understanding of both the functional roles of HSPs and the regulation of HSFs. Manipulations of the HS-response in transgenic plants have the potential to improve common abiotic stress tolerance and this may have a significant impact on the exploitation of the inherent genetic potential of agronomically important plants. Transgenic overexpression of certain HSFs and HSF-fusion proteins results in an expression of HSPs at normal temperature. The increased acquired thermotolerance of transgenic lines is attributed to the higher levels of HSP chaperones. It has also been demonstrated that tomato *MT-sHSP* has a molecular chaperone function *in vitro* and recently it has been demonstrated that *MT-sHSP* gene exhibits thermotolerance in transformed tobacco with the tomato *MT-sHSP* gene (Sanmiya *et al*., 2004) at the plant level. Experimental data obtained from transgenic, reverse-genetics and mutation approaches in non-cereal species confirm causal involvement of HSPs in thermotolerance in plants (Queitsch *et al*., 2000). However, the cellular targets of HSPs are still unknown.

In summary, transformation technology for improving plant stress tolerance is still at its infancy, and the success to date represents only a beginning. Advancements in marker technology and genetic transformation are expected to contribute significantly to the development of plants with tolerance to high temperatures in future. With the current transformation technology, it is becoming possible to transfer multiple genes, which may act synergistically and additively to improve plant stress tolerance. Future knowledge of tolerance components and the identification and cloning of responsible genes may allow transformation of plants with multiple genes and production of highly stress tolerant transgenic plants. In addition, there is no report to date of any studies testing the performance of transgenic plants under field stress conditions. Therefore, much more work is needed to gain a clearer understanding of the genetics, biochemical and physiological basis of plant heat tolerance.

Induction of heat tolerance

Although genetic approaches may be beneficial in the production of heat-tolerant plants, it is likely that the newly produced plants are low yielding compared to near-isogenic heat sensitive plants. Thus, considerable attention has been devoted to the induction of heat tolerance in existing high-yielding cultivars. Among the various methods to achieve this goal, foliar application of, or pre-sowing seed treatment with, low concentrations of inorganic salts, osmoprotectants, signaling molecules (e.g., growth hormones) and oxidants (e.g., H_2O_2) as well as preconditioning of plants are common approaches. High-temperature preconditioning has been shown to drastically reduce the heat-induced damage to black spruce seedlings at moderately high temperature. Similarly, heat acclimated, compared to non-acclimated, turfgrass leaves manifested higher thermostability, lower lipid peroxidation product malondialdehyde (MDA) and lower damage to chloroplast upon exposure to heat stress (Xu *et al.*, 2006).

In pearl millet, pre-sowing hardening of the seed at high temperature (42°C) resulted in plants tolerant to overheating and dehydration and showing higher levels of water-soluble proteins and lower amounts of amide-N in leaves compared to non-hardened plants. The researchers proposed that the higher heat tolerance was due to enhanced glutathione synthase activity, promoting binding of the ammonia accumulated during exposure to high temperature. In tomato, it was demonstrated that heat treatment administered to plants prior to chilling stress resulted in reduced incidence and severity of chilling injury in fruit and other organs. In some cool season grasses, under heat stress, Ca^{2+} is required for maintenance of antioxidant activity and not for osmotic adjustment (Jiang and Haung, 2001). Under heat stress, Ca^{2+} requirement for growth is high to mitigate adverse effects of the stress (Kleinhenz and Palta, 2002). It has been shown that exogenous application of Ca^{2+} promotes plant's heat tolerance. Application of Ca^{2+} in the form of $CaCl_2$ prior to the stress treatment elevated the content of lipid peroxidation product, MDA and stimulated the activities of guaiacol peroxidase, SOD and catalase, which could be the reasons for the induction of heat tolerance (Kolupaev *et al.*, 2005).

Among the low molecular weight organic compounds, glycinebetaine and polyamines have been successfully applied to induce heat tolerance in various plant species. For example, barley seeds pre-treated with glycinebetaine led to plants with lower membrane damage, better photosynthetic rate, improved leaf water potential and greater shoot dry mass, compared to untreated seeds (Wahid and Shabbir, 2005). Also, exogenous application of 4mM spermidine improved tomato heat resistance by improving chlorophyll fluorescence properties, hardening and higher resistance to thermal damage of the pigment-protein complexes structure, and the activity of PSII during linear increase in temperature (Murkowski, 2001). Thus, to improve plant heat tolerance, alternative approaches to genetic means would include pre-treatment of plants or seeds with heat stress or certain mineral or organaic compounds. The success of such approach, however, depends on plant species and gentoypes and must be studied on case basis.

Conclusions

Plants exhibit a variety of responses to high temperatures, which are depicted by symptomatic and quantitative changes in growth and morphology. The ability of the plant to cope with or adjust to the heat stress varies across and within species as well as at different developmental stages. Although high temperatures affect plant growth at all developmental stages, later phenological stages, in particular anthesis and grain filling are generally more susceptible. Pollen viability, patterns of assimilate partitioning, and growth and development of seed/grain are highly adversely affected. Other notable heat stress effects include structural changes in tissues and cell organelles, disorganization of cell membranes, disturbance of leaf water relations, and impedance of photosynthesis via effects on photochemical and biochemical reactions and photosynthetic membranes. Lipid peroxidation via the production of ROS and changes in antioxidant enzymes and altered pattern of synthesis of primary and secondary metabolites are also of considerable importance.

In response to heat stress, plants manifest numerous adaptive changes. The induction of signaling cascades leading to profound changes in specific gene expression is considered an important heat-stress adaptation. Although various signaling molecules are synthesized under heat stress, the role of Ca^{2+} remains critical. A fundamental heat-stress response ubiquitous to plants is the expression of HSPs, which range from low (10 kDa) to high (100 kDa) molecular mass in different species. Evidence on synthesis and accumulation of some other stress-related proteins is also available. Such stress proteins are thought to function as molecular chaperones, helping in folding and unfolding of essential proteins under stress, and ensuring three-dimensional structure of membrane proteins for sustained cellular functions and survival under heat stress.

In addition to genetic means to developing plants with improved heat tolerance, attempts have been made to induce heat tolerance in a range of plant species using different approaches. These include preconditioning of plants to heat stress and exogenous applications of osmoprotectants or plant growth-regulating compounds on seeds or whole plants. Results from such applications are promising and further research is forthcoming. Also, while some notable progress has been reported as to the development of crop plants with improved heat tolerance via traditional breeding, the prospect for engineering plants with heat tolerance is also good considering accumulating molecular information on the mechanisms of tolerance and contributing factors.

Although physiological mechanisms of heat tolerance are relatively well understood, further studies are essential to determine physiological basis of assimilate partitioning from source to sink, plant phenotypic flexibility which leads to heat tolerance, and factors that modulate plant heat-stress response. Furthermore, applications of genomics, proteomics and trascriptomics approaches to a better understanding of the molecular basis of plant response to heat stress as well as plant heat tolerance are imperative. As in the case of most other abiotic stresses, foliar plant parts are more directly impinged upon by high temperatures than roots. However, an understanding of root responses to heat stress, most likely involving root–shoot signaling, is crucial and warrants further exploration. Molecular knowledge of response and tolerance mechanisms will pave the way for engineering plants that can tolerate heat stress and could be the basis for production of crops which can produce economic yield under heat-stress conditions.

References

Adams, S.R., Cockshull, K.E., Cave, C.R.J., 2001. Effect of temperature on the growth and development of tomato fruits. *Ann. Bot.* 88: 869–877.

Albertini, E., Marconi, G., Reale, L., Barcaccia, G., Porceddu, A., Ferranti, F., Falcinelli, M., 2005. SERK and APOSTART: candudate genes for apomixisin in *Poa partensis*. *Plant Physiol.* 138: 2185–2199.

Allakhverdieva, Y.M., Mamedov, M.D., Gasanov, R.A., 2001. The effect of glycinebetaine on the heat stability of photosynthetic membranes. *Turk. J.Bot.* 25: 11–17.

Allan, A.C., Maddumage, R., Simons, J.L., Neill, S.O., Ferguson, I.B., 2006. Heat-induced oxidative activity protects suspension-cultured plant cells from low temperature damage. *Funct. Plant Biol.* 33: 67–76.

Anon, S., Fernandez, J.A., Franco, J.A., Torrecillas, A., Alarc´on, J.J., S´anchez-Blanco, M.J., 2004. Effects of water stress and night temperature preconditioning on water relations and morphological and anatomical changes of *Lotus creticus plants*. *Sci. Hortic.* 101: 333–342.

Arshad, M., Frankenberger, W.T.J., 2002. Ethylene, Agricultural Sources and Applications. Kluwer Academic/Plenum Publishers, New York.

Barlow, K. M., Christy, B. P. And O'Leary, G. J. 2015. Riffkin, Nuttall Simulating the impact of extreme heat and frost events on wheat crop production: a review. *Field Crops Res*, 171:109-119.

Barua, D., Downs, C.A., Heckathorn, S.A., 2003. Variation in chloroplast small heat-shock protein function is a major determinant of variation in thermotolerance of photosynthetic electron transport among ecotypes of *Chenopodium album*. *Funct. Plant Biol.* 30: 1071–1079.

Blum, A., Klueva, N., Nguyen, H.T., 2001. Wheat cellular thermotolerance is related to yield under heat stress. *Euphytica* 117: 117–123.

Bowen, J., Michael, L.-Y., Plummer, K.I.M., Ferguson, I.A.N., 2002. The heat shock response is involved in thermotolerance in suspension-cultured apple fruit cells. *J. Plant Physiol.* 159: 599–606.

Burke, J.J., 2001. Identification of genetic diversity and mutations in higher plant acquired thermotolerance. *Physiol. Plant.* 112:167–170.

Camejo, D., Jim´enez, A., Alarc´on, J.J., Torres, W., G´omez, J.M., Sevilla, F., 2006. Changes in photosynthetic parameters and antioxidant activities following heat-shock treatment in tomato plants. Funct. *Plant Biol.* 33: 177–187.

Charng, Y., Liu, H., Liu, N., Hsu, F., Ko, S., 2006. Arabidopsis Hsa32, a novel heat shock protein, is essential for acquired thermotolerance during long term recovery after acclimation. *Plant Physiol.* 140: 1297–1305.

Chinnusamy,V., Schumaker, K., Zhu, J.K., 2004. Molecular genetic perspectives on cross-talk and specificity in abiotic stress signaling in plants. *J. Exp. Bot.* 55: 225–236.

Crafts-Brander, C., Salvucci, M.E., 2002. Sensitivity to photosynthesis in the C4 plant, maize to heat stress. *Plant Cell* 12: 54–68.

Crafts-Brandner, S.J., Salvucci, M.E., 2000. Rubisco activase constrains the photosynthetic potential of leaves at high temperature and CO2. *Proc. Natl. Acad. Sci. USA.* 97: 13430–13435.

Dat, J., Vandenbeele, S., Vranova, E., Van Montagu, M., Inz′e, D., Van Breusegem, F., 2000. Dual action of the active oxygen species during plant stress responses. *Cell. Mol. Life Sci.* 57: 779–795.

De Block, M., Verduyn, C., De Brouwer, D., Cornelissen, M., 2005. Poly(ADPribose) polymerase in plants affects energy homeostasis, cell death and stress tolerance. *Plant J.* 41: 95–106.

De Ronde, J.A.D., Cress, W.A., Kruger, G.H.J., Strasser, R.J., Staden, J.V., 2004. Photosynthetic response of transgenic soybean plants containing an *Arabidopsis P5CR* gene, during heat and drought stress. *J. Plant Physiol.* 61: 1211–1244.

Diamant, S., Eliahu, N., Rosenthal, D., Goloubinoff, P., 2001. Chemical chaperones regulate molecular chaperones in vitro and in cells under combined salt and heat stresses. *J. Biol. Chem.* 276: 39586–39591.

Dubey, R.S., 2005. Photosynthesis in plants under stressful conditions. In: Pessarakli, M. (Ed.), Handbook of Photosynthesis. CRC Press, Boca Raton, Florida, pp. 717–737.

Funk, J.L., Mak, J.E., Lerdau, M.T., 2004. Stress-induced changes in carbon sources for isoprene production in *Populus deltoides*. *Plant Cell Environ.* 27: 747–755.

Goyal, K., Walton, L.J., Tunnacliffe, A., 2005. LEA proteins prevent protein aggregation due to water stress. *Biochem. J.* 388: 151–157.

Hall, A.E., 2001. Crop Responses to Environment.CRCPress LLC, Boca Raton, Florida.

Hatfield, J. L., Boote, K. J., Kimball, B.A., Ziska, L., Izaurralde, R.C., Ort, D., Thomson, A.M. and Wolfe, D.A. 2011. Climate impacts on agriculture:Implications for crop production. Agron. J. **xx**:xx–xx.

Hatfield, J. L., Boote, K. J., Fay, P., Hahn, L., Izaurralde, C., Kimball, B. A., Mader, T., Morgan, J., Ort, D., Polley, W., Thomson, A. and Wolfe, D. 2008. Agriculture. In The effects of climate change on agriculture, landresources, water resources, and biodiversity in the United States. *U.S.Climate Change Science Program and the Subcommittee on GlobalChange Res., Washington, DC.*

Horton, P., 2002. Crop improvement through alteration in the photosynthetic membrane., ISB News Report. Virginia Tech, Blacksburg, VA. Horv′ath, G., Arellano, J.B., Droppa, M., Bar′on, M., 1998. Alterations in Photosystem II electron transport as revealed by thermoluminescence of Cu-poisoned chloroplasts. *Photosyn. Res.* 57: 175–181.

Howarth, C.J., 2005. Genetic improvements of tolerance to high temperature. In: Ashraf, M., Harris, P.J.C. (Eds.), Abiotic Stresses: Plant Resistance Through Breeding and Molecular Approaches. Howarth Press Inc., New York.

Jiang, Y., Haung, B., 2001. Plants and the environment. Effects of calcium on antioxidant activities and water relations associated with heat tolerance in two cool-season grasses. *J. Exp. Bot.* 52: 341–349.

Kaur, N., Gupta, A.K., 2005. Signal transduction pathways under abiotic stresses in plants. *Curr. Sci.* 88: 1771–1780.

Kavi Kishore, P.B., Sangam, S., Amrutha, R.N., Laxmi, P.S., Naidu, K.R., Rao, K.R.S.S., Rao, S., Reddy, K.J., Theriappan, P., Sreenivasulu, N., 2005. Regulation of proline biosynthesis, degradation, uptake and transport in higher plants: its implications in plant growth and abiotic stress tolerance. *Curr. Sci.* 88: 424–438.

Kepova, K.D., Holzer, R., Stoilova, L.S., Feller, U., 2005. Heat stress effects on ribulose-1,5-bisphosphate carboxylase/oxygenase, Rubisco bindind protein and Rubisco activase in wheat leaves. *Biol. Plant.* 49: 521–525.

Klein, J.D., Dong, L., Zhou, H.W., Lurie, S., 2001. Ripeness of shaded and sun-exposed apples (*Malus domestica* L.). *Acta Hortic.* 553: 95–98.

Kleinhenz, M.D., Palta, J.P., 2002. Root zone calcium modulates the response of potato plants to heat stress. *Physiol. Plant.* 115: 111–118.

Kolupaev, Y., Akinina, G., Mokrousov, A., 2005. Induction of heat tolerance in wheat coleoptiles by calcium ions and its relation to oxidative stress. *Russ. J. Plant Physiol.* 52: 199–204.

Korotaeva, N.E., Antipina, A.I., Grabelynch, O.I., Varakina, N.N., Borovskii, G.B.,Voinikov,V.K., 2001. Mitochondrial low-molecular-weight heat shock proteins and tolerance of crop plant's mitochondria to hyperthermia. *Fiziol. Biokhim Kul'turn. Rasten.* 29: 271–276.

Larkindale, J., Huang, B., 2004. Thermotolerance and antioxidant systems in *Agrostis stolonifera*: involvement of salicylic acid, abscisic acid, calcium, hydrogen peroxide, and ethylene. *J. Plant Physiol.* 161: 405–413.

Larkindale, J., Huang, B., 2005. Effects of abscisic acid, salicylic acid, ethylene and hydrogen peroxide in thermotolerance and recovery for creeping bentgrass. *Plant Growth Regul.* 47: 17–28.

Larkindale, J., Knight, M.R., 2002. Protection against heat stress-induced oxidative damage in *Arabidopsis* involves calcium, abscisic acid, ethylene and salicylic acid. *Plant Physiol.* 128: 682–695.

Liu, H.-T., Li, B., Shang, Z.-L., Li, X.-Z., Mu, R.-L., Sun, D.-Y., Zhou, R.-G., 2003. Calmodulin is involved in heat shock signal transduction in wheat. *Plant Physiol.* 132: 1186–1195.

Liu, X., Huang, B., 2000. Heat stress injury in relation to membrane lipid peroxidation in creeping bent grass. *Crop Sci.* 40: 503–510.

Maestri, E., Klueva, N., Perrotta, C., Gulli, M., Nguyen, H.T., Marmiroli, N., 2002. Molecular genetics of heat tolerance and heat shock proteins in cereals. *Plant Mol. Biol.* 48: 667–681.

Majoul, T., Bancel, E., Triboi, E., Ben Hamida, J., Branlard, G., 2003. Proteomic analysis of the effect of heat stress on hexaploid wheat gra: characterization of heat-responsive proteins from total endosperm. *Proteomics* 3: 175–183.

Mazorra, L.M., Nunez, M., Echerarria, E., Coll, F., S´anchez-Blanco, M.J., 2002. Influence of brassinosteriods and antioxidant enzymes activity in tomato under different temperatures. *Plant Biol.* 45: 593–596.

Meehl, G. A., Stocker, T. F., Collins, W. D. and Friedlingstein, P. 2007. Global climate projections. In: Solomon S, Qin D, Manning M, Chen Z and others (eds) Climate change 2007. The physical science basis. Contribution of Working Group I to the Fourth Assessment Report of the Inter - governmental Panel on Climate Change. *Cambridge University Press, Cambridge*, 749–844

Miller, P., Lanier, W., Brandt, S., 2001. Using Growing Degree Days to Predict Plant Stages. Ag/ Extension Communications Coordinator, Communications Services, Montana State University-Bozeman, Bozeman, MO.

Moriarty, T., West, R., Small, G., Rao, D., Ristic, Z., 2002. Heterologous expression of maize chloroplast protein synthesis elongation factor(EFTU) enhances *Escherichia coli* viability under heat stress. *Plant Sci.* 163: 1075–1082.

Munne-Bosch, S., Lopez-Carbonell, M., Alegre, L.A., Van Onckelen, H.A., 2002. Effect of drought and high solar radiation on 1-aminocyclopropane-1- carboxylic acid and abscisic acid in *Rosmarinus officinalis* plants. *Physiol. Plant.* 114: 380–386.

Murkowski, A., 2001. Heat stress and spermidine: effect on chlorophyll fluorescence in tomato plants. *Biol. Plant.* 44: 53–57.

Neta-Sharir, I., Isaacson, T., Lurie, S., Weiss, D., 2005. Dual role for tomato heat shock protein 21: protecting photosystem ii from oxidative stress and promoting color changes during fruit maturation. *Plant Cell* 17: 1829–1838.

Nollen, E.A.A., Morimoto, R.I., 2002. Chaperoning signaling pathways: molecular chaperones as stress-sensing 'heat shock' proteins. *J. Cell Sci.* 115: 2809–2816.

Ono, K., Hibino, T., Kohinata, T., Suzuki, S., Tanaka, Y., Nakamura, T., Takabe, T., Takabe, T., 2001. Overexpression of DnaK from a halotolerant cyanobacterium *Aphanothece halophytica* enhances the high-temperatue tolerance of tobacco during germination and early growth. *Plant Sci.* 160: 455–461.

Ortiz, C., Cardemil, L., 2001. Heat-shock responses in two leguminous plants: a comparative study. *J. Exp. Bot.* 52: 1711–1719.

Panchuk, I.I., Volkov, R.A., Sch¨offl, F., 2002. Heat stress- and heat shock transcription factor-dependent expression and activity of ascorbate peroxidase in *Arabidopsis*. *Plant Physiol.* 129: 838–853.

Porter, J.R., 2005. Rising temperatures are likely to reduce crop yields. *Nature* 436: 174.

Quan, R., Shang, M., Zhang, H., Zhao, Y., Zhang, J., 2004. Engineering of enhanced glycine betaine synthesis improves drought tolerance in maize. *Plant Biotech. J.* 2: 477–486.

Queitsch, C., Hong, S.W., Vierling, E., Lindquest, S., 2000. Heat shock protein 101 plays a crucial role in thermotolerance in *Arabidopsis*. *Plant Cell* 12: 479–492.

Rivero, R.M., Ruiz, J.M., Garcia, P.C., Lopez-Lefebre, L.R., Sanchez, E., Romero, L., 2001. Resistance to cold and heat stress: accumulation of phenolic compounds in tomato and watermelon plants. *Plant Sci.* 160: 315– 321.

Rizhsky, L., Hongjian, L., Mittler, R., 2002. The combined effect of drought stress and heat shock on gene expression in tobacco. *Plant Physiol.* 130: 1143–1151.

Sairam, R.K., Tyagi, A., 2004. Physiology and molecular biology of salinity stress tolerance in plants. *Curr. Sci.* 86: 407–421.

Salvucci, M.E., Crafts-Brandner, S.J., 2004. Relationship between the heat tolerance of photosynthesis and the thermal stability of rubisco activase in plants from contrasting thermal environments. *Plant Physiol.* 134: 1460–1470.

Sangwan, V., Dhindsa, R.S., 2002. In vivo and in vitro activation of temperatureresponsive plant map kinases. *FEBS Lett.* 531: 561–564.

Sch¨offl, F., 2005. The role of heat shock proteins in abiotic stress response and the development of plants. Universitat Tubingen, ZMBP, Allgemeine Genetik.

Setimela, P.S., Andrews, D.J., Partridge, J., Eskridge, K.M., 2005. Screening sorghum seedlings for heat tolerance using a laboratory method. *Eur. J. Agron.* 23: 103–107.

Shah, F., Huang, J., Cui, K., Nie, L., Shah, T., Chen, C. and Wang, K. 2011. Impact of high-temperature stress on rice plant and its traits related to tolerance. *J. Agric. Sci.,* 149: 545-556.

Sharkey, T.D., Badger, M.R., Von-Caemmerer, S., Andrews, T.J., 2001. Increased heat sensitivity of photosynthesis in tobacco plants with reduced Rubisco activase. *Photosyn. Res.* 67: 147–156.

Shi, Q., Bao, Z., Zhu, Z.,Ying, Q., Qian, Q., 2006. Effects of different treatments of salicylic acid on heat tolerance, chlorophyll fluorescence, and antioxidant enzyme activity in seedlings of *Cucumis sativa* L. *Plant Growth Regul.* 48: 127–135.

Singh ,V., Nguyen, C. T., vanosterom, E. J., Chapman, S. C., Jordan, D. R. and Hammer, G. L. 2015. Sorghum genotypes differ in high temperature responses for seed set. *Field Crops Res.*, 171: 32-40

Stone, P., 2001. The effects of heat stress on cereal yield and quality. In: Basra, A.S. (Ed.), Crop Responses and Adaptation to Temperature Stress. Food Products Press, Binghamton, NY, pp. 243–291.

Sun, A., Yi, S., Yang, J., Zhao, C., Liu, J., 2006. Identification and characterization of a heat-inducible ftsH gene from tomato (*Lycopersicon esculentum* Mill.). *Plant Sci.* 170: 551–562.

Sung, D.-Y., Kaplan, F., Lee, K.-J., Guy, C.L., 2003. Acquired tolerance to temperature extremes. *Trends Plant Sci.* 8: 179–187.

Suzuki, N., Mittler, R., 2006. Reactive oxygen species and temperature stresses: a delicate balance between signaling and destruction. *Physiol. Plant.* 126: 45–51.

Taiz, L., Zeiger, E., 2006. Plant Physiology. Sinauer Associates Inc. Publishers, Massachusetts.

Wahid, A., 2007. Physiological implications of metabolites biosynthesis in net assimilation and heat stress tolerance of sugarcane sprouts. *J. Plant Res.* 120: 219–228.

Wahid, A., Shabbir, A., 2005. Induction of heat stress tolerance in barley seedlings by pre-sowing seed treatment with glycinebetaine. *Plant Growth Reg.* 46: 133–141.

Wang, L.-J., Li, S.H., 2006a. Thermotolerance and related antioxidant enzyme activities induced by heat acclimation and salicylic acid in grape (*Vitis vinifera* L.) leaves. *Plant Growth Regul.* 48: 137–144.

Wang, L.-J., Li, S.-L., 2006b. Salicylic acid-induced heat or cold tolerance in relation to Ca2+ homeostasis and antioxidant systems in young grape plants. *Plant Sci.* 170: 685–694.

Wang, W., Vinocur, B., Shoseyov, O., Altman, A., 2004. Role of plant heat-shock proteins and molecular chaperones in the abiotic stress response. *Trends Plant Sci.* 9: 244–252.

Xiong, L., Lee, H., Ishitani, M., Zhu, J.-K., 2002. Regulation of osmotic stress responsive gene expression by LOS6/ABA1 locus in *Arabidopsis*. *J. Biol. Chem.* 277: 8588–8596.

Xu, S., Li, J., Zhang, X., Wei, H., Cui, L., 2006. Effects of heat acclimation pretreatment on changes of membrane lipid peroxidation, antioxidant metabolites, and ultrastructure of chloroplasts in two cool-season turfgrass species under heat stress. *Environ. Exp. Bot.* 56: 274–285.

Yang, K.A., Lim, C.J., Hong, J.K., Park, C.Y., Cheong, Y.H., Chung, W.S., Lee, K.O., Lee, S.Y., Cho, M.J., Lim, C.O., 2006. Identification of cell wall genes modified by a permissive high temperature in Chinese cabbage. *Plant Sci.* 171: 175–182.

Yang, X., Liang, Z., Lu, C., 2005. Genetic engineering of the biosynthesis of glycinebetaine enhances photosynthesis against high temperature stress in transgenic tobacco plants. *Plant Physiol.* 138: 2299–2309.

Young, T.E., Ling, J., Geisler-Lee, C.J., Tanguay, R.L., Caldwell, C., Gallie, D.R., 2001. Developmental and thermal regulation of maize heat shock protein, HSP101. *Plant Physiol.* 127: 777–791.

Zhang, J.-H., Huang, W.-D., Liu, Y.-P., Pan, Q.-H., 2005. Effects of temperature acclimation pretreatment on the ultrastructure of mesophyll cells in young grape plants (*Vitis vinifera* L. cv. Jingxiu) under cross-temperature stresses. *J. Integr. Plant Biol.* 47: 959–970.

Zidenga, T., 2005. Improving Stress Tolerance through Energy Homeostasis in Plants. Department of Plant Cellular and Molecular Biology, Ohio State University.

4

Low Temperature Stress Induced Changes in Plants and Their Management

Ashok, Shantappa, T., Vithal Navi , Ratnakar M. Shet, Shivanand Hongal and Athani, S.I.

Food productivity is decreasing due to detrimental effects of various biotic and abiotic stresses; therefore minimizing these losses is a major area of concern to ensure food security under changing climate. Environmental abiotic stresses, such as drought, extreme temperature, cold, heavy metals, or high salinity, severely impair plant growth and productivity worldwide. Drought, being the most important environmental stress, severely impairs plant growth and development, limits plant production and the performance of crop plants, more than any other environmental factor (Shao *et al.,* 2009). Plant experiences drought stress either when the water supply to roots becomes difficult or when the transpiration rate becomes very high. Available water resources for successful crop production have been decreasing in recent years. Furthermore, in view of various climatic change models scientists suggested that in many regions of world, crop losses due to increasing water shortage will further aggravate its impacts. Drought impacts include growth, yield, membrane integrity, pigment content, osmotic adjustment water relations, and photosynthetic activity (Praba *et al.,* 2009). Drought stress is affected by climatic, edaphic and agronomic factors. The susceptibility of plants to drought stress varies in dependence of stress degree, different accompanying stress factors, plant species, and their developmental stages. Acclimation of plants to water deficit is the result of different events, which lead to adaptive changes in plant growth and physio-biochemical processes, such as changes in plant structure, growth rate, tissue osmotic potential and antioxidant defenses (Duan *et al.,* 2007 and Shakeel *et al.,* 2011).

Plant growth, productivity and distribution were greatly affected by environmental stresses such as high and low temperature, drought and high salinity. Different plant species have different optimal growth temperatures. The temperatures considerably lower than the optimal growth temperatures result in low temperature stress in the plants. In majority of the plants this temperature is usually below 10 °C. The low temperature i.e. lower than 10 °C can be further subdivided into two main categories – the "chilling" and "freezing" temperatures. This is decided by the fact whether there is formation of ice within the tissues or not. Both these conditions also cause water deficit inside the plant and damage membranes and macromolecules. In response to abiotic stresses, plants undergo a variety of changes at the molecular level (gene expression) leading to physiological adaptation (Mantri *et al.,* 2012). The situation has become more serious with concerns of global climate change. Therefore, studies on abiotic stress tolerance have become one of the main areas of research worldwide. Recent advances in this area include unraveling the physiological, biochemical, and molecular mechanism of abiotic stress tolerance and corresponding development of tolerant cultivars through transgenic technology or molecular breeding (Niu *et al.,* 2012). Desired tolerant genotypes/varieties so developed need to be screened at laboratory as well as field levels for functional validation. Numerous physio-biochemical indicators for tolerance screening and indirect selection using molecular markers linked to desired loci are being deployed for accelerating the production of stress-tolerant varieties (Ashraf and Foolad 2013 and 2007).

During the drought stress, in the plants transcriptional regulatory networks such as the induction of a large range of genes encoding for enzymes involved in osmotic adjustments, osmoprotection, wax biosynthesis and changes in fatty acid composition. Adjustment of osmotic pressure allows the plant to take up more soil water and maintain turgor and cell function for a longer time under drought (Geoffrey and Kerstin 2016).

In a condition where temperature ranges between 10 °C to 0 °C, but there is no ice formation, solute leakages and loss occur in cellular compartments indicating loss in membrane permeability and integrity. With freezing temperatures i.e. below 0 °C the formation of ice within the extracellular space or apoplast takes place. Intracellular ice formation is usually lethal. Winter rye and other over wintering cereals also secrete antifreeze proteins into the leaf apoplasts that have the ability to bind to the surface of ice and modify its growth (Pihakaski-Maunsbach *et al.,* 2003). Angiosperms have the ability to survive under dessicated states and most temperate zone woody deciduous trees, including apple require a certain degree of chilling to break endodormancy before active shoot growth in spring, a phenomenon referred to as chilling requirement. Chilling stress induces oxidative processes in plant cells. These processes are initiated by reactive oxygen species, which arise from disturbed operation of electron transport chains and bring about various manifestations of chilling damage. These stresses affect the functions of cell membranes as the primary site of freezing injury, elevate membrane viscosity, and promote lipid transition from a liquid crystalline to a gel phase. As a result, the cells either adapt to these changes or perish. The fluidity of membranes is important for maintaining the barrier properties and for the activation and functioning of certain membrane bound enzymes. Plants acclimate to environmental low temperature primarily due to the shifts in cell metabolism determined by differential gene expression, which results in changes in the membrane composition and accumulation of cryoprotectants and antioxidants.

Types of low temperature injury

A plant can sustain two types of injuries through exposure to low temperatures. The first one is chilling injury that occurs from approximately 20 to 0 °C. The resultant injuries may include a variety of physiological disruptions in germination, flower and fruit development, yield and storage life. Minor chilling stress at non-lethal temperatures is normally reversible. Exposure to gradually decreasing temperatures above the critical range can also result in hardening of plants that may reduce or eliminate injury during subsequent exposure to low temperatures. The second type of injury occurs when the external temperature drops below the freezing point of water which is called as freezing injury.

Many plants that are native to cold climates can survive extremely low temperatures without injury. Plants may experience intracellular freezing and / or extracellular freezing. Intracellular freezing damages the protoplasmic structure and the ice crystals kill the cell once they grow large enough to be detected microscopically. In extracellular freezing, the protoplasm of the plant becomes dehydrated because a water vapour deficit is created as cellular water is transferred to ice crystals forming in the intercellular spaces. In some cases, water can remain liquid as low as -47 °C without nucleating and forming ice. When nucleation of this supercooled water does occurs, intracellular ice forms suddenly resulting in death of the plant.

Types of low temperature tolerance plant

There are three main classes of plant grouped according to their low temperature tolerance. The first group includes frost tender plants that are sensitive to chilling injury and can be killed by short periods of exposure to temperatures just below freezing. They cannot tolerate ice in their tissues and readily exhibit frost injury symptoms that include water soaked flaccid appearance with loss of turgor followed by rapid drying upon exposure to warm temperatures. Beans, corns, rice and tomatoes are examples of plants in this category. Second group of plants allows them to tolerate the presence of extracellular ice in their tissues. Their frost resistance ranges from the broad-leafed summer annuals, which are killed at temperatures slightly below freezing, to perennial grasses that can survive exposure to -40 °C. As temperatures decrease the outward migration of intracellular water to the growing extracellular ice crystal causes dehydration stress that will eventually result in irreversible damage to the plasma membrane, which is the primary site of low temperature injury. If ice nucleation does not occur at -3 to -5 °C, supercooling may result in intracellular freezing and death of individual cells. The third and final group is made up of very cold hardy plants that are predominantly temperate woody species. Like the plants in the previous group, their lower limits of cold tolerance are dependent on the stage of acclimation, the rate and degree of temperature decline, and the genetic capability of tissues to accommodate extracellular freezing and the accompanying dehydration stress. Deep supercooling allows certain tissues in plants from group to survice low temperatures without the formation of extracellular ice. However, the most cold hardy species do not rely in supercooling and can withstand temperatures of -196 °C.

Plant chilling stress and its repercussions

Both tropical and subtropical origin crops are sensitive to chilling temperatures. This limits production areas and causes potential damage to during storage if they are exposed to low temperatures. The temperatures below which chill injury can occur varies with species and regions of origin, ranging from 0 to 4 °C for temperate fruits, 8 °C for subtropical fruits and about 12 °C for tropical fruits such as banana. Amongst the highest volume world crops, maize (*Zea mays*) and rice (*Oryza sativa*) are sensitive to chilling temperatures. Their growth and development can be adversely effected by temperatures below 10 °C resulting in yield loss or crop failure. Chilling during the seedling stage in cotton can reduce plant height, delay flowering and adversely affect yield and lint quality. Seedlings can also suffer water stress and leaf desiccation at chilling temperatures, floral initiation is inhibited at 7 °C and seed set is inhibited at 15 °C. Other crops suffering stand loss, delayed maturity and reduced yield as a result of chilling after planting include soybean, lima bean, cucurbits, tomato, pepper, egg plant, okra and various cereal crops.

The influence of chilling sensitivity can also be noticed at physiological age, seedling development and pre-harvest climate. Freshly imbibed seeds of chill sensitive species tend to be very sensitive as does the pollen development stage. Fruits maturing at high temperature are more susceptible than those maturing at lower temperatures. Post-harvest storage at lower temperatures is commonly used to extend the storage life of fruits and vegetables.

Both tropical and subtropical plants are often subject to physiological damage and loss of quality due to chill injury under these storage conditions. The severity of injury to chill-sensitive tissues tends to increase with decreasing temperatures and with length of low-temperature exposure. Chilling has been found to change the entire metabolic system of the cell with some processes recovering quickly and others only slowly. Chilling affects the entire internal environment of each cell and each molecule within the cells. Enzymatic reactions, substrate diffusion rates and membrane transport properties are all affected. Chilling injury is therefore likely a direct consequence of these effects.

Cold acclimation in woody plants

Low temperature acclimation is an ability of plants to cold acclimate when exposed to gradually decreasing temperatures below a specific threshold. This is the most common mechanism that plants have evolved for adapting to low temperature stress and examples of plants with the capacity to cold harden can be found in most species. Many plants adapted to temperate and cold climates, when exposed to low temperatures (LT) (0-10 °C) develop physiological and biochemical responses which increase their freezing tolerance. This process is known as cold acclimation (Levitt, 1980). In annual herbaceous species ample knowledge has been acquired about the nature of the genes and the mechanisms responsible for freezing tolerance as well as sensing and regulatory mechanisms that activate the cold acclimation-response (Yamaguchi-Shinozaki and Shinozaki, 2005). Much less research has been carried out concerning this process in woody plants, despite their adaptation to two different types of cold acclimation: acclimation to temperature fluctuations during the growing season, and the seasonal acclimation for overwintering (Li *et al.*, 2004). Forest species in cold and temperate climates are regularly exposed to freezing temperatures during winter months. Their ability to survive is based on adaptative mechanisms by which plants enter a state of dormancy and develop freezing tolerance. The onset of winter deep dormancy (endodormancy) is preceded by a stage of ecodormancy. Endodormancy is caused by plant endogenous factors and, once established, no growth can be achieved until a chilling requirement has been satisfied. In order for bud break to occur, plants need to be exposed to LT for a cumulative number of hours (chilling requirement). In contrast, during ecodormancy, growth is arrested by adverse environment and resumes when conditions become favourable (Howe *et al.*, 1999). The onset of endodormancy is one of the most frequently studied photoperiodic phenomena. Some important endodormancy-related traits such as growth cessation, bud set and the initial stages of cold acclimation can be induced in many tree species by a short day photoperiod (SD) (Thomas and Vince-Prue, 1997). Other changes, including leaf senescence and abscission, are not induced by SD alone but require exposure to LT. These stages coincide with sequential phases of cold acclimation of which the first is initiated by SD and the second by LT and freezing temperatures. When fully hardened, the trees can tolerate exposure to extreme temperatures of –50 to –100 °C. There appears to be a third stage of acclimation in hardy woody species which is induced by low temperatures (–30 to –50 °C), that may not commonly be achieved in nature. Thus, dormancy in a woody plant is superimposed on a seasonal development of cold hardiness. This makes it difficult to distinguish physiological and molecular changes associated with dormancy regulation from those underlying the seasonality of cold hardiness (Arora *et al.*, 2003 and Gomez *et al.*, 2005).

Cold acclimation induction by Short Day Photoperiod and Low Temperature

Photoperiod is the most important environmental cue controlling the onset of dormancy in perennial plants. By responding to short day photoperiod (SD), plants are able to synchronize cold acclimation and dormancy induction with the end of the growing season and the onset of low temperatures in the fall. Because the length of the growing season varies latitudinally, photoperiodic responses often differ between northern and southern populations of the same species. In the northern hemisphere, trees from southern locations usually require shorter days to induce bud set than do northern trees. Differences have also been observed in the critical photoperiod length (the longest photoperiod inducing growth cessation) between ecotypes of different elevations. This behavior has been described in different forest species of several genera, such as *Betula, Picea, Populus, Salix*. Bud phenology is often found to be under strong genetic control. Quantitative trait locus (QTL) mapping experiments have shown that three major genes may be involved in the control of bud set in *Populus.* Two of these map respectively next to *PHYB1* gene, involved in photoperception of the photoperiod, and to *ABI1B* gene, related to the activity of abscisic acid (ABA) (Frewen *et al.*, 2000). However, other QTL analyses suggest that environmental factors other than day-length significantly influence genetic differences in the timing of bud set in the field (Howe *et al.*, 1999).

As the temperature is lowered in chilling-sensitive plants, lipids in cellular membranes solidify (Crystallize) at a critical temperature that is determined by the ratio of saturated to unsaturated fatty acids. This critical temperature for a phase transition from liquid to crystalline often proves to be equivalent to the temperature that causes chilling damage. Development of tolerance to chilling temperatures in chilling sensitive plants apparently involves changes in this ratio. An increase in the proportion of unsaturated fatty acids or in the quantity of sterols causes the membranes to remain functional at lower temperature.

Cold stress, especially the chilling stress in cereal crops, is one major form of stress which affects the crop growth and yield (Hasanuzzaman *et al.,* 2013). Cold stress-induced tissue dehydration further leads to membrane disintegration, reduced growth and development of plants in maize which was due to the accumulation of MDA content as a result of lipid peroxidation in membranes (Yadav, 2010). According to Yordanova and Popova (2007), exposure of wheat plants to low temperature (3°C) for 48 and 72hr resulted in decreased levels of chlorophyll, CO_2 assimilation, transpirations rates and photosynthesis due to the reduced activities of ATPsynthase, which further restricted RuBisCo regeneration and limited photophosphorylation. Physio-biochemical responses to cold stress in tetraploid and hexaploid wheat were studied where, the elevated levels of electrolyte leakage index, H_2O_2 and MDA content were observed in stressed plants (Nejadsadeghi *et al.,* 2014). According to some previous reports, oxidative stress as a result of chilling stress has been observed in some other crops also (Turan and Ekmekci, 2011). Cold stress adversely affected membrane properties and enzymatic activities leading to plant and tissue necrosis, as observed in banana (*Musa* spp.) (Chinnusamy *et al.,* 2007). Some other crops, which are chilling-sensitive and have been studied for the adverse effects on growth and development include Coffee plant (*Coffea arabica*), tomato (*Lycopersicum esculentum*) and its wild varieties, potato (*Solanum* spp.), Citrus plant, muskmelons (*Cucumis melon*), cotton (*Gossipium hirusutum*), and sugarcane (*Saccharum officinarum* L.).

Conventional techniques for abiotic stress tolerance at field level

Earlier, screening the breeding population/developed varieties for abiotic stress tolerance at field level was based on visual symptoms and/or shoot/root biomass reduction on exposure to the stress (es). Further, to quantitate the tolerance level, necrosis scores on stress exposure relative to unstressed controls were proposed for tolerance screening (Mantri *et al.,* 2010). Moses *et al.* (2008) screened 600 accessions of chickpea (*Cicer arietinum* L.) for salt tolerance under greenhouse conditions based on necrosis scores and shoot biomass reduction compared to unstressed controls at harvest stage. The results indicated wide variation in salinity tolerance determined by both measures. In addition, increase in grain yield on exposure to stress has been commonly used to screen for tolerance in the field. However, these measures of screening for tolerance are laborious, destructive, and time consuming, and the results are subject to environmental variation. Therefore, nondestructive biomass measurement techniques based on satellite remote sensing have been recently developed (Masuka *et al.*, 2012).

Chilling stress induces oxidative processes in plant cells. These processes are initiated by reactive oxygen species, which arise from disturbed operation of electron transport chains and bring about various manifestations of chilling damage. These stresses affect the functions of cell membranes as the primary site of freezing injury, elevate membrane viscosity, and promote lipid transition from a liquid crystalline to a gel phase. As a result, the cells either adapt to these changes or perish. The fluidity of membranes is important for maintaining the barrier properties and for the activation and functioning of certain membrane bound enzymes. Plants (including chickpea) acclimate to environmental low temperature primarily due to the shifts in cell metabolism determined by differential gene expression, which results in changes in the membrane composition and accumulation of cryoprotectants and antioxidants (Leila *et al.*, 2011).

In cold acclimation, plants acquires tress tolerance on prior exposure to sub optimal, low and non-freezing temperatures however; various plant species differ in their ability to face cold stress, which is governed by appropriate changes in gene expression to alter their metabolism, physiology and growth (Chinnusamy *et al.*, 2010). Plant species acclimate during cold stress, by synthesis of cryoprotective molecules such as soluble sugars (saccharose, raffinose, stachyose, trehalose), sugar alcohols (sorbitol, ribitol, inositol) and low molecular weight nitrogenous compounds (proline, glycine betaine) (Janska *et al.*, 2009). These molecules stabilize both membrane phospholipids and proteins, and cytoplasmic proteins in conjunction with dehydrin proteins (DHNs), cold regulated proteins (CORs) and heat shock proteins (HSPs). Cryoprotective solutes are also involved in maintenance of hydrophobic interactions, homeostasis of ions, protection of the plasma membrane from adhesion of ice, scavenging ROS and consequent damage to cells (Iba, 2002). Also, the increased activity of the antioxidative enzymes such as super oxide dismutase, glutathione peroxidase, glutathione reductase, ascorbate peroxidase and catalase, as well as the presence of a series of non-enzymatic antioxidants, such as tripeptidthiol, glutathione, ascorbicacid (vitamin C) and alpha- tocopherol (vitaminE) play important role in cold acclimation and maintenance of cellular redox homeostasis(Chen and Li, 2002). Cold acclimation also affects cell lipid composition by Temperature stress and oxidative responses in crops increasing the proportion of unsaturated fatty acids making up the phospholipids, which is necessary for the maintenance of plasma membrane functionality (Rajashekar, 2000). Cold acclimation induced chilling tolerance in chickpea was found to be associated with marked increase in endogenous ABA, cryoprotective solutes, antioxidative enzymes like ascorbate, glutathione, superoxide dismutase and catalase, relative growth rate of roots and significant decrease in electrolyte leakage and oxidative damage. Some previous observations on this aspect also related higher chilling tolerance imparted by cold acclimation to elevated endogenous ABA, calcium, carbohydrates and proline (Rashmi *et al., 2015).* During cold acclimation, changes in H_2O_2 concentrations and GSH/ GSSG ratio alter the redox state of cells and activate special defense mechanisms through redox signaling chain. H_2O_2 generated by NADPH oxidase in the apoplast of plant cells plays a crucial role in cold acclimation induced chilling tolerance in tomato (*Lycopersicon esculentum*). Some plants modulate their antifreeze activity by Ca2+, which is either released from pectin or bound to specific proteins and enhance the synthesis of proteins that inhibit the activity of ice nucleators in response to cold stress. An altered ratio of abscisic acid (ABA) to gibberellins content, in favor of ABA, results in the retardation of growth required for cold acclimation. Gibberellin content is regulated by a family of nuclear growth repressing proteins called DELLAs, and these are components of the C-repeat (CRT) binding factor1 (CBF1)-mediated cold stress response. However, the degradation of DELLAs is stimulated by gibberellins.

Physiological symptoms attributed to low temperature stress

Leakage of solutes, discoloration and lesions on leaves are some physiological symptoms attributed to low temperature stress (Fig. 1). Other physiological parameters include leaf widening, thickening and changes in foliage. The crown of the plant, consisting of shoot apex and young leaves, is much more cold hardy and can tolerate low temperatures than the more mature older parts. In general, roots much less cold hardened than aerial parts due to the buffering action of soil. This differential ability to cold acclimation indicates that the phenomenon is under genetic and developmental control involving multiple genes. Low temperature stress also manifests increased oxidative damage in plants. There is enhancement of antioxidant defense mechanism of the plant under low temperature stress. A major response of the plant under low temperature stress involves enhanced activities and contents of each of the scavenging enzymes and non-enzymatic antioxidants. The plant's ability to withstand chilling which is a condition very similar to photo-oxidative damage strongly depends upon its existing antioxidant reserves and the ability to add/ enhance the overall antioxidant defense under stress. An enhancement in the antioxidant capacity has been elicited as a response to oxidative stress in chloroplasts which would be due to the oxidative damage to photosynthetic apparatus, due to low temperature stress. The antioxidant action is a concerted effect of the accumulation of conjugated polyamines and phenolic compounds in the extracellular fluid that act as free scavengers.

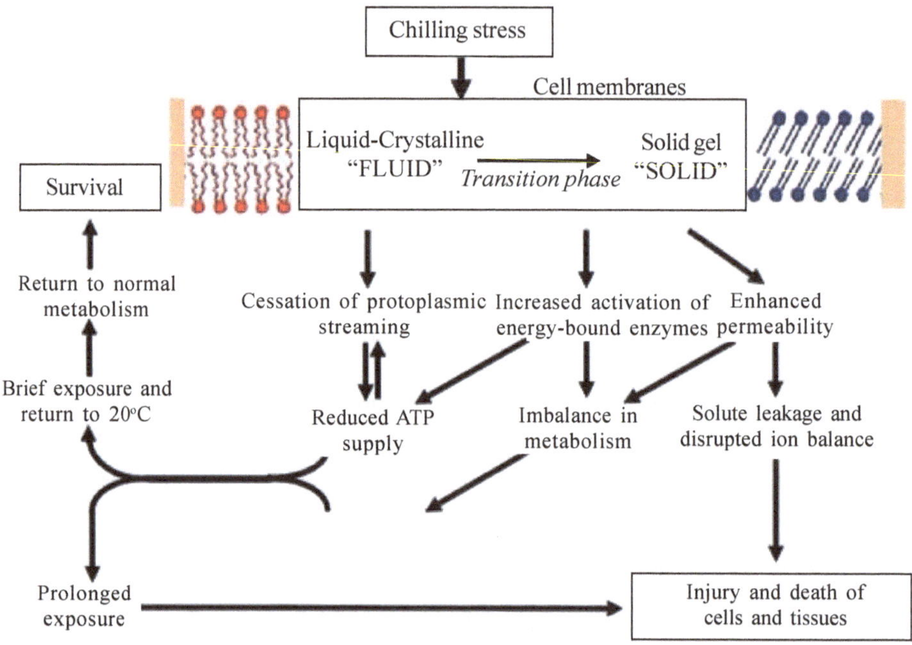

Fig. 1: Symptoms attributed to chilling stress in plants

Molecular responses to low temperature

Molecular mechanisms by which trees respond to cold stress remain poorly understood despite their biological and practical importance (Arora *et al.*, 2003). By far, more cold hardiness research has been carried out on herbaceous annuals like Arabidopsis than on woody perennials. However, two aspects of tree response to low temperature must be underlined. The induction of protective proteins and 2. Change in carbohydrate metabolism. Dehydrins (DHNs) are plant proteins belonging to the group 11 of late embryogenesis-abundant (LEA) proteins that are believed to play a protective role during cellular dehydration. It has been suggested that DHNs stabilize membranes and rescue hydrolytic enzyme function under dehydrative conditions. Cryoprotective and antifreeze activity has also been reported for a peach DHN (Wisniewski *et al.*, 1999). The overexpression of multiple dehydrin genes enhances tolerance in freezing stress in Arabidopsis. Woody plants accumulate DHN during periods of cold acclimation in leaves, buds and bark, and the presence of these proteins is correlated with increased freeze tolerance. In addition, seasonal dehydrin fluctuations have been observed in the xylem of multiple woody species. Different groups have carried out a series of experiments in an attempt to separate cold acclimation/ deacclimation and dormancy transitions and evaluate changes in dehydrin levels in association with each phenomenon separately. Altogether, the results obtained indicate that different genes of DNH family behave differently, the response of the majority would be more closely associated with cold acclimation, whereas that of some others would be with dormancy status (Karlson *et al.,* 2003). Members of another group of proteins called heat shock proteins (HSP) have been identified in herbaceous plants as being responsive to LT stress. The association of HSP, especially belonging to small HSP (sHSP) family, with cold acclimation and endodormancy (López-Matas *et al.*, 2004) has also been shown in some tree species. Changes in tree carbohydrate metabolism in response to SD or LT, or during endodormancy has been observed (Renaut *et al.*, 2004). With the onset of autumn, starch is broken down while at the same time there is a build-up of oligosaccharides, predominantly sucrose, trehalose, raffinose and stachyose. The rise in sucrose levels occurs in direct response to SD, while raffinose and stachyose levels rise later in autumn, in response to temperature drop. Under chilling conditions, sucrose, and trehalose accumulated rapidly, while raffinose content increased after one week at 4 °C (Renaut *et al.*, 2004). The protecting role suggested for these oligosaccharides in cold acclimation of herbaceous plants can be important in woody plants. Indeed, sucrose and trehalose stabilize proteins and membranes during freezing and the accumulation of raffinose may supplement sucrose in membrane stabilisation and might help to prevent sucrose crystallisation during glassy state.

Amelioration of chilling injury

Planting dates can be altered to avoid chilling injury though this is often difficult because of its effect on later development of the plant. To overcome this problem, cultivars have been bred for early vigour and maturity. Maintenance of storage appropriate storage temperatures is essential to avoid chilling injury. Low temperature hardening allowing tolerance to chilling temperatures appears to have little effect although some sensitivity to slight chilling can be reduced by exposure to temperatures slightly above the chilling range. It also appears that chilling injury to stored fruits and vegetables can be ameliorated by warm temperatures if they are imposed before tissue degeneration becomes advanced. Other treatments such as waxing, fungicides, hormones and antioxidants have produced variable results that have been dependent upon the species and treatment conditions. Ultrastructural chilling injury increases with time and with prolonged exposure the injury becomes irreversible. It is therefore important to minimize the time of chilling temperature exposure. Apart from the above, high relative humidity has been found to protect chloroplasts from chill injury, an effect that is enhanced by darkness.

Effect of chilling injury on physiological processes

Chilling injury causes an imbalance in plant physiological processes. Chilling was found to affect O_2 evolution, organic acids, sugars, polyphenols, phospholipids, protein and ATP. Research indicates that chilling stress in sensitive plants changes most chemical entities. There is evidence of accumulation of toxins such as ethanol and acetaldehyde. It is difficult to separate the critical chilling-sensitive metabolic processes from those that are byproducts of metabolic disruptions or of ultrastructural breakdown. Ion leakage due to membrane permeability changes has often been reported in chill sensitive plants. Phase transition of the lipid portion of the cellular membranes has also received considerable attention as the primary response to chilling temperatures.

The ultrastructural symptoms are very similar across species. Ultrastructural symptoms of chilling injury become evident before obvious physical symptoms are visible. These include changes to chloroplasts, mitochondria and membranes associated with these organelles and the vacuoles. The symptoms include swelling and disorganization of the chloroplasts and mitochondria, reduced size and number of starch granules, dilation of thylakoids and unstacking of grana, formation of small vesicles of chloroplast peripheral reticulum. Lipid droplet accumulates in chloroplasts and condensation of chromatin in the nucleus. Chloroplasts are the first and most severely affected organelle. Irradiance during chilling greatly exacerbates the resulting injury. Chilled plants in darkness have been found to remain green and except for starch depletion, chloroplasts appear normal. In the presence of light, however, chlorophyll becomes bleached, lipid droplets accumulate and thylakoids degenerate. Mitochondria appear more resistant to chilling temperature but an immediate effect of low temperature on chilling-sensitive species is a suppression of mitochondrial activity. Electron micrographs of chilled sweet potato roots revealed that the mitochondria had a swollen appearance due to the release of phospholipids from the inner and outer membranes during storage at chilling temperatures. The capacity to bind phospholipids was also greatly decreased.

Solute leakage or ion permeability has provided evidence of increased membrane permeability in response to chilling. The plasma membrane is often considered the primary site of freezing injury and electrolyte leakage. Early work indicated that plants originating in warm climates tend to have more saturated fatty acids in their membranes has shown that membranes do undergo a physical phase transition from a flexible liquid-crystalline to a solid-gel structure at 10 to 12 °C, which coincides with temperature sensitivity range of species of tropical origin. Fruits of several apple cultivars have been observed to undergo phase transition in the 3 to 10 °C range suggesting the same mechanism of chilling injury as found in tropical species. The correlation between fatty acid composition and temperature induced phase transition is however, not precise. It may be that other membrane components such as sterols also play a role. It is possible that the phase transition of cellular membranes could account for the entire range of physiological and metabolic changes associated with chilling injury. Increased membrane permeability could lead to an altered ion balance and also to the ion leakage observed from chilling of sensitive tissues.

Chilling injury can be manipulated by modulating levels of unsaturation of fatty acids by the action of acyl-lipid desaturases and glycerol-3-phosphate acyltransferase while temperature induced phase transition of membrane lipids may play a primary role in chilling sensitivity of plants. Continued exposure to chilling temperatures would result in phase separated membranes becoming incapable of maintaining ionic gradients resulting in metabolic disruption and eventual cell death. A positive correlation has been found between chilling sensitivity of herbaceous plants and the level of saturated and trans-monounsaturated molecular species of phosphatidylglycerol in thylakoid membranes. Growth at low temperature generally increases the degree of unsaturation of membrane lipids, which compensates for the decrease in fluidity caused by the lower temperature. This increased unsaturation is also correlated with the sustained activity of membrane-bound enzymes at low temperature. The unsaturation of membrane lipids is therefore considered critical for the functioning of biological membranes and the survival of plant cells at low temperature. Recently the role of unsaturation of membrane lipids in chilling tolerance and in response to low temperature has been re-examined using mutant and transgenic lines. In this way unsaturated fatty acids can be manipulated independent of temperature so that their individual effects can be evaluated. Tobacco was transformed with squash and Arabidopsis phosphatidylglycerol (PG) species found in thylakoid membranes. Squash has low levels of *cis*-unsaturated PG while Arabidopsis has relatively high levels of *cis*-unsaturated PG. It was found that tobacco transformed with squash PG was more chilling sensitive and tobacco transformed with Arabidopsis PG was the most chilling resistant, as measured by photosynthesis at 1 °C under strong illumination. These results indicate that chilling sensitivity can be manipulated by altering the level of unsaturated PG in the chloroplasts. These and other experiments have shown that unsaturation of membrane lipids protect the photosystem II complex from low temperature photoinhibition by accelerating recovery from the photoinhibited state. However, it is likely that other factors such as accumulation of polyols and amino acids, or their derivatives, contribute to chilling sensitivity in plants. Some specific proteins may also be responsible for chilling tolerance.

Intracellular pH was actively controlled by H^+-transport from the cytoplasm to the vacuole catalysed by H^+-ATPase located on the vacuolar membrane in mung bean (*Vigna radiata* L.), which is a very chilling-sensitive species. The vacuolar H^+-ATPase is extremely sensitive to low temperatures and is preferentially inactivated upon exposure to chilling temperatures. This inactivation occurs much earlier than the symptoms of cell injury and the decrease in enzyme activity associated with plasma membranes, endoplasmic reticulum and mitochondria. Cold induced inactivation of H^+-ATPase also occurs in chilling sensitive rice. Cold induced suppression of proton transport disrupts cytoplasmic homeostasis and cause a change in the pH. The chilling sensitivity of cultured mung bean cells changed markedly during the growth cycle and a close relationship was found between sensitivity of the cells and of H^+-ATPase to the cold. Cold induced inactivation of the vacuolar H^+-ATPase was closely linked to acidification of the cytoplasm and the corresponding alkalization of the vacuoles suggesting a passive release of H^+-ions across the vacuolar membrane. The susceptibility of vacuolar H^+-ATPase to low temperature in vivo was found to be markedly different between chilling-sensitive and chilling-resistant species. In contrast to the H^+-ATPases of chilling-sensitive species like mung bean and kidney bean, the H^+-ATPases of the chilling-tolerant species such as pea and broad bean were very stable over long periods of low-temperature exposure. The molecular structures of the 16kDa proteolipids from the two types of H^+-ATPase appeared to be very different. Low temperature-induced pH reduction of the cytoplasm caused by inactivation of vacuolar H^+-ATPase may therefore be the cause of extreme chilling-sensitivity.

Chilling/cold stress limits plant growth and causes significant crop loss. Stress responses are influenced by duration of exposure, species and the stage of plant development. Plants native to temperate regions exhibit varying degrees of cold tolerance and can acquire freezing tolerance after cold acclimation, while many tropical and subtropical plants are sensitive to chilling at 0–15 °C (M. To cope with low temperatures, plants often upregulate the expression of some cold related genes (*CORs*). At the molecular level, a C-repeat (*CRT*) or dehydration responsive element (*DRE*) is commonly found in the promoter regions of *COR*, and is bound by CRT binding factor (CBF)/DRE binding (DREB) protein. Expression of *VrDREB2A* in mungbean seedlings was markedly induced by drought, high-salt stress and ABA treatment, but only slightly by cold stress (Chen *et al.*, 2016). Although the CBF pathway has been one of the dominant signal mechanisms mediating cold acclimation and is widely conserved in higher plants, it may function differently in mungbean plants. Different mungbean cultivars vary in their physio-biochemical responses to UV-B, heat stress and bruchid damage. Genome size of different mungbean varieties ranged from 494 to 554 Mb and at least 52,739 genes have been annotated. Transcriptomic comparison between bruchid resistant and susceptible mungbean lines identified nucleotide variations caused by differential expressed genes and sequence-changed-protein genes of mungbean and transposon elements, besides bruchid-resistant (*Br*) genes, as putative modifier factors for bruchid resistance. Nevertheless, little is known about the molecular and physiological mechanisms underlying the intrinsic susceptibility of mungbean to chilling stress. We previously isolated 1198 mungbean expressed sequence tags (ESTs) informative to early seedling development and chilling response, and showed that variety NM94 maintained better membrane integrity than VC1973A under chilling/cold stress (Chen *et al.,* 2008 and 2017).

Effect of freezing stress on physiological and agronomic aspects

Generally, plants have adapted two mechanisms to protect themselves from damage due to below freezing temperatures. Supercooling is a low-temperature tolerance mechanism that is usually associated with acclimated xylem parenchyma cells of moderately hardy woody plants. When sources of ice nucleation are absent, pure water can supercool or remain unfrozen to its homogeneous nucleation point of approximately -40°C. The initiation of freezing at the limit of supercooling occurs suddenly and is accompanied by an exotherm that can be detected by thermal analyses of plant tissues. Plant tissues suffer irreversible damage once ice nucleation of supercooled water occurs and the distribution in nature of tree species with the ability to deep supercool is normally restricted to regions where winter temperatures are warmer than -40°C. Second mechanism is acclimation. Low temperature acclimation is a gradual process during which there are changes in just about every measurable morphological, physiological and biochemical characteristic of the plant.

Cold acclimation of winter wheat plants begins once fall temperatures drop below approximately 9°C. A translocatable substance that promotes cold acclimation is not produced when winter wheat plants are exposed to acclimating temperatures. Consequently, the cold-hardiness level of different plant parts, such as leaves, crowns and roots, is dependent upon the temperature to which each part has been exposed. Because the crown contains tissues that are necessary for plant survival, it is the soil temperature at crown depth that determines critical cold-acclimation rates. Plant growth slows considerably at temperatures that promote cold acclimation. In the field, soil temperatures gradually decrease as winter approaches and four to seven weeks at temperatures below 9 °C is usually required to fully cold-harden plants. Cold acclimation during this period is dependent upon crown temperatures and the rate of acclimation increases dramatically as temperature drop from 9 to 0°C. Exposure of winter wheat crowns to soil temperatures above 9°C during this period results in a rapid loss of cold hardiness. The rate of dehardening is dependent upon the temperature to which the crown is exposed. At this stage, plants that have been exposed to crown temperatures above 9°C will resume cold acclimation once they return to temperatures below 9°C. Once verbalization saturation is complete and the plant enters the reproductive stage, it loses its ability to cold acclimate and it will start to deharden at temperatures warmer than approximately -4°C. This means that winter wheat will eventually completely deharden once plant growth resumes in the spring. Growth rate and rate of dehardening are both temperature dependent and because frozen soils warm slowly in the spring, several weeks of warm air temperatures are required to re-establish and completely deharden winter cereal plants that have survived without serious winter damage.

Plant tolerance to cold stress

Cold stress occurs at temperatures less than 20°C and varies with the degree of temperature duration and plant type. Chilling (<20°C) or freezing (<0°C) temperatures can trigger the formation of ice in plant tissues, which causes cellular dehydration. Ultimately, cold stress reduces plasma membrane (PM) integrity, causing leakage of intracellular solutes. Cold stress severely affects plant growth and survival, and leads to substantial crop losses in temperate climatic regions and hilly areas of the tropics and subtropics. In rice, for instance, losses due to cold stress can range from 0.5 to 2.5 t/ha and grain yields can drop by up to 26%, especially when low temperatures occur during the reproductive stage. To cope with this adverse condition, plants adapt several strategies such as producing more energy by activation of primary metabolisms, raising the level of anti-oxidants and chaperones, and maintaining osmotic balance by altering cell membrane structure. These mechanisms of plant response to cold stress are closely similar to that of heat stress. However, the difference lies in the fact that membrane rigidification occurs in cold stress as opposed to heat stress. Thus, membrane rigidification is the upstream trigger for the induction cytosolic Ca^{2+} signatures leading to a transient increase in cytosolic Ca^{2+}levels. It is assumed that dimethyl sulfoxide (DMSO) mediates the perception of membrane rigidification by mechanosensitive Ca^{2+}channels. Other upstream factors such as changes in the metabolic reactions and metabolite concentrations, protein and nucleic acid conformation could contribute to enhance perception of cold stress. These factors as well, either directly or indirectly, induce an increase in cytosolic Ca^{2+}, which is a well-known upstream second messenger, regulating cold regulated (*COR*) gene expression.

Cold-stress-induced cytosolic Ca^{2+} signals can be decoded by different pathways. More recently, Ca^{2+}signal were reported to be transduced directly into the nucleus. The concentration of nuclear Ca^{2+} is monitored by a chimera protein formed by the fusion of aequorin to nucleaoplasmin, which is also transiently increased after cold shock. Aequorin possesses several EF-hand-type binding sites for Ca^{2+} ions. The binding of Ca^{2+} to these sites causes a conformational change in aequorin which enables the monitoring of Ca^{2+} concentration. It has been reported that nuclear Ca^{2+} concentration peaks at about 5–10s later than the cytosolic Ca^{2+}. In the cytoplasm, a range of Ca^{2+} sensors including calmodulin (*CaM*), CaM-like (*CMLs*), Ca^{2+}-dependent protein kinases (*CDPKs*), Ca^{2+}-and Ca^{2+}/CaM-dependent protein kinase (*CCaMK*), CaM-binding transcription activator (*CAMTA*), calcineurin B-like proteins (*CBLs*) and *CBL*-interacting protein kinases (*CIPKs*). Some of the sensors work as negative regulators of cold tolerance in plants, e.g., calmodulin3, a SOS3-like or a CBL calcium-binding protein and a protein phosphatase 2C (*AtPP2CA*). The positive regulators, e.g., CDPKs and probably some *CBLs*, relay the Ca^{2+} signal by interacting with and regulating the family of *CIPKs*. For instance, *CBL1* has been shown to regulate cold response by interacting with *CIPK7*, whereas *CAMTA3* has been identified as a positive regulator of *CBF2/DREB1C* through binding to a regulatory element (*CG-1, vCGCGb*) in its promoter. Although *CBF2/DREB1C* was negatively regulate *CBF1/DREB1B* and *CBF3/DREB1A*, its expression appears to be necessary for integrating cold-inducible calcium signaling with gene expression, but under transient and tight control to avoid repression of freezing tolerance. Both *CBF1 / DREB1B* and *CBF3/DREB1A* are required for constitutive expression of cold-inducible genes in *Arabidopsis,* and play an important role in cold acclimation. Ca^{2+} influx into the cytoplasm also apparently activates phospholipase C (PLC) and D (PLD), which are precursors for IP_3 and PA, respectively. IP_3 activates IP_3-gated Ca^{2+} channels that can amplify Ca^{2+}signatures in the cytoplasm, leading to higher induction of *COR* genes and CBFs.

The chloroplast may also play a role in sensing low temperature. Cold stress is considered to cause excess photosystem II (PSII) excitation pressure, as a result of the imbalance between the capacity for harvesting light energy and the capacity to consume this energy on metabolic activity in the leaves, which probably leads to ROS generation. The damaging effect of ROS on the photosynthetic apparatus presumably leads to photo-inhibition, which occurs even under relatively low irradiance and is apparently a mechanism of cold acclimation or freezing tolerance. ROS also acts as the second messenger and may reprogramme transcriptome changes through induction of Ca^{2+} signatures and activation of MAPKs and redox-responsive TFs. The MAPK cascades in *Arabidopsis*, including AtMEKK1/ANP1 (MAPKKK)–AtMKK2 (MAPKK)–AtMPK4/6 (MAPK), positively regulate cold acclimation in plants.

Downstream of these TFs are *COR* genes, which are mainly linked to the onset of tolerance mechanisms and ultimately lead to acclimation. Genes encoding for annexin; hyper-sensitive-induced response (HIR) protein families (e.g., prohibitins and stomatins); dehydrins (e.g., 25 kDa dehydrin-like protein, *ERD14*, and *cold acclimation-specific protein 15* (*CAS15*)); anti-oxidants (e.g., superoxide dismutase, catalase and ascorbate peroxidase); *HSPs* (e.g., *HSP70* family being the most abundant); defence-related proteins such as protein disulfide isomerase; disease resistance response proteins, peptidylprolyl isomerase *Cyp2* and cysteine proteinase; amino acids, polyamines and polyols; and cellulose synthesis, such as UDP-glucose pyrophosphorylase, are commonly reported in expression studies. Several metabolism-associated proteins, including carbohydrate metabolism enzymes, such as phosphogluconate dehydrogenase, NADP-specific isocitrate dehydrogenase, fructokinase, cytoplasmic malate dehydrogenase, pyruvate orthophosphate dikinase precursors (PPDK), aconitate hydratase, glycine dehydrogenase and enolase, have activated during cold stress. Thus, several genes and the corresponding proteins are associated with the regulation of the metabolic pathways operating under cold stress (Geoffrey and Kerstin, 2016).

Nitrogen Assimilation and Low Temperature Stress

Plants accumulate certain compatible nitrogenous compounds specific to a particular stress. These compounds help the plants to combat that particular stress condition (Choudhary and Singh, 2000). The most common nitrogenous compound which accumulates upon chilling injury is proline and glycine betaine. In barley and radish, substantial amounts of proline and amino acids serine, glycine and alanine accumulate on exposure to low temperatures. Accumulation of proline is most common in most of the plants as this amine acid does not interfere with the normal cell functioning and serves as N source specially during stress recovery. Accumulation of the osmolytes proline and glycine betaine decreases the cytosolic-free water potential, which restricts the loss of water. Further, these osmolytes protect protein complexes in organelles and cytosol against dehydration damage by keeping them hydrated. Transgenic tobacco plants expressing the enzyme pyrroline-5-carboxylate synthetase, an enzyme of proline biosynthetic pathway, from *Vigna aconitifolia* synthesize and accumulate about 10 fold more proline than control plants and show greater tolerance to freezing (Srivastava, 2002). Dufeu *et al.* (2003) reported that there existed a flexibility of polyamine and aromatic amine metabolism in the seedling of *Pringlea antiscorbutica,* in relation to temperature. The cold-cultivated seedlings of *Pringlea* maintained high levels of the polyamine spermidine which perhaps plays a crucial role in plant development under cold conditions.

Low temperature reduce NO_3 uptake by the roots and affect the partitioning of N within the whole plant. Short term exposure of roots of *Lolium multiflorum* grass, *Cicer arietinum* and barley to low temperatures caused reduced NO_3 uptake by the roots with a concomitant decrease in total N content of the plants. Soybean plants when exposed to 15 °C for 4 days showed about 52 to 61% decrease in N partitioning in the young shoots compared to plants grown at 25 °C. About 22% remobilization of N from older leaves to the young shoots was observed in such plants suggesting that cold tolerant cultivars have increased N partitioning in shoots. It is suggested that tolerance to low temperature can be increased by increasing N supply to young shoots (Walsh and Layzell, 1986).

The level of soluble proteins during the cold-hardening was enhanced which was accompanied with increased activity of enzymes of nitrate reduction, nitrate reductase (NR) and glutamine synthetase (GS) in many plant species including wheat, barley, sugar beet, rose, sunflower, etc. The prime enzyme of nitrate reduction, nitrate reductase (NR) shows varying behaviours depending on the plant species when plants are exposed to low temperatures. Barley and maize seedlings, when grown at 20°C for 7 days, showed a drastic reduction NR activity compared to the seedlings grown at 28°C. Low temperature stress along with low soil pH influenced nitrate reduction by strongly decreasing NR activity in growing plants as NR is a substrate inducible enzyme, its activity is dependent on the level of NO_3 in the active pool. Reduced NO_3 uptake under low temperature would thus lead to decrease in NR activity. However, in N_2 fixing plant species like black alder (*Alnus glutinosa*), exposure of seedlings to chilling temperatures of -1 to 4 °C for 2 h during the night led to an increased NR activity in both roots and shoots of the plants. In such plants the apparent increase in NR activity is due to constitutive NR which is reported to be present in many N_2O fixing plants (Dubey and Pessaralki, 2002).

Low Temperatures and Photosynthesis

Exposure of plants to low temperatures leads to changes in membrane fluidity. Chloroplast membranes are also susceptible to these changes. A strong reduction in chloroplast membrane content and its disorganization has been reported in spinach leaves under chilling. Chilling injury causes discoloration of leaves and disorganization of thylakoid membranes. Swelling of chloroplast and accumulation of starch grains within the chloroplast has also been reported in plants upon exposure to low temperature. A decrease in the photosynthetic capacity has been observed in many plants upon exposure of leaves to chilling temperatures (below 10 °C, but above freezing point). The site of inactivation has been identified in the oxygen-evolving system of photosystem II. This inactivation is irreversible by nature. PSI is much more susceptible in aerobic photoinhibition at chilling temperature. Low temperature enhanced the distribution of excitation energy of PSI and altered the amount transferred from Chl b to Chl a in maize. The photoinhibition of PSI is irreversible involving inactivation of Fe-S centers and the reaction center subunit (Psa B protein).

Pyruvate orthophosphate dikinase enzyme plays an important role in decreased photosynthetic capacity resulting from chilling injury. The activity of this enzyme is lowered at temperatures below 11°C, the enzyme reversibly dissociates to dimeric and monomeric forms which are less active. The activity of carboxylating enzyme RUBISCO significantly decreases in plant species like *Zea mays* which are sensitive to chilling. At 14^{0}C a 75% decrease in RUBISCO activity and 50% decrease in the activity of NADP-malate dehydrogenase (C_4) with no change in PEP case was reported in *Zea mays*. However, an increase in the activities of RUBISCO, NADP-MDH and many other enzymes of carbon metabolism in spinach subjected to 10^{0}C for 10 days. A lack of an NAD(P)H dependent cyclic electron flow around PSI in tobacco plants attributed to chilling induced photoinhibition and inhibited to maintain normal C metabolism under adverse conditions of low temperatures. Overall photosynthetic-CO_2 assimilation decreased to 50% indicating a significant inhibition of PS II at 10 °C (Nie *et al.*, 1992).

The most commonly accumulated sugar in plants exposed to low temperatures is sucrose. It gets accumulated to about 10-fold in spinach leaves with an increase in sucrose phosphate synthase activity upon exposure to cold temperatures. In winter wheat, activities of enzymes of sucrose biosynthesis sucrose phosphate synthase and acid invertase increase several fold on exposure to low temperatures (Srivastava, 2002). Sucrose acts as a cryoprotectant and helps in accumulation of other photosynthate such as starch. Chilling also leads to an increase in amount of glucose and fructose.

Low Temperatures and Oxidative Stress

Low temperatures lead to production of reactive oxygen species, because the light harvesting reactions continue to function, while biochemical reactions are severely effected. Over-expression of SOD gene in alfalfa could ameliorate oxidative stress caused due to ROS and protect the plants against freezing injury. Plasmalemmae of plants cells, under chilling stress produces oxidative bursts of H_2O_2 overwhelming the antioxidative defense machinery of the cells. Low temperature stress responses include oxidative stress responses which are also observed under water deficit conditions in higher plants. One such trait is radical scavenging during stress. Transgenic tobacco plants that over-express SOD are more tolerant than untransformed control to a superoxide-generating herbicide and ozone. Increased SOD activity in response to stresses has been shown to confer increased protection from oxidative damage (Shah *et al.*, 2001). Catalase is another key enzyme that helps in the removal of toxic peroxides including H_2O_2. This is most universal oxidoreductase that scavenges H_2O_2 via a two electron transfer producing O_2 and H_2O. A decline in catalase activity is regarded as a general response to several stresses including low temperature stress. It is attributed to inhibitions of enzyme synthesis or an alternation in enzyme sub units under low temperatures.

Synthesis of cold-related proteins

Cold acclimation involves switching on of multiple genes which in turn induce the synthesis of cold-related proteins also known as cold acclimation proteins in plants. These proteins have been given names after the inducing agent e.g. early-dehydration-inducible (ERD), low temperature induced (LTI), low temperature responsive or cold responsive (COR), cold induced (KIN) or after the developmental stage at which they were first recognized e.g. late embryogenesis abundant (LEA) or they are named after the labeling pattern of their encoded product e.g. early methionine-labeled (Em). Cold acclimation also involves synthesis/induction of enzymes of lipid metabolism and synthesis of antifreeze proteins (AFPs) in some plants. LEA proteins and other hydrophilic proteins which have been shown to be synthesized under low temperature stress appear to confer freezing tolerance in rice plants transformed with the Harley HVAI cDNA, encoding a LEA III protein. Transgenic Arabidopsis plants expressing COR15a encoding a plastidic polypeptide, show increased freezing tolerance of chloroplasts *in situ* (Srivastava, 2002).

Membrane integral proteins (MIPs) which facilitate water movement by forming small pores across membranes are termed as aquaporins. The aquaporin genes are reported to be up regulated under water stress in pea and Arabidopsis and down regulated in *M. crystallinum* showing their role in chilling injury. A family of proteins associated with cell dehydration commonly known as dehydrins assist cells in tolerating dehydration caused within the cell due to environmental factors. Recent findings demonstrate that a low temperature induced dehydrin from peach (PCA60) and cold-induceed dehydrin from barley possess cryoprotective and antifreeze activity (Bravo *et al.,* 2003). The upregulation of these proteins by cold suggest their role in stabilizing cell structures during freeze imposed dehydration of cytoplasm.

Synthesis of AFPs occurs in a wide range of plant species on exposure to freezing temperatures. The synthesis of AFPs is more common in winter cereals during cold acclimation. After synthesis these AFPs are escorted to the extra cellular space and deposited in the apoplast. Plant AFPs exhibit thermal hysteresis and their main role lies in preventing the growth of ice crystals by directly interacting with ice by adsorption on to the surface of ice crystals thereby inhibiting further addition of water molecules to already frozen surface. AFPs isolated from winter rye and other cereals belongs to certain classes of chitinasaes, glucanases and thaumatin like proteins and thus cereal AFPs presumably appear to be isozymes of pathogenesis related proteins (Srivastava, 2002). In winter rye these AFPs have been shown to be gene products, localized in cell walls and that the corresponding genes are expressed in all living leaf cells under cold acclimation (Pihakaski-Maunsbach *et al.,* 2003). Plants also produce an array of small defence peptides in response to plan stresses, including defensins, thionins and non-specific lipid transfer proteins (LTPs). These compounds exhibit ability to arrest growth of plant pathogens and levels of certain LTPs have been shown to increase in winter barley following exposure to low, above-freezing, hardening temperatures (2°C) both in the field and under controlled environment conditions.

Low temperature and membrane fluidity

Membrane integrity compromise both Chilling as well as freezing,. The maintenance of membrane functional integrity necessarily involves the appropriate membrane fluidity. Membrane fluidity is important to ensure selective permeability, transport of ions and membrane associated electron transport. Freezing injury to membranes involves loss of activity of intrinsic membrane proteins and enzymes, loss of permeability, leakage of electrolytes, etc., many workers have reported desaturation of membrane lipids upon acclimation of plants to low temperatures, allowing the maintenance of correct fluidity at these temperatures. The desaturation is mediated by desaturases enzymes that introduce double bonds at specific locations in fatty acid chains.

Chilling tolerant plant species have been shown to possess a large ratio of unsaturated/saturated fatty acids in the membrane lipids, when compared to the chilling sensitive plant varieties. The FADS, desaturase gene of Arabidopsis has been shown to be induced specifically under cold acclimation, thereby increasing the relative ratio of phospholipids to sterols in membranes which enhances membrane fluidity (Srivastava, 2002). The relative proportions of polyunsaturated fatty acids as well as amounts of linoleic acid (C18:2) or linolenic acid (C18:3) increase in plants on acclimation to cold. This is also elicited by the presence of specific isoforms of the enzyme like glycerol-3-phosphate acyltransferase (GPAT) responsible for transfer of a fatty acid chain to glycerol-3-phpsphate in the plastids or cytosol. The level of this specific GPAT isoforms has been shown to be elevated in chill- resistant Arabidopsis where it adds unsaturated fatty acids. In chill sensitive varieties however, the isoform leads to addition of saturated fatty acid.

In plants exposed to freezing temperatures, ice formation takes place within the cells that leads to complete loss of membrane integrity and fluidity thereby ultimately resulting in cell-death. Withdrawal of intracellular water leads to dessication and developed strain on cell-wall constituents. Freezing also destroys the intracellular compartmentation, again proving lethal for the cells. Severe damage occurs when ice crystals grow and puncture into the cytoplasm. Ice crystal formation begins at -3 °C to -5 °C in the cell walls. In many plants, the temperature of water initially falls below its freezing temperature but still remains in fluid state termed as supercooling, then the transition to solid phase takes place. While becoming solid, ice gives off heat and the temperature rises but when water in the cell gets frozen, the temperature drops again and freezing injury occurs. Many plants tolerate subfreezing temperatures in winter by supercooling. In such plants, as supercooled water still remains liquid, the dehydrative damage to membranes is minimal (Srivastava, 2002). It is believed that temperature changes leading to cold acclimation trigger desaturate enzymes and activate other genes by physical phase transitions that permit certain sensor proteins to undergo conformational changes on transduction of the temperature signal.

Conclusions

Low temperature stress severely impairs plant growth and development, limits plant production and the performance of crop plants. Numerous physio-biochemical indicators for tolerance screening and indirect selection using molecular markers linked to desire loci are being deployed for accelerating the production of stress-tolerant varieties. Plants acclimate to environmental low temperature primarily due to the shifts in cell metabolism determined by differential gene expression, which results in changes in the membrane composition and accumulation of cryoprotectants and antioxidants. Post-harvest storage at lower temperatures is commonly used to extend the storage life of fruits and vegetables.

Photoperiod is the most important environmental cue controlling the onset of dormancy in perennial plants. By responding to short day photoperiod, plants are able to synchronize cold acclimation and dormancy induction with the end of the growing season. The plant's ability to withstand chilling which is a condition very similar to photo-oxidative damage strongly depends upon its existing antioxidant reserves and the ability to add/ enhance the overall antioxidant defense under stress. An enhancement in the antioxidant capacity has been elicited as a response to oxidative stress in chloroplasts which would be due to the oxidative damage to photosynthetic apparatus, due to low temperature stress. The antioxidant action is a concerted effect of the accumulation of conjugated polyamines and phenolic compounds in the extracellular fluid that act as free scavengers.

Increased membrane permeability could lead to an altered ion balance and also to the ion leakage observed from chilling of sensitive tissues. The chloroplast may also play a role in sensing low temperature. Cold stress is considered to cause excess photosystem II (PSII) excitation pressure, as a result of the imbalance between the capacity for harvesting light energy and the capacity to consume this energy on metabolic activity in the leaves, which probably leads to ROS generation. The damaging effect of ROS on the photosynthetic apparatus presumably leads to photo-inhibition, which occurs even under relatively low irradiance and is apparently a mechanism of cold acclimation or freezing tolerance. ROS also acts as the second messenger and may reprogramme transcriptome changes through induction of Ca^{2+} signatures and activation of MAPKs and redox-responsive TFs. The MAPK cascades in *Arabidopsis* , including AtMEKK1/ANP1 (MAPKKK)–AtMKK2 (MAPKK)–AtMPK4/6 (MAPK), positively regulate cold acclimation in plants. Cold acclimation involves switching on of multiple genes which in turn induce the synthesis of cold-related proteins also known as cold acclimation proteins in plants. These proteins have been given names after the inducing agent e.g. early-dehydration-inducible (ERD), low temperature induced (LTI), low temperature responsive or cold responsive (COR), cold induced (KIN) or after the developmental stage at which they were first recognized e.g. late embryogenesis abundant (LEA) or they are named after the labeling pattern of their encoded product e.g. early methionine-labeled (Em). Cold acclimation also involves synthesis/induction of enzymes of lipid metabolism and synthesis of antifreeze proteins (AFPs) in some plants. LEA proteins and other hydrophilic proteins which have been shown to be synthesized under low temperature stress appear to confer freezing tolerance in rice plants transformed with the Harley HVAI cDNA, encoding a LEA III protein. Transgenic Arabidopsis plants expressing COR15a encoding a plastidic polypeptide, show increased freezing tolerance of chloroplasts in situ.

References

Arora, R., Rowland, L. J. and Tanino, K. 2003. Induction and release of bud dormancy in woody perennials: a science come of age. *Hort. Science,* 38: 911-921.

Ashraf, M. and Foolad, M.R. 2013. Crop breeding for salt tolerance in the era of molecular markers and marker-assisted selection. *Plant Breeding,* 132(1): 10–20.

Ashraf, M. and Foolad, M.R. 2007. Roles of glycine betaine and proline in improving plant abiotic stress resistance. *Environ. Exp. Bot.,* 59: 206-216.

Bravo, L.A., Gallardo, J., Navarrete, A., Olave, N., Martinez, J., Alberdi, M., Close, T.J. and Corcuera, L.J. 2003. Cryoprotective activity of a cold induced dehydrin purified from barley. *Physiol. Plant.,* 118: 262-269.

Chen, W. P. and Li, P. H. 2002. "Attenuation of reactive oxygen production during chilling in ABA-treated maize cultured cells" In: *Plant Cold Hardiness*, eds C.Li and E. T. Palva (Dordrecht: Kluwer Academic Publishers), 223–233.

Chen, L.R., Markhart, A.H., Shanmugasundaram, S. and Lin, T.Y. 2008. Early developmental and stress responsive ESTs from mungbean, *Vigna radiata* (L.) Wilczek, seedlings. *Plant Cell Rep.,* 27: 535–552.

Chen, H., Liu, L., Wang, L., Wang, S. and Cheng, X. 2016. VrDREB2A, a DREB-binding transcription factor from *Vigna radiata,* increased drought and high-salt tolerance in transgenic *Arabidopsis thaliana. J. Plant Res.,* 129: 263–273.

Chen, Li Ru., Chia Yun Ko, William R. Folk and Tsai Yun Lin. 2017. Chilling susceptibility in mungbean varieties is associated with their differentiallyexpressed genes. *Bot Stud.,* 58 (7): 1-9.

Chinnusamy,V., Zhu, J. and Zhu, J.K. 2007. Cold stress regulation of gene expression in plants. *Trends Plant Sci.,* 12: 10-16.

Chinnusamy,V., Zhu, J. K. and Sunkar, R. 2010. Gene regulation during cold stress acclimation in plants methods. *Mol.Biol.* 639: 39–55.

Choudhary, A. and Singh, R. P. 2000. Cadmium induced changes in diamine oxidase activity and polyamine levels in Vigna radiate Wilczek seedlings. *J. Plant Physiol.*, 156: 704-710.

Duan, B., Yang, Y., Lu, Y., Korpelainen, H., Berninger, F. and Li, C. 2007. Interactions between drought stress, ABA and genotypes in *Picea asperata. J. Exp. Bot.*, 58: 3025-3036.

Dubey, R.S. and Pessarakli, M. 2002. Physiological mechanisms of nitrogen absorption and assimilation in plants under stressful conditions. In: *Handbook of Plant and Crop Physiology,* Marcel Dekker Inc, N.Y. 2nd Edition, pp 637-656.

Dufeu, M., Tanguy, J.M. and Hennion, F. 2003. Temperature-dependent changes of amine levels durinf early seedling development of the cold-adapted subantarctic crucifer *Pringlea antiscorbutica. Physiol. Plant.*, 118: 164-172.

Frewen, B. E., Chen, T. H., Howe, G. T., Davis, J., Rohde, A., Boerjan, W. and Bradshaw, H. D., 2000. Quantitative trait loci and candidate gene mapping of bud set and bud flush in *Populus*. *Genetics*, 154: 837-845.

Geoffrey Onaga and Kerstin Wydra. 2016. Advances in plant tolerance to abiotic stresses. In: *Plant Genomic,* Book Edited by Ibrokhim Y Abdurakhmonor.

Gómez, L. , Allona, I., Ramos, A., Núñez, P., Ibáñez, C., Casado, R. and Aragoncillo, C. 2005. Molecular responses to thermal stress in woody plants. *Invest Agrar: Sist. Recur. For.*, 14(3): 307-317.

Hasanuzzaman,M., Nahar,K. and Fujita, M. 2013. Extreme temperatures, oxidative stress and antioxidant defense in plants, in *Abiotic Stress—Plant Responses and Applications in Agriculture*, eds K.Vahdati and C.Leslie (Rijeka: In Tech), 169–205.

Howe, G. T., Davis, J., Jeknic, Z., Chen, T. H. H., Frewen, B., Bradshaw, H. D. and Saruul, P. 1999. Physiological and genetic approaches to studying endodormancy-related traits in *Populus*. *Hort. Science*, 34: 1174-1184.

Iba, K. 2002. Acclimative response to temperature stress in higher plants: approaches of gene engineering for temperature tolerance. *Annu. Rev. Plant Biol.*, 53: 225-245.

Janska, A., Mars, P., Zelenkova, S. and Ovesna, J. 2009. Cold stress and acclimation–what is important for metabolic adjustment? *Plant Biol.*, 12: 395–405.

Leila Heidarvand, Reza Maali Amiri, Mohammad Reza Naghavi, Yadollah Farayedi, Behzad Sadeghzadeh and Khoshnood Alizadeh. 2011. Physiological and morphological characteristics of chickpea accessions under low temperature stress. *Russian Journal of Plant Physiology*. 58 (1): 1-7.

Li, C., Junttila, O., Heino, P. and Palva, E. T. 2004. Low temperature sensing in silver birch *(Betula pendula* Roth) ecotypes. *Plant Sci.*, 167: 165-171.

Levitt, J. 1980. Responses of plants to environmental stresses. New York: Academic Press.

López-Matas, M. A., Nuñez, P., Soto, A., Allona, I., Casado, R., Collada, C., Guevara, M. A., Aragoncillo, C. and Gomez, L. 2004. Protein cryoprotective activity of a cytosolic small heat shock protein that accumulates constitutively in chestnut stems and is up-regulated by low and high temperatures. *Plant Physiol.*, 134: 1708-1717.

Karlson, D. T., Zeng, Y., Stirm, V. E., Joly, R. J. and Ashworth, E. N. 2003. Photoperiodic regulation of a 24-kD dehydrin-like protein in red-osier dogwood *(Cornus sericea* L.) in relation to freeze-tolerance. *Plant Cell Physiol.*, 44: 25-34.

Mantri, N., Ford, R., Coram, T.E. and Pang, E.C.K. 2010. Evidence of unique and shared responses to major biotic and abiotic stresses in chickpea. *Environ. Exp. Bot.*, 69(3): 286–292.

Mantri, N., Patade, V., Penna, S., Ford, R. and Pang, E.C.K. 2012. Abiotic stress responses in plants - present and future. In: Ahmad, P. and Prasad, M.N.V. (eds) Abiotic stress responses in plants: metabolism to productivity. Springer, Science + Business Media NY, USA, pp 1–19.

Masuka, B., Araus, J.L., Das, B., Sonder, K. and Cairns, J.E. 2012. Phenotyping for abiotic stress tolerance in maize. *J. Integr. Plant Biol.*, 54(4): 238–249.

Moses, F.A., McNeil, M.D., Redden, B., Kollmorgen, J.F. and Pittock, C. 2008. Sampling strategies and screening of chickpea (*Cicer arietinum* L.) germplasm for salt tolerance. *Genet. Resour. Crop Evol.,* 55: 53–63.

Nejadsadeghi, L., Maali-Amiri, R., Zeinali, H., Ramezanpour, S. and Sadeghzade, B. 2014. Comparative analysis of physio-biochemical responses to cold stress in tetraploid and hexaploid wheat. *Cell Biochem.Biophys.,* 70: 399–408.

Nie, G.Y., Long, S.P. and Baker, N.R. 1992. In: *Handbook ofplant and Crop Physiology*. Marcel Dekker Inc, N.Y. 2nd Edition, pp 637-656.

Niu, C. F., Wei, W., Zhou, Q.Y., Tian, A, G., Hao, Y. J., Zhang, W. K., Ma, B., Lin, Q., Zhang, Z. B., Zhang, J. S. and Chen, S.Y. 2012. Wheat WRKY genes TaWRKY2 and TaWRKY19 regulate abiotic stress tolerance in transgenic Arabidopsis plants. *Plant Cell Environ.,* 35(6): 1156–1170.

Pihakaski-Maunsbach, K., Tamminen, I., Pietiainen, M. and Griffith, M. 2003. Antifreeze proteins are secreted by winter rye cells in suspension culture. *Physiol. Plant.,* 118: 390-398.

Praba, M. L., Cairns, J. E., Babu, R. C. and Lafitte, H. R. 2009. Identification of physiological traits underlying cultivar differences in drought tolerance in rice and wheat. *J. Agron. Crop Sci.,* 195: 30-46.

Rajashekar, C. B. 2000."Cold response and freezing tolerance in plants" in *Plant Environ. Interactions,2nd Edn.*, ed R.E.Wilkinson (NewYork, NY: Marcel Dekker, Inc.), 321–341.

Rashmi Awasthi, Kalpna Bhandari and Harsh Nayyar. 2015. Temperature stress and redox homeostasis in agricultural crops. *Frontiers Env. Sci.* 3(11): 1-24.

Renaut, J., Lutts, S., Hoffmann, L. and Hausman, J. F. 2004. Responses of poplar to chilling temperatures: proteomic and physiological aspects. *Plant Biol.,* 6: 1-10.

Shah, K., Kumar, R.G., Verma, S. and Dubey, R.S. 2001. Effect of cadmium on lipid peroxidation, superoxide anion generation and activities of antioxidant enzymes in growing rice seedlings. *Plant Sci.,* 161: 1135-1144.

Shakeel Ahmad Anjum, Xiao-yu Xie, Long-chang Wang, Muhammad Farrukh Saleem, Chen Man and Wang Lei. 2011. Morphological, physiological and biochemical responses of plants to drought stress. *African Journal of Agricultural Research,* 6(9): 2026-2032.

Shao, H. B., Chu, L. Y., Jaleel, C. A., Manivannan, P., Panneerselvam, R. and Shao, M. A. 2009. Understanding water deficit stress-induced changes in the basic metabolism of higher plants-biotechnologically and sustainably improving agriculture and the ecoenvironment in arid regions of the globe. *Crit. Rev. Biotechnol.,* 29: 131-151.

Srivastava, L.M. 2002. Abscisic acid and stress tolerance in plants. In: *Plant Growth and Development of Hormones and Environment* (ed. Srivastava, L. M.). Academic Press, New York, pp. 381-408.

Thomas, B. and Vince-Prue, D. 1997. Photoperiodism in Plants. 2nd Ed. London: Academic Press.

Turan, O. and Ekmekci, Y. 2011. Activities of photosystem II and antioxidant enzymes in chickpea (*Cicer arietinum* L.) cultivars exposed to chilling temperatures. *Acta Physiol. Plant,* 33: 67–78.

Walsh, B. and Layzell, B. 1986. Carbon and nitrogen assimilation and partitioning in soybeans exposed to low root temperatures. *Plant Physiol.,* 80: 249-255.

Wisniewski, M., Webb, R., Balsamo, R., Close, T. J., Yu, X. and Griffith, M. 1999. Purification, Immune localization, cryoprotective and antifreeze activity of PCA60: A dehydrin from peach *(Prunuspersica)*. *Physiol Plant,* 105: 600-608.

Yadav, S. K. 2010. Cold stress tolerance mechanisms in plants. *Agron. Sustain. Dev.,* 30:515–527.

Yamaguchi-shinozaki, K. and Shinozaki, K. 2005. Organization of cis-acting regulatory elements in osmotic and cold-stress-responsive promoters. *Trends Plant Sci.,* 10: 88-94.

Yordanova, R. and Popova, L. 2007. Effect of exogenous treatment with salicylic acid on photosynthetic activity and antioxidant capacity of chilled wheat plants. *Gen. Appl. Plant Physiol.,* 33: 155–170.

5

Biotechnological Approaches to Improve Abiotic Stress Tolerance-I

Asha Rani, Monika, Jyoti Taunk, Neelam R Yadav and Ram C. Yadav

Crop yields are affected by a combination of abiotic stresses, biotic stresses, and nutritional factors but abiotic stresses (drought, heat, cold or salinity) are the major factors that prevent crops from realizing their full yield potential (Edmeades, 2009). In traditional approach, breeders grow and cross varieties and then evaluate how the progenies vary in their ability to deal with stresses. The best-adapted plants are then selected for growing in fields exposed to stresses. Biotechnologists have taken advantage of recent advances in biotechnology and functional genomics to genetically engineer crops which can give better yield in adverse conditions than the unmodified ones (Manavalan *et al.*, 2009; Umezawa *et al.*, 2006). Drought, extreme temperatures (high or low), high salinity and cold are the major abiotic stresses that affect plant growth and result in significant yield losses. Although plants have evolved a wide spectrum of programs for adaptation to changing environment, the current understanding of the mechanisms associated with the ability of crops to maintain yield under abiotic stress are poorly understood (Witcombe *et al.*, 2008; Munns *et al.*, 2008; Bartels *et al.*, 2005). New advances in 'omic' technologies are providing opportunities for identification of transcriptional, translational and post-translational mechanisms and signaling pathways that regulate the plant response. This chapter describers several examples of how modern crop technologies may be applied to broaden crop tolerance of various abiotic stresses and to increase total biomass.

Genetic engineering for abiotic stress tolerance

To overcome the food insecurity in developing countries due to abiotic stresses, researchers are trying to design crops that could tolerate these stresses with optimum yield levels. The development of tolerant crops by genetic engineering requires the identification of key genetic determinants underlying stress tolerance in plants and introducing these genes into crops. Molecular control mechanisms for abiotic stresses are based on the activation and regulation of stress specific genes. These genes are involved in whole sequence of stress response, such as transcriptional control, signaling, free radical scavenging and protection of membrane and proteins. Furthermore, several reproductive barriers limit the transfer of favourable alleles from intergeneric and interspecific sources. Transgenic development is another straight forward technology to improve crop yield against abiotic stresses (Roy and Basu, 2009). Molecular change by genetic engineering takes less time as here desired genes can be transferred whereas, conventional method of breeding approach is associated with simultaneous transfer of undesired genes. Genetically modified (GM) crops have been grown commercially for more than 15 years, with a continuous increase of their cultivation area-from 1.7million ha in 1996 to 181million hectares in 2014 (ISAAA, 2014). Practically, only three major food-crops-soybean, maize and rapeseed-plus cotton are used for cultivation. This 'first generation' of GM plants was designed to improve some agronomic traits such as herbicide tolerance (HT) and insect resistance (IR), and also, to a much lesser extent, virus resistance (VR). The present trend is to combine several transgenes in the same plant ('stacked traits'). It is clear that these genetically modified crops have allowed for obtaining higher average yields. Importantly, 2014 comprehensive global meta-analysis, on 147 published biotech crop studies over the last 20 years (1995 to 2014) worldwide confirmed the significant and multiple benefits of biotech crops. GM technology adoption has reduced chemical pesticide use by 37%, increased crop yields by 22%, and increased farmer profits by 68%. These findings corroborate consistent results from other annual global studies which estimated increases in crop productivity valued at US$133.3 billion over 1996-2013 (ISAAA, 2014).

Abiotic stress tolerance mechanism

The isolation and characterization of genes conferring stress tolerance by expression in GM crops requires in-depth understanding of the mechanisms that plants use as a response to stress, which together with the academic interest of the topic has stimulated the study of these mechanisms over the last two decades. These studies have revealed a series of basic and conserved stress response pathways, apparently used by all plants-tolerant as well as sensitive-which are activated at the cellular level in response to different abiotic stresses. These include: i) the control of water transport, ion transport and ion homeostasis, to prevent cellular dehydration and to maintain osmotic balance, including the synthesis and accumulation of compatible solutes or 'osmolytes' in the cytosol. These osmolytes have additional functions as 'osmoprotectants', directly stabilizing proteins and cellular structures under dehydration conditions and protecting the cell against oxidative stress as scavengers of 'reactive oxygen species' (ROS); ii) synthesis of specific protective proteins, such as heat-shock proteins, 'late-embryogenesis abundant' (LEA) proteins, osmotine etc. and iii) activation of enzymatic antioxidant systems (ascorbate peroxidase, superoxide dismutase, glutathione peroxidase, catalase, glutathione reductase, etc.), induced in response to oxidative stress generated either directly (e.g., by ozone or high UV irradiation) or secondarily by other stressful environmental conditions (Zhu, 2001; Munns, 2002; Wang *et al.*, 2003; Vinocur and Altman, 2005; Ashraf, 2009; Hussain *et al.*, 2008). It was expected that overexpression of genes involved in these response mechanisms would increase the stress tolerance of transgenic plants. In fact, some positive results were obtained by expression of genes encoding ion transporters, enzymes of osmolyte biosynthesis, antioxidant enzymes, which conferred variable levels of tolerance to drought, high temperatures, salinity and/or other abiotic stresses.

Drought limits crop yields in many parts of the world, and research has identified many genes that may enhance plant performance under drought stress condition. Seven independent genes involved in drought tolerance were examined in a transgenic rice study (Xiao *et al.*, 2008). The involved genes were transcription factors, genes for abscisic acid (ABA) biosynthesis and genes involved in oxygen-radical detoxification. Transgenic plants carrying these genes yielded more than the wild-type rice under drought conditions. The first biotech drought tolerant crop (DroughtGard ™ maize) has been commercially launched by Monsanto in collaboration with BASF. They developed a GM maize variety transformed with bacterial genes encoding RNA chaperones (Castiglioni *et al.*, 2008). After going through all the regulatory process and field trials, the company obtained approval in USA and Canada, in 2012 and were grown in the more drought-prone U.S. states of Nebraska and Kansas. This crop were planted in the US in 2013 on 50,000 hectares increased over 5 fold to 275,000 hectares in 2014 reflecting high acceptance by US farmers. However, the expected increments in yield were very modest (no more than 10%); some improvement can be expected with more advanced 'versions' of the crop and by introducing the trait in other, more drought-tolerant cultivars obtained by classical breeding. It is expected that drought tolerant maize will probably be available for Sub Saharan Africa by 2017. Promising results have also been obtained in field trials of drought tolerant wheat in Australia, with the best GM lines yielding 20% more than their conventional counterparts (GMO Safety, 2008).

ABA signaling and stress responsive transcription factors

The plant hormone ABA regulates the plant's adaptive response to environmental stresses (drought, salinity, and chilling) via diverse physiological and developmental processes. One biotechnological target for improvement of these stress tolerance is the genetic manipulation of the stress response to the hormone abscisic acid (ABA). During water stress, ABA levels in plant greatly increase resulting in closure of stomata, thereby reducing the level of water loss through transpiration from leaves and activate responsive genes. Several transcriptional responses to abiotic stresses have been well characterized and are classified as being ABA-independent, ABA-dependent, or both. ABA induces expression of many genes under drought, cold, and salinity stress when applied exogenously (Shinozaki and Yamaguchi-Shinozaki 1996). ABA-dependent transcription responses are of two types. First is the "direct" ABA-dependent transcription pathway which involves cis-acting ABA-responsive elements (ABREs), that are directly activated by binding with TFs such as basic-domain leucine zipper (bZIP)-type DNA binding proteins (Shinozaki and Yamaguchi-Shinozaki 1996; Kobayashi *et al*., 2008). Second is the "indirect" pathway involves other cis-acting elements, such as MYC and MYB. Such transcription factor elements are activated through binding with ABA- or drought-inducible TFs, such as basic helix–loop– helix (bHLH)-related protein AtMYC2 and an MYB-related protein, AtMYB2 (Abe *et al*., 2003). Some genes are induced by drought stress but are ABA-independent e.g. *rd29a* (also known as *lti78* and *cor78*). A dehydration-responsive element (DRE) were identified in the promoter region of *rd29a* and the DRE-binding (DREB) protein transcription pathway has since been explored for its important roles in drought, cold, and salinity stress (Shinozaki and Yamaguchi- Shinozaki 1996; Qin *et al*., 2007). Many C-repeat (CRT) binding factor (CBF)/ DREB proteins have now been identified from the promoter regions of other stress inducible Arabidopsis genes (e.g. *cor15a*, *kin1*, *cor6.6* and *cor47/rd17*), and the CBF/DREB pathway has been shown to be conserved across species (Benedict *et al*., 2006; Pasquali *et al*., 2008). CBF/DREB1 and DREB2, belong to the ethylene responsive element/apetela 2 (ERE/AP2) TF family express during cold or drought stress.DREB2A overexpression in Arabidopsis confers significant drought tolerance (Sakuma *et al*., 2006a, b). DREB genes have been used in transformation of many crops, including rice and wheat, in attempts to increase drought tolerance (Chen *et al*., 2008; Kobayashi *et al*., 2008). DREB1A, a transcription factor that recognizes dehydration response elements, has been shown in *Arbidopsis thaliana* that promote the expression of drought tolerant genes (Pellegrineschi *et al*., 2003, 2004). Sakuma *et al*. (2006a) provided evidence that DREB2A has direct role in heat stress responsive gene expression in Arabidopsis. Many cold regulated genes are under the control of a primary master regulator, CBF/DREB1. Over expression of CBF4 from barley has been shown to confer salinity, drought and low temperature tolerance in transgenic rice (Oh *et al*., 2007). The NAC gene family members encode one of the largest families of plant specific TFs that expressed in various developmental stages, environmental factors. NAC gene *SNAC1* were isolated and chacterized by Hu *et al.* (2006). Transgenic rice plants having *SNAC1* showed significantly improved drought resistance under field conditions and strong tolerance to salt stress (Hu *et al.*, 2006). The phosphoinositide pathway implicated in plant responses to stress.Drought stress studies revealed, surprisingly, that transgenic plants containing inositol polyphosphate 5-phosphatase (InsP 5-ptase) lost less water and exhibited increased drought tolerance.Transcript profiling revealed that the drought-inducible ABA-independent transcription factor DREB2A and a subset of DREB2A-regulated genes were basally upregulated in the InsP 5-ptase plants (Perera *et al.*, 2008). In a separate study, over-expression of the transcription factor Nuclear Factor Y B subunit conferred protection against drought in *Arabidopsis thaliana* (L.) (Nelson *et al.*, 2007).

ERA1, a gene identified in Arabidopsis, encodes β-subunit of a farenesyl-transfarase that involves in ABA signaling. It was shown that down-regulation of farnesyl transferase β-subunit (FTB) in Arabidopsis, using either anti-sense or RNA interference, resulted in a drought-tolerant phenotype without the negative effects of the full knock-out (Wang *et al.*, 2005). The primary physiological mechanism underlying this response is increased in sensitivity to ABA signal produced under drought-stress which results in more-rapid stomatal closure, increased water retention in the plant, and increased seed yield. This research was subsequently extended to the farnesyl transferase α-subunit. Down-regulation of FTA also resulted in a drought-tolerant phenotype in Arabidopsis (Wang *et al.*, 2009). Down-regulation of FTB or FTA in canola (*Brassica napus* L.) has been shown to confer protection to plants growing in the field over several growing seasons in western Canada (Wang *et al.*, 2005; Wang *et al.*, 2009) and about 26% increase in yield were observed in transgenic canola over the wild-type growing under dryland conditions. This approach of down-regulating FTA /FTB is currently being extended to several other important crops, with the aim of protecting their yields under drought stress conditions. Transgenic Arabidopsis plants containing the zeaxanthin epoxidase gene, *AtZEP*, which encodes an enzyme required for an initial step in ABA synthesis from isopentyl diphosphate (IPP) and β-carotene (Schwartz *et al.*, 2003) showed increased tolerance to drought and salinity stress. The increased drought stress tolerance was attributed to increased leaf and lateral root development, higher fresh weight, and increased survival than control plants. In transgenics, the levels of ABA were higher, the expression of stress responsive genes such as *Rd29a* was much higher, and stomatal aperture was smaller under salt and/or drought stress.

Compatible organic solutes

Overproduction of different types of compatible organic solutes are also the most common plant responses to abiotic stresses (Ashraf and Foolad, 2007). The most common organic solutes that play an active role in the stress tolerance mechanism include proline, sucrose, trehalose, polyols and quaternary ammonium compounds such as glycinebetaine, prolinebetaine, hydroxyprolinebetaine, alaninebetaine, pipecolatebetaine, and choline O-sulfate (Rhodes and Hanson, 1993). Compatible solutes accumulate under osmotic stress and their primary function is to maintain cell turgour and thus the driving gradient for water uptake. Studies also indicated that compatible solutes can act as free radical scavengers by directly stabilizing membrane (Diamant *et al*., 2001). Although increased accumulation of such organic solutes is widely reported under salt stress, some plant species accumulate very low quantity of these compounds, while some others naturally do not synthesize them under non-stress or stress conditions (Ashraf and Foolad, 2007). For example, glycinebetaine, one of the most common quaternary ammonium compounds, naturally accumulate in response to salt stress in many plant species, including sorghum (*Sorghum bicolor*), sugar beet (*Beta vulgaris*), barley (*Hordeum vulgare*), spinach (*Spinacea oleracea*), and wheat (*T. aestivum*) (Yang *et al*., 2003; Ashraf and Foolad, 2007). However, some plant species such as tobacco (*N. tabacum*), rice (*Oryza sativa*), mustard (Brassica spp.), and Arabidopsis (*Arabidopsis thaliana*) are not capable of synthesizing GB under optimal or stress conditions (Ashraf and Foolad, 2007). Whether the plants produce low amount of an organic solute, engineering for overproduction of such organic compound seems to be a plausible approach. Accumulation of proline has been demonstrated to be associated with abiotic stress. The gene for Ä^{1}-pyrroline-5-carboxylate synthetase (*P5CS*) has been overexpressed in rice (Su and Wu, 2004), potato (Hmida-Sayari *et al*., 2005), wheat (Sawahel and Hassan, 2002), and tobacco (Hong *et al*., 2000). Indeed, all transgenic lines accumulated prolinemany fold higher than as compared to wild type plants and showed improved growth under saline conditions. Transgenic soybean plants were developed by transferring *P5CR* (L-"'-pyrroline-5-carboxylate reductase) gene that showed increased proline accumulation, leading to higher water stress tolerance (De Ronde *et al*., 2000). In plants, polyamine accumulatess under several abiotic stresses. It has been suggested that increase in concentration of polyamine could be considered as indicator of plant stress. It is now possible to manipulate polyamine content using sense and antisense constructs of polyamine biosynthesis gene such as *ADC* (encodes for arginine decarboxylase) in transgenic plants. Engineering of polyamine biosynthesis pathway has concentrated mostly on two species, tobacco and rice (Capell and Christou, 2004). Capell and Christou (2004) developed a diverse rice germplasm with altered polyamine content. Transgenic rice plant expressing

S- adenosylmethionine decarboxylase (SAMDC) DNA accumulated 2-3 fold higher levels of spermidine and apermine as compared to wild type plants under drought stress. Transcript levels for rice SAMDC reached at maximum levels at 6 days after stress induction. Turhan (2005) produced transgenic potato plants expressing the oxalate oxidase enzyme using Agrobacterium-mediated genetic transformation. His findings revealed a relatively higher salt tolerance ability of transgenic than the non-transgenic plants. Manitol as an osmoprotectory compound found in microbes that help the cells to lower their osmotic potential and to draw water from the outside medium.*mt1D* gene were introduced into upland rice (*Oryza sativa* var. japonica) using microprojectile bombardment by Li *et al.* (2004). Transgenic plants growth rate was significantly higher than the control on MS medium containing 1% NaCl whereas Non-transgenic plants died after 35 days. Long-term acclimation to the cold is strongly correlated with the recovery of photosynthesis through upregulation of sucrose biosynthesis. Strand *et al.* (2003) compared the acclimation responses of wild type Arabidopsis with transgenic plants overexpressing sucrose phosphate synthase or with antisense repression of either cytosolic fructose-1,6-bisphosphatase or sucrose phosphate synthase. Plants overexpressing sucrose phosphate synthase showed improved photosynthesis and increased sucrose concentration when shifted to 5°C, whereas both antisense lines showed reduced soluble sugars relative to WT plants. Fructans are synthesized in vacuoles from sucrose by the action of two or more different fructosyltransferases. These include sucrose:sucrose 1-fructosyltransferase (1-SST), sucrose:fructan 6-fructosyltransferase (6-SFT), fructan:fructan 1-fructosyltransferase (1-FFT), and fructan:fructan 6G-fructosyltransferase (6G-FFT; Vijn and Smeekens 1999). Kawakami *et al.* (2008) used rice (sensitive to chilling temperatures), to study the effect of fructan biosynthesis against water stress. Two wheat fructan-synthesizing enzymes, 1-SST, encoded by wft2, or 6-SFT, encoded by wft1, were transferred into rice plants. The transgenic seedlings with wft2 showed higher concentrations of oligo- and polysaccharides than nontransgenic rice seedlings, and exhibited enhanced chilling tolerance (11-day exposure to 5°C). Parvanova *et al.* (2004) transformed tobacco to accumulate different compatible solutes (proline, fructans or glycine-betaine) to improve tolerance to low temperature. Transgenic plants successfully survived againsed freezing stress. To increase glycinebetaine levels, Yang *et al.* (2005) overexpressed betaine aldehyde dehydrogenase protein from spinach into tobacco plants. Tobacco transformants showed increased thermotolerance in terms of growth of young seedlings as well as CO_2 assimilation rates. Trehalose is a non-reducing disaccharide of glucose that functions as a protectant in the stabilization of biological structure under abiotic stress. Zhang *et al.* (2005) transferred trehalose synthetase (*Tsase*) gene in tobacco plants for manipulating abiotic stress tolerance. They reported higher trehalose accumulation and enhanced tolerance to drought and salt stresses in transgenic plants than non-transgenics. By using agrobacterium mediated genetic transformation, Almeida *et al.* (2004) transformed *Nicotiana tabaccum* with *Arabidopsis thaliana* gene (AtTPS1), which is involved in trehalose biosynthesis. They observed high germination rates at higher levels of mannitol than wild type plants. Another trehalose phosphate synthetase (*TPS1*) gene was introduced from yeast into tobacco chloroplast (Lee *et al.*, 2003). Stable integration of this gene into tobacco chloroplast genome was confirmed by PCR and Southern blots analysis. Transgenic chloroplast thylakoid membranes were found with high integrity under osmotic stress as evidenced by retention of chlorophyll when grown in 6% PEG6000, whereas the untransformed plants chloroplast were bleached. Thus they suggested that trehlose functions by biological membrane protection rather than regulating water potential. Abiotic stresses induce Late-embryogenesis-abundant (LEA) proteins in plants which are synthesized during the late embryogenesis and induced by stress. It has been observed that LEA proteins act as water binding molecules, in ion sequestration and membrane stabilization. Rohilla *et al.* (2002) transformed Pusa Basmati 1 with *HVA1* to increase abiotic stress tolerance. Transgenic lines showed increased stress tolerance in terms of cell integrity and growth after imposed water stress. They observed that accumulation of *LEA3* in leaves of transgenic Pusa Basmati 1 rice plants might have conferred the significant increase in drought stress tolerance. A wheat dehydrin, *DHN-5*, was ectopically overexpressed in Arabidopsis. Transgenic plants displayed superior growth, seed germination rate, water retention, ion accumulation, and higher proline contents than wild type plants under salt and/or drought stress (Brini *et al.*, 2007a). Houde *et al.* (2004) transferred the wheat *Wcor410a* acidic dehydrin gene to strawberry. Freezing tests showed that cold-acclimated transgenic strawberry leaves had a 5°C improvement in freezing tolerance as compared to wild type leaves. Lal *et al.* (2008) observed the effects of overexpression of the *HVA1* gene in mulberry under a constitutive promoter. *HVA1* is a group 3 LEA protein isolated from barley aleurone layers that has been found to be inducible by ABA.Transgenic plants showed better less photooxidative damage, cellular membrane stability, photosynthetic yield, and better water use efficiency than nontransgenic plants under both salinity and drought stress.

Heat shock protein factors

Many abiotic stress responsive proteins, particularly Heat Shock Proteins (HSPs) have been shown to act as molecular chaperones. They are responsible for protein synthesis, targeting, maturation and degradation in a broad array of normal cellular process. Furthermore, function of molecular chaperones is to stabilize the proteins and membranes and to assist protein folding under stress conditions. In a research, DcHSP17.7 gene (a carrot heat shock protein gene encoding HSP17.1) were fused to a 6XHistidine (His) tag to distinguish the engineered protein from endogenous potato proteins and it was introduced into the potato cultivar Desiree under the control of the cauliflower mosaic virus (CaMV) 35S promoter (Ahn and Zimmerman, 2006).The integration was confirmed by Western Blot. They observed improved cellular membrane stability at high temperature, as compared to wild type and vector controlled plants. Mishra *et al.* (2002) over-expressed tomato *hsfA1* gene in tomato plants that showed increased thermotolerance. In various studies, it was observed that plant heat shock proteins are not only expressed in response to heat shock, but also under salt, water, oxidative stress and at low temperature. During cold acclimation, several stress proteins i.e. chaperones and membrane stabilizers during freeze dehydration are expressed in the cytosol (Puhakainen *et al.*, 2004). Li *et al.* (2003) suggested that Hsf (heat shock factor) gene may play a pivotal role in heat-shock-induced chilling tolerance. They developed transgenic tomato by transferring *Arabidopsis thaliana Hsf1b* (*AtHsf1a*) gene. The transgenic tomato plants harbouring this gene were observed with increased chilling tolerance. A sweet pepper cDNA clone, CaHSP26 encoding the chloroplast (CP)-sHSP were characterized with regard to its sequence, response to various temperatures, and function in transgenic tobacco plants by Guo *et al.* (2007). *CaHSP26* gene expression showed that the mRNA accumulation of *CaHSP26* was induced by heat stress. However, the expression of the *CaHSP26* gene was not induced by chilling stress (4°C) in the absence of HS (heat schok). The transcripts were detected at 48 h at 4°C after HS while not at 25°C. The photochemical efficiency of PSII and the oxidizable P700 in transgenic tobacco overexpressing *CaHSP26* were higher than that in wild type during chilling stress. These results suggest that the CP sHSP protein have an important role in the protection of PSI and PSII during chilling stress.

Detoxifying genes

Reactive oxygen species (ROS) are produced at high level in cells under stress conditions (Alscher *et al.*, 2002).Electron transport chain systems in mitochondria and chloroplasts are the major source for the production of the O_2 radical. Component of cell detoxification mechanisms have been employed in specific experiments to change thermotolerance response. Chen *et al.* (2004) overexpressed gene encoding for glutathione peroxidase from tomato in tobacco which protected transgenic leaves from salt and heat stress. Overexpression of Cu/Zn superoxide dismutase were also noted to protect plants from high temperature stress (Tang *et al.*, 2006). Glutathione peroxidase (GPX), as an antioxidant enzyme reduces hydroperoxides in the presence of glutathione and protect the cells from oxidative damage, including lipid peroxidation (Maiorino *et al.*, 1995). Gaber *et al.* (2006) generated transgenic Arabidopsis plants containing *GPX- 2* genes in cytosol (*AcGPX2*) and chloroplasts (*ApGPX2*). It was seen that both transgenic lines showed increased tolerance to oxidative damage caused by the treatment with H_2O_2, Fe ions, and environmental stress conditions, such as chilling with high light intensity, high salinity or drought. The degree of tolerance to all types of stress was correlated with the levels of lipid peroxide suppressed by the overexpression of the *GPX-2* genes. SOD is the first enzyme in the enzymatic antioxidative pathway. Halophytic plants, such as mangroves, reported to have a high level of SOD activity. SOD have an important role in defending mangrove species against severe abiotic stresses. Prashanth *et al.* (2008) further characterized the Sod1 cDNA (a cDNA encoding a cytosolic copper/zinc SOD) by transforming it into rice from the mangrove plant (*Avicennia marina*). Transgenic plants were more tolerant to oxidative stress than wild type.

Salt tolerant genes (Ion transporter and antiporter genes)

There are another large number of genes found to be instrumental for engineering tolerance to salinity. In a research it was found that STGs (salt tolerant grasses) were a potent source of salt tolerant genes, capable of surviving at increasing salt stress by utilizing different mechanisms that include vacuolization of toxic Na+ and Cl" in mature or senescing leaves, secretion of excess salts by salt glands, accumulation of osmolytes like proline and glycine betaine, and scavenging of ROS by antioxidative enzymes. Roy and Chakraborty (2014) transformed these genes into cereal crops and found enhanced growth and reproducibility at increasing salinity stress. Ion transporters selectively transport ions and maintain them at physiologically relevant concentrations. Na+/H+ antiporters play a crucial role in maintaining cellular ion homeostasis and permit plant survival and growth under saline conditions. This type of antiporters catalyze the exchange of Na+ for cytoplasmic pH, sodium levels and cell turgour (Serrano *et al.*, 1998). Na+/H+ antiporters are found in every biological kingdom, from bacteria to higher plants (Padan *et al.*, 2001). There are many different families of Na+/H+ antiporters including NhaA, NhaB, NhaC, NhaD, and NapA in prokaryotes, SOD2 and Nha1 in fungi, and AtNHX1 and SOS1 in Arabidopsis (Wu *et al.*, 2005). Plant engineering for overexpression of genes for different types of antiporters has an effective approach for controlling the uptake of toxic ions and hence enhanced plant salt tolerance. In a study, it was seen that transgenic *Brassica juncea* plants over expressing *pgNHX1* withstand 300 mM salt stress till the seed setting stage and exhibited normal growth phenotype without much loss in seed yield (Rajgopal *et al.*, 2007). Xue *et al.* (2004) generated transgenic wheat by transferring a vacuolar Na+/H+ antiport gene *AtNHX1*. The transgenic lines exhibited improved biomass production.The *Escherchia colinahA* gene encodes a Na+/H+ antiporter, having critical role in ion homeostasis. This gene was transferred into rice (*Oryza sativa* L. sp. Japonica) by Wu *et al.* (2005). Increasing the abundance of vacuolar Na+/H+ antiporters (NHX), involved in the synthesis of compatible solutes (such as proline and glycinebetaine) had successful in improving crop salinity tolerance . The transgenic plants showed better germination rate than control. Bao-Yan *et al.* (2008) and Gu *et al.* (2008) have generated two transgenic lines of Arabidopsis using *MsNHX1* from alfalfa (*M. sativa*) and *ZmOPR1* from maize (*Zea mays*), respectively with considerable improvement in seed germination of both transgenic lines under saline conditions. The introgression of high affinity potassium transporter gene from *T. monococcum* (*TmHKT1;5-A*) into the durum wheat, Tamaroi, resulted in a significant improvement in grain yield in salt stressed, field grown durum by increasing its ion exclusion (James *et al.*, 2012).

With the advent of genetic engineering, plant breeding has got a new dimension to produce crop varieties with more desirable characters i.e. QTLs mapping, Association mapping, Marker assisted selection (MAS) and other genomics approaches.

QTL mapping

The identification of genes that are responsible for important agricultural traits has been mostly conducted by traditional molecular genetics (forward and reverse genetic screens) for discrete traits and by quantitative trait locus (QTL) mapping for complex traits (Ashikari *et al.*, 2006). A striking advancement in DNA marker technology has revolutionized the field of genetic studies. The advent of high-density molecular maps coupled with the rapid development of PCR-based techniques has created opportunities for the application of DNA markers both in genetic and breeding research. The inheritance of quantitative traits is classically thought to be controlled by a large number of genes with small effect scattered throughout different chromosomes. Currently, tremendous progress has been made in mapping many agriculturally important genes with DNA markers in several crop plants (Koundal *et al.*, 2008). The identification of DNA markers linked to desirable genes or QTL affecting target traits is a prerequisite for MAS. In recent years, the use of molecular markers has facilitated breeding of crop plants including breeding for biotic and abiotic stresses. A good progress has been made in soybean, Brassica and groundnut in development of molecular markers and genome maps, mapping and tagging QTLs and their application in MAS. DNA markers, particularly anonymous markers such as RFLP, RAPD, AFLP and SSR, have been widely used for genome mapping and tagging of many agronomically important traits in plants. SNPs, on the other hand, have been found to be more abundant and can be used for germplasm fingerprinting, marker-assisted breeding and can potentially be used to create high-density genetic maps. Commonly used markers for *Brassica juncea* such as restriction fragment length polymorphisms (RFLP), amplified fragment length polymorphism (AFLP) and random amplified polymorphic DNA (RAPD) (Mohapatra *et al.*, 2002) are widely used alongwith combination of markers like AFLP, RFLP and ISSR, SSR, RAPD (Kalita *et al.*, 2007).

QTL mapping principle

The term QTL (Quantitative Trait Loci: QTLs) is a term first coined by Gelderman (1975). QTL are genomic regions associated with complex quantitative traits governed by several large effect as well as smaller effect genes. Special statistical software is needed to identify the locations and effects associated with these regions. Conceptually a QTL can be a single gene or it may be a cluster of linked gene that affect the trait. QTL mapping is used to offer direct mean to investigate the number of genes influencing the trait, to find out the location of the gene and to know the effect of dosage of these genes on variation of the trait. QTL analysis is based on the principle of detecting an association between phenotype and the genotype of markers. The markers are used to partition the mapping population in to different genotypic classes based on genotypes at the marker locus, correlative statistics is then used to determine whether the individual of one genotype differ significantly with the individuals of other genotype with respect to the trait of interest. Significant differences between phenotypic means of the two / more groups depending on the marker system and type of population indicates that the marker locus being used to partition the mapping population is linked to a QTL controlling the trait (Veeresha *et al.*, 2015).

'QTL mapping' is actually based on the principle that genes and markers segregate via chromosome recombination (called crossing-over) during meiosis (i.e. sexual reproduction), thus allowing their analysis in the progeny. Recent technical progress in the area of molecular biology and genomics have made possible the molecular dissection of major loci (Quantitative Trait Loci: QTLs) responsible for the major quantitative traits in crops (Tuberosa, and Salvi, 2005). QTL mapping usually begins with the collection of genotypic (based on molecular markers) and phenotypic data from a segregating population, followed by a statistical analysis to reveal all possible marker loci where the allelic state correlates with the phenotype. In general, QTLs are identified by correlating genetic variation with trait variation; a significant correlation between genotype and phenotype suggests that DNA status helps determine trait expression. The complexity of these phenotypic traits, particularly of those involved in adaptation, probably arises from segregation of alleles at many interacting loci (quantitative trait loci, or QTL), the effects of which are sensitive to the environments (Lynch & Walsh, 1997). QTL are controlled by several genes and the phenotype observed is the combined effect of all the alleles at all the loci, influenced by environmental conditions, which can then be untangled with certain statistical determination. Identifying existing genetic variation by QTL mapping is important because, unlike molecular genetic approaches, this can provide a basis for crop improvement through conventional breeding approaches (Mackay, 2001). QTL mapping relies on statistical linkage analyses among quantitative traits of interest and genetic markers, using a population that carries genetic mosaics derived from parental varieties, such as second generation (F_2) plants or recombinant inbred lines (RILs). Parental varieties are used for the preparation of RILs for mapping QTL. At each genetic locus, phenotypic variance for plants with the different genotypes is scored. It should be noted that RILs, rather than the F_2 or F_3 population, are better to evaluate genotype-by-environment interactions. The MB approach involves first identifying quantitative trait loci (QTLs) for traits of interest, such as tolerance to drought and salinity. Until recently, QTLs were identified by linkage mapping, but now a days association genetics has started to supplement these efforts in several economically important crops (Hall, 2010). The feasibility of QTL mapping depends on the availability of genetic markers and various populations of segregating individuals showing measurable phenotypic variation, whereas positional cloning of QTL genes is facilitated by sequence information and/or a physical map of the genome. Using various markers, genes affecting quantitative trait loci (QTL) have been mapped in several species of crop plants. Rice is the leading species among cereal crops, because its complete genome sequence is available. As the genome structure is conserved between rice and other cereals, such as maize, barley, wheat and sorghum, sequence information from the rice genome is also useful for comparative mapping in these monocotyledonous (monocot) crops. Marker-assisted selection increases the efficiency and flexibility of a breeding program by selecting for marker linked to target genes or quantitative trait loci (QTL). Linkage maps are constructed from the analysis of many segregating markers. The three main steps of linkage map construction are: (1) production of a mapping population; (2) identification of polymorphism and (3) linkage analysis of markers.

Steps in QTL mapping

Developing of mapping population

A suitable mapping population generated from phenotypically contrasting parents (Highly resistant and susceptible lines) is the main prerequisite for QTL mapping. The parental lines used for the development of mapping population should be genetically diverse, because it enhances the possibility of identifying a large set of polymorphic markers that are well distributed across the genome. The ability to detect QTL in F_2 or F_2 derived populations and RILs are relatively higher than other mapping population. The $F_{2:3}$ families have the advantage that it is possible to measure the effects of additive and dominant gene actions at specific loci. The advantage with RILs is that the experiments can be performed at several locations in multiple years. The size of the mapping population for QTL analysis depends on several factors viz., type of mapping population used for QTL analysis, genetic nature of the target trait and objective of the study. In general size of the mapping population is around 200-300 individuals.

Generating saturated linkage map

Mapping means placing the markers in order, indicating the relative genetic distance between them and assaying them to their linkage groups on the basis of recombination values from all pair wise combination between the markers. Linkage map indicates the position and relative genetic distance between markers along chromosomes. A variety of molecular markers viz., RFLPs, RAPD, SSRs, AFLP, and SNPs etc have been used to identify individual QTLs and to find out effects and position of these QTLs.

Phenotyping of mapping population

The target quantitative traits have to be measured as precisely as possible. The data is taken over location and replications to obtain a single quantitative value for the line. It is also necessary to measure the target traits in experiments conducted in multiple location to have better understanding of the QTL x Environment interaction.

QTL detection

The basic purpose of QTL mapping is to detect QTL, while minimizing the occurrence of false positive. QTL detection is done by following methods.

a. Single Marker Analysis (SMA)

It is also referred as single point analysis. It is the simplest method for detecting QTL associated with single markers. This method does not require complete linkage map and can be performed with basic statistical software programs. However the major disadvantage is that the further QTL is from a marker, the less likely it will be detected. This is because recombination may occur between the marker and the QTL in this analysis.

b. Simple Interval Mapping (SIM)

Simple Interval Mapping was first proposed by Lander and Botstein in 1989. SIM method uses linkage maps and analysis intervals between adjacent pairs of linked markers along the chromosomes, simultaneously, instead of analyzing single markers. Many researchers have used Mapmaker/QTL (Lincoln *et al.*, 1993) and QGene (Nelson, 1997) to construct SIM.

c. Composite Interval Mapping (CIM)

Composite Interval Mapping is one of the popular methods used to detect QTLs. CIM was developed by Zeng (1994). This method combines internal mapping with linear regression. It considers a marker interval plus a few other well-chosen single markers in each analysis. The main advantage of CIM is that it is more precise and effective at mapping QTLs compared to SMA and SIM, especially when linked QTL are involved.

d. Multiple Interval Mapping (MIM)

Most recently MIM has become popular for mapping QTLs. MIM is the extension of internal mapping to multiple to multiple QTLs, just as multiple regression extends analysis of variance. MIM allows one to infer the location of QTLs to position between markers and it makes proper allowance for missing genotype data and can allow interaction between QTLs.

e. AB-QTL analysis

Advanced backcross QTL (AB-QTL) analysis was proposed by Tanskley and Nelson (1996). It is another important approach for QTL mapping, which aims at simultaneous detection and transfer of useful QTLs from the wild relatives to a popular cultivar for improvement of various traits. Using this method QTL analysis has been done in various crops like Wheat (Kunert *et al.*, 2007) and Barley (Li *et al.*, 2006) and Rice (Cheema *et al.*, 2008).

Distinguished examples of QTL or linkage mapping in crops

QTL mapping is used to offer direct mean to investigate the number of genes influencing the trait, to find out the location of the gene and to know the effect of dosage of these genes on variation of the trait. Genetic mapping is the first step to map based cloning. QTL maps based on linkage studies and marker trait association can be effectively utilized for gene pyramiding, germplasm screening of diversified material for abiotic (salinity, cold, salt, drought) and biotic stresses (disease, pest) etc. The identification and location of specific genes mediating quantitative characters is having great importance in plant breeding. QTLs so identified for diverse traits in different crops have been met in crop improvement especially to enhance the yield and to develop disease resistant elite lines. The introgression of QTLs into elite lines / germplasm, and maker-aided selection (MAS) for QTLs in crop improvement has to be undertaken in some of the crop like Maize, rice, and wheat. Among crop species, sorghum has been studied as a model for drought resistance because of its adaptation to hot and dry environments. A particularly relevant trait conferring drought tolerance towards improvement of sorghum is the 'stay green' trait. This is characterized by delayed leaf senescence during grain ripening under water-limited conditions, which ensures better grain filling and is often associated with resistance to charcoal rot and lodging. Consistent with this, drought tolerance is enhanced by delaying leaf senescence when an isopentenyltransferase gene for cytokinin synthesis is expressed under the control of a stress- and maturation induced promoter (Sanchez *et al*., 2002). A particularly relevant trait conferring drought tolerance QTL has been towards improvement of sorghum is 'stay green' trait. Several QTL-mapping studies for stay green identified four major QTLs, designated Stg1, Stg2, Stg3, and Stg4, which account for approximately 20%, 30%, 16% and 10% of the phenotypic variance, respectively (Harris *et al*., 2007). In maize, a major QTL designated root-ABA1 is involved in root architecture, ABA concentration and other traits according to water availability (Landi *et al*., 2007) recently, major QTLs with large effects (accounting for 32–33% of the genetic variation) for grain yield under drought stress conditions in upland and lowland rice have been reported (Kumar *et al*., 2007). However, the genes responsible have not yet been identified in any of these studies. Whether the genes underlying these QTLs prove to be novel or identical to previously identified genes, there are promising opportunities for improving drought tolerance in various crops. A major QTL, Submergence1 (Sub1), is linked to the submergence tolerance of FR13A65 (Xu *et al*., 2000). This locus contains a cluster of three genes (Sub1A, Sub1B, and Sub1C) that encode putative ethylene response factors (Gutterson & Reuber, 2004). The gene for submergence tolerance has been identified as Sub1A. Moreover, introgression of the Sub1 genes into the widely grown Indian variety Swarna, which lacks Sub1A, confers strong submergence tolerance. In durum wheat, the QTL Nax1 has been studied as a genetic component that confers lower Na+ and higher K+ concentrations in the leaf blade (Lindsay *et al*., 2004). In rice, several research groups have dissected the QTLs responsible for variations in Na+ and K+ content (Sahi *et al*., 2006). Using the salt tolerant indica rice variety Nona Bokra and the salt sensitive japonica variety Koshihikari, Lin *et al*. mapped a major QTL for shoot K+ content in seedlings, Skc1, which was revealed to encode a HKT-type transporter, SKC1 (Ren *et al*., 2005). The development of molecular markers and identification of QTL controlling various traits such as oil content (Mahmood *et al*., 2003), disease resistance (Zhao and Meng, 2003), flowering time (Camargo and Osborn, 1996), and fertility restoration (Jean *et al*., 1997) are positive steps toward marker-assisted breeding of Brassica and other oilseed crops. Many QTL conferring resistance to downy mildew of maize (Nair *et al.*, 2005) have been identified in IARI. During the last decade considerable progress has been made in terms of positional cloning of several QTL's including Brix 9-2-5, fw 2.2, Hd 1, Hd6, FRI in tomato, rice, and Arabidopsis (Salvi *et al.*, 2002).

Mapping and tagging of seed coat colour and the identification of microsatellite markers for marker-assisted manipulation of the trait in *Brassica juncea* was carried out (Padmaja *et al.*, 2005). QTLs for six yield-related traits in oilseed rape (*Brassica napus*) using DH and immortalized F_2 populations were detected. A genetic linkage map was constructed using 208 SSR and 189 SRAP markers for the DH population. Using composite interval mapping analyses, 30 and 22 significant QTL were repeatedly detected across environments for the six traits in the DH and IF_2 populations, respectively. The results provided a better understanding of the genetic factors controlling yield related traits in rapeseed (Chen *et al.*, 2007). A genetic linkage map of *Brassica carinata* (BBCC; 2n = 34) has been constructed with 23 RAPD, 29 ISSR and 17 SSR markers using an F_2 hybrid population of 150 individuals (Priyamedha *et al.*, 2012). A genetic map was constructed with PCR-based markers, and a total of 212 loci, which covered 1,703 cM, were assigned to eight linkage groups in the B genome and nine linkage groups in the C genome, which allowed comparison with genetic maps of other important Brassica species that contain the B/C genome(s). Loci for two Mendelian-inherited traits related to pigmentation (petal and anther tip colour) and one quantitative trait (seed coat colour) were identified using the linkage map (Shaomin *et al.*, 2012).

A pioneering study for crop yield using QTL or linkage mapping was the identification of a gene controlling fruit size in tomato by QTL mapping (Frary *et al.*, 2000). This gene fw2.2, the first QTL gene to be positionally cloned in plants is responsible for 30% of the difference in fruit mass between wild and cultured tomato. fw2.2 encodes a protein which acts as a negative regulator of cell division. Subsequent studies showed that the accumulation of fw2.2 transcripts varied in terms of timing and quantity between the two accessions (Cong *et al.*, 2002) and that expression levels are negatively correlated to fruit mass in an artificial gene dosage series. (Liu *et al.*, 2003). In cereal crops, many QTLs affecting yield and their map positions have been listed in the Gramene database. Recently, rice genes controlling grain mass have been identified by QTL mapping. gs3 was identified as a major QTL for grain length and a minor QTL for grain width and thickness (Fan *et al.*, 2006), gs3 encodes a putative transmembrane protein, which might function as a negative regulator of the growth of grains. Another QTL controlling grain width and weight, gw2, has also been identified. gw2 (Song *et al.*, 2007). encodes a RING-type ubiquitin

E3 ligase, suggesting that it functions through degradation of an unknown target protein by the 26S proteasome. The encoded protein is suggested to be a negative regulator of cell division in spikelet hulls, as loss of function increases cell division and thereby grain mass. In terms of branch numbers, Gn1a has been identified as a major QTL affecting the number of secondary and tertiary branches in rice, and thereby grain numbers per panicle (Ashikari *et al*., 2005).

Association mapping

Association mapping, also called linkage disequilibrium (LD) mapping, generally relies on correlation between a genetic marker and a phenotype among collections of diverse germplasm (Yu and Buckler, 2006). Thus, large numbers of alleles can be tested in it in contrast to QTL mapping, in which only parental alleles are tested. Association mapping also allows easier fine mapping than the classical approaches, in which a large number of segregants are often needed. This is because hundreds or thousands of generations are likely to have passed since the establishment of association between a marker and a linked allele, allowing time for a large numbers of recombination events during meiosis. Such events break down association between a causal variant and a genetic maker that is not tightly linked to the trait. In association mapping, the marker density and experimental design are determined by linkage disequilibrium (LD) patterns (Flint-Garcia *et al.,* 2003) which reflect recombination history and demographic factors, including population history and inbreeding. Using large marker sets, association mapping can potentially be performed across the whole genome, an approach that is currently being used in human genetics (Hirschhorn & Daly, 2005). The feasibility of genome-wide association mapping in *A. thaliana* was recently tested using 95 accessions in (Zhao *et al.,* 2007) a search for associations with flowering time and pathogen resistance, and major genes (*FRI*, *Rpm1*, *Rps2* and *Rps5*) were successfully identified. Among cereal crops, maize is most suitable for association analysis because it is an outcrossing species exhibiting a high recombination rate and rapid decay of LD (<1 kb) (Yu & Buckler, 2006). In both maize and other crop species including rice, barley, sorghum and soybean, which are self-crossing diverse Germplasm or genotype panels, are also being established for whole-genome association mapping (Buckler & Gore, 2007). In self-crossing crops, however, the scale of LD is relatively large, leading to lower mapping resolution.

Phenotypes are scored for plants that have also been genotyped. Association scanning is performed by comparing phenotypic scores respective to each haplotype. Data sets are analysed using statistical methods, which have been designed to deal with the population structure. Nested association mapping, which combines the advantages of linkage analysis and association mapping in a single unified mapping population, is nowadays being used for the genome-wide dissection of complex traits in many crops. Association mapping, compared with linkage mapping, is a high-resolution and relatively less expensive approach. In the near future, it is likely to be routinely used for identifying traits associated with abiotic stresses, particularly given the availability of high-throughput marker genotyping platforms. After identifying the markers associated with QTLs or genes for traits of interest, the candidate QTLs or genes can be introgressed in elite lines through marker-assisted backcrossing (MABC). One of the main difficulties of developing superior genotypes for abiotic stresses such as drought or heat is that these traits are generally controlled by several small effect QTLs or several epistatic QTLs.

Marker assisted selection

Conformist plant breeding is primarily based on phenotypic selection of superior individuals among segregating progenies. Although significant strides have been made in crop improvement through phenotypic selections for desirable traits, considerable difficulties are often encountered due to genotype-environment interactions. Besides, testing procedures may be unreliable or expensive due to the nature of the target traits or the target environment and requires many years to achieve the goal. Most of the traits are quantitative in nature, i.e. they are controlled by many genes together with environmental factors and the underlying genes have small effects on the phenotype observed. The regions within genomes having genes associated with a particular quantitative trait are known as quantitative trait loci (QTLs). The identification of QTLs was initiated by the development of DNA (or molecular) markers in the 1980s. In agricultural research, the main use of DNA markers has been in the construction of linkage maps for diverse crop species. Linkage maps have been utilized to identify chromosomal regions that contain genes controlling simple traits (controlled by a single gene) and quantitative traits through QTL mapping (Collard *et al.*, 2005).

DNA markers that are tightly linked to desired genes (called gene 'tagging') may be used as molecular tools for marker-assisted selection (MAS) in plant breeding (Friedt and Ordon, 2007). MAS involves the presence/absence of a marker as a substitute to assist in phenotypic selection, which make it more effective, efficient, reliable and cost-effective as compared to the more conventional plant breeding methodology (Choudhary *et al.*, 2008).Although, MAS has proven successful in case of simple trait controlled by major genes, the availability of genetic maps and molecular markers make it feasible also for QTL (Torres *et al.*, 2010). Theoretically, all markers tightly linked to QTLs could be used for MAS. However, due to the cost of utilizing several QTLs, only markers that are tightly linked to no more than three QTLs are typically used (Ribaut and Betran, 1999), although there have been reports of introgression of up to 5 QTLs into tomato using MAS (Lecomte *et al.*, 2004). Even a single QTL selection via MAS can be beneficial for crop improvement strategies; such QTL should account for the largest proportion of phenotypic variation for the specific trait. Furthermore, all QTLs that are selected for MAS should be stable across the environments (Hittalmani *et al.*, 2002). Several successful stories were reported for MAS in many crops. Using marker-assisted selection, HAU and ICRISAT have successfully demonstrated the use of RFLP markers in the transfer of additional downy mildew resistance into a parental line of popular pearl millet hybrid "HHB 67" and thus a new version of this hybrid, "HHB 67 Improved" based on the improved parental line, was released for commercial cultivation in 2005. This was the first public-bred product of marker-assisted breeding in India (Hash *et al.*, 2006; Khairwal *et al.*, 2007). MAS 946-1 was the 1st drought tolerant aerobic rice variety released in India in 2007 using marker assisted selection. To develop this new variety, scientists at the University of Agricultural Sciences (UAS), Bangalore, crossed a deep rooted upland japonica rice variety from the Philippines with a high yielding indica variety which consumed upto 60% less water than traditional varieties (Gandhi *et al.*, 2007).

Transgenic plants developed through genetic engineering are facing several legislative constraints due to biosafety concerns and ethical consideration while introduction of varieties obtained through MAS is not a subject to such restrictions. As MAS does not necessarily involve genetic engineering the thought that it will not be subject to public distrust like GMOs is justified (Muller-Rober *et al.*, 2007). Organizations that criticize genetic engineering for GMOs seem to accept MAS to a large extent (Then, 2005).

Marker assisted backcrossing

Backcrossing is a plant breeding method to incorporate one or a few genes into an adapted or elite variety. Mostly, the parent used for backcrossing has a large number of desirable attributes but is deficient in few characteristics (Allard 1999). This method was described in 1922 and was widely used between the 1930s and 1960s (Stoskopf *et al.*, 1993). The DNA markers use in backcrossing programme greatly increases the efficiency of selection. Three general levels of marker-assisted backcrossing are described (Holland 2004). The first level is referred as 'foreground selection' (Hospital & Charcosset 1997) or referred as positive selection (Takeuchi *et al.*, 2006). In this, the markers can be used in combination with or to replace screening for the target gene or QTL. This may be particularly useful for traits that have laborious phenotypic screening procedures. It can also be used for reproductive-stage traits selection in the seedling stage that allows the best plants to be identified for backcrossing. Furthermore, recessive alleles can be selected, which is difficult to do via conventional methods of breeding. The second level involves selecting BC (backcross) progeny with the target gene and recombination events between the target loci and linked flanking markers. This is referred as 'recombinant selection'. The purpose of this type of selection is to reduce the size of the donor chromosome segment containing the target locus. This is important because the rate of reduction of this donor fragment is slow than unlinked regions and many undesirable genes may be linked to the target gene from the donor parent. This is referred to as 'linkage drag' (Hospital 2005). This donor segment can remain very large even with many BC generations using conventional breeding methods (Salina *et al.*, 2003). Linkage drag can be minimized by using markers that flank a target gene (e.g. less than 5 cM on either side) (Hospital, 2011). The third level of MAB is 'background selection' which involves selecting BC progeny with the greatest proportion of recurrent parent (RP) genome, using markers that are unlinked to the target locus (Collard and Mackill, 2007). This was also called negative selection (Takeuchi *et al.*, 2006). This is extremely useful because the RP genome recovery can be greatly accelerated as compared to conventional backcrossing where it takes a minimum of six BC generations to recover the RP and there may still be several donor chromosome fragments unlinked to the target gene. Using MAB, it can be achieved by BC_4, BC_3 or even BC_2 (Morris *et al.*, 2003; Collard *et al.*, 2007), thus saving two to four BC generations. The background selection use during MAB to accelerate the development of a recurrent parent with additional genes has been referred as 'complete line conversion' (Ribaut *et al.*, 2002).

The use of this powerful approach for MABC was first reported by Chen *et al.* (2000) for rice introducing resistance to bacterial blight disease into Chinese hybrid parents. At IRRI, it was also described for submergence tolerance using the Sub1 gene (Neeraja *et al.* 2007). By using MABC breeding programmes, submergence sub1 gene at the locus RM23805 derived from IR64 was incorporated into susceptible variety OM1490 (Lang *et al.*, 2011). Sub1A gene has been introgressed into a popular high-yielding variety from India, Swarna using MABC procedure within 2 years (Neeraja *et al.* 2007). Through MABC, successful introgression of sub1 derived from donor rice variety IR64, has been done in Vietnam into popular rice variety AS996 (Cuc *et al.*, 2012). The sub1 introgression was confirmed using ART5 and SC3. Parental diversity was carried out with 460 markers, out of which 53 polymorphic markers were used for assessment on BC_1F_1, BC_2F_1 and BC_3F_1 generations. RP genome recovery was observed 87.5%, 93.75% and 96.15%, respectively in BC_1F_1, BC_2F_1 and

BC_3F_1 generations. There was the highest genetic background i.e. 100% in BC_4F_1 generation. Saltol QTL derived from FL478 was introgressed in genetic background of Bacthom 7 cultivar of rice. The background analysis in the introgression line revealed the recovery of recurrent parent genome up to 96.8%-100% based on the screened markers after three generations (Vu *et al.*, 2012). QTLs for drought tolerant traits were transferred successfully into Thai variety of rice KMDL105 by following MABC programme (Siangliwa *et al.*, 2007). Near isogenic lines (NILs) were developed through MABC by introgression of three root QTLs from CT9993, an upland japonica into IR20, a lowland indica cultivar. Among the NILs, variation in grain yield and drought response under rain-fed condition in target populations was monitored. Out of 41 NILs, 5 showed high yield permanence in both rain-fed and irrigated conditions compared to the IR20 (Suji *et. al.*, 2012).

The development of molecular genetics and associated technology like MAS has led to the emergence of a new field in plant breeding-Gene pyramiding. Pyramiding deals with stacking multiple genes leading to the simultaneous expression of more than one gene in a variety to develop durable resistance expression. Gene pyramiding is now a days gaining considerable importance as it would improve the efficiency of plant breeding leading to the precise development of broad spectrum resistance crops. The success of gene pyramiding is dependent upon several critical factors, including the number of genes to be transferred to the crop, the distance between the target genes and flanking markers, the total number of genotype being selected in each breeding generation, the nature of germplasm etc. Innovative tools such as DNA chips, micro arrays and SNPs are making rapid strides that are aiming towards assessing the gene functions through genome wide approaches. The power and efficiency of genotyping are expected to improve in the forthcoming decades by selecting the marker linked to the gene of interest. MAS have become a reality with development and availability of an array of molecular markers and dense molecular genetic maps in crop plants (Joshi and Nayak, 2010). Gene pyramiding offers greater prospects to attain durable resistance against biotic and abiotic stresses in crop plants. It could facilitate in pyramiding of genes effectively into a single genetic background. In a gene pyramiding scheme, strategy is to cumulate into a single genotype, genes that have been identified in multiple parents. The use of DNA markers, which permits complete gene identification of the progeny at each generation, increases the speed of pyramiding process. In general, the gene pyramiding aims at the derivation of an ideal genotype that is homozygous for the favorable alleles. The gene pyramiding scheme can be distinguished into two parts. The first part is called a pedigree, which aims at cumulating of all target genes in a single genotype called the root genotype. The second part is called the fixation step which aims at fixing the target genes into a homozygous state i.e. to derive the ideal genotype from the one single genotype. Each node of the tree is called an intermediate genotype and has two parents. Each of this intermediate genotype variety can resist. Moreover, pyramiding can also improve a parent in the next cross. The intermediate genotypes are not just an arbitrary offspring of a given cross but it is a particular genotype selected from among the offspring in which all parental target genes are present. Although the pedigree step may be common, several different procedures can be used to undergo fixation in gene pyramiding. Historically, long-term cultivation of varieties carrying single resistance gene has resulted in a significant shift in pathogen or insect race frequency and consequent breakdown of resistance. MAS is in contrast to genetic engineering which involves the artificial insertion of such individuals genes from one organism into the genetic material of another (Wamanda and Jonah, 2006).One solution to resistance breakdown is pyramiding of multiple resistance genes in the background of modern high yielding varieties. The main use of gene pyramiding is to improve an existing elite cultivar through introgression of a few genes of large effects from other crop sources. Gene pyramiding is difficult to achieve using conventional breeding alone because of linkage with some undesirable traits that is very difficult to break even after repeated backcrossing (Suresh and Malathi, 2013). Pyramiding of two or more resistance genes through Marker Assisted Selection (MAS) has been a successful approach to ensure multiple and durable host plant resistance against a wide spectrum of biotic stresses. The successful effort on gene pyramiding in rice includes resistance to blight, blast, gall midge etc. In abiotic stresses, many economically important traits such as yield, quality and tolerance are controlled by a relatively large number of loci termed as QTLs.MAS was used to transfer four QTL alleles for deeper roots from 'Azucena', a japonica upland cultivar (well adapted to rainfed conditions) into 'IR64', a rice cultivar with shallow root system. The MAS-generated lines, when grown in drought-stressed field conditions, were characterized by a root mass greater than that of 'IR64' (Courtois *et al.*, 2003).In 2006, strong submergence tolerance lines were developed by introgressing a locus conferring submergence tolerance from cultivar 'FR13A' into the variety 'Swarna' (Xu *et al.*, 2006). QTL analysis has been carried out on a segregating population derived from a cross between a high salt-tolerant indica variety and a susceptible elite japonica variety (Lin *et al.*, 2004). Two major QTLs, on chromosome 7 for shoot Na+ concentration (called as qSNC-7) and on chromosome 1 for shoot K+ concentration (called as qSKC-1), explained 48.5 and 40.1% of the total phenotypic variance respectively. In three F_3 lines of the segregating population, the alleles of the salt-tolerant variety at a discrete number of QTLs (three to four, including qSNC-7 and qSKC-1) for physiological traits related to salt tolerance were pyramided. These lines showed enhanced level of seedling survival under salt stress. MABC was also successfully used to pyramid a submergence tolerant QTL allele together with three genes for different biotic stresses tolerance to KDML105, an indica cultivar widely grown in Thailand (Toojinda *et al.*, 2005).

These results would indicate that QTLs pyramiding breeding method could be applied for the development of varieties with high level of stress tolerance. In the future, MAS pyramiding could facilitate the combination of QTLs for several abiotic stress tolerances. With the help of MAS based gene pyramiding, it is possible for the researchers to conduct many rounds of selections in a year. New developments in technology such as automation, allele-specific diagnostics and diversity array technology (Jaccoud *et al.*, 2001) will make MAS based gene pyramiding more powerful and effective technology. Needless to point out that the current advances in tissue-derived techniques, genetic transformation and MAS, together with the advances in powerful new omics technologies offer a great potential to improve abiotic stress conditions.

Moreover plant responses to abiotic stress are complex, so there is need to bridge genotype-phenotype gap so as to understand and exploit the mechanisms for increasing yield stability under stressful conditions which can also be accomplished through integration of omics scale studies involving transcriptomics, proteomics and metabolomics which is crucial for rapid progress in stress biology research.

Transcriptomic analysis

Transcriptome is the total set of transcript in a given organism or specific subset of transcript present in a given cell type. The responses of plants to abiotic stresses involve major changes in transcriptome composition. To have an unbiased insight into how abiotic stress effects the expression of genes, understanding transcriptome dynamics is desirable. This can be achieved by the use of microarray or related high throughput profiling technologies. The development of next generation sequencing technologies like Illumina, RNA-Seq etc. with reduced costs have greatly expedited the precise RNA transcript profiling. There are two methods of inferring transcriptomes. One approach reads sequence maps onto a reference genome, the other approach which involves de novo transcriptome assembly, uses software to infer transcripts directly from short sequence reads. A generalised strategic approach for transcript profiling using next generation sequencing involves RNA extraction and apparent c-DNA library preparation. Reads from each library then needs to be assembled and clean reads filtered out. Transcriptome sequences are assembled into distinct contigs. The reads are realigned to the contig sequences and paired-end-relationships between the contigs are used to construct scaffolds. Unigene sets are constructed and to assess sequence coverage of the transcriptomic assemblies, Unigene sequences are searched against the databases. For functional annotation, accessions of best fit are retrieved and Gene Ontology (GO) accessions are mapped to GO terms according to molecular function, biological process and cellular component ontologies. Following the same approach Qui *et al.* (2011) have done genome scale transcriptome analysis of desert poplar. In their study, de novo assembly generated 86,777 Unigenes using Solexa sequence data and they found that 27% of total Unigenes generated were differentially expressed in response to salt stress. They also found that numerous putative genes involved in abscisic acid (ABA) regulation and biosynthesis were also differentially regulated.

To help characterize the cellular mechanisms underlying the toxicity of Aluminium (Al) to plants Kumari *et al.* (2008) presented large scale transcriptomic analysis of Arabidopsis root responses to Al using microarray, which represent approximately 93% of predicted genes in the genome of Arabidopsis. They found that exposure to Al have triggered changes in the transcript levels of several genes related to oxidative stress pathway, membrane transporters, cell wall, energy, and polysaccharide metabolism. Al exposures induced differential abundance of transcripts for several ribosomal proteins, peptidases and protein phosphatases. They also detected increased abundance of transcripts for several membrane receptor kinases and non-membrane calcium response kinases involved in signal transduction. Among Al responsive transcription factors, the most predominant families identified were AP2/EREBP, MYB and bHLH. Similarly, to obtain better insights into the effects of Selenium (Se) on plant metabolism, Hoewyk *et al.*, (2007) have undergone transcriptome analysis. They found that many genes involved in sulfur (S) uptake and assimilation were upregulated. Se treatment enhanced sulfate levels in plants, but the quantity of organic S metabolites decreased. Transcripts regulating the synthesis and signaling of ethylene and jasmonic acid were also upregulated by Se. Down regulation of genes involved in cell wall synthesis and auxin-regulated proteins in their study indicated that Se repress plant development.

Affymetrix GeneChip® Wheat Genome Array hybridization was used to reveal differences in global expression profiles of drought tolerant and sensitive wild emmer wheat genotypes to compare transcriptomes in root and leaf tissues (Ergen *et al.*, 2009). The comparison reveal several unique genes like differential usage of IP3-dependent signal transduction pathways, ethylene- and abscisic acid (ABA)-dependent signaling, and faster induction of ABA-dependent transcription factors by the drought tolerant genotype, which is indicative of distinctive stress response pathways (Efgen *et al.*, 2009). Similarly, Zheng *et al.* (2009) used Affymetrix Maize Genome Array to understand the transcriptome dynamics during drought stress in maize seedlings. Genome-wide gene expression profiling was compared between the drought-tolerant line Han21 and drought-sensitive line Ye478. They found that differential expression levels of cell wall-related and transporter genes may contribute to the different tolerances of the two lines.

To understand the molecular basis of differences in salinity tolerance in salt-sensitive indica rice IR64 and its salt-tolerant relative Pokkali, Kumari *et al.* (2008) have generated transcriptome map for seedling stage specific salinity stress response in rice. They indicated a specific set of genes like CaMBP, GST, LEA, V-ATPase, OSAP1 zinc finger protein, and transcription factor HBP1B, that were expressed at high levels in Pokkali even in the absence of stress as candidate for saline stress. Few genes mapped on major quantitative trait loci Saltol – on chromosome 1 were found to be differentially regulated in the two contrasting genotypes. Zeller *et al.*, (2009) have used whole genome tilling arrays to analyse the effects of salt, osmotic, cold, heat and application of ABA in Arabidopsis which enabled them to analyse more than 9000 genes that are not represented on the standard ATH1 array. They used entropy based detection of genes with a specific stress response. Many stress-responsive genes like NCED3, ATHB-7, RD29b, a PP2C and diverse LEA genes, were strongly induced preferentially in response to salt and osmotic stress, ABA treatment and after prolonged exposure to high temperatures. They observed that some transposable elements and pseudogenes are more strongly expressed under conditions of stress. They also employed tilling array data for detection of transcriptionally activated regions (TARs). In several cases they found that where novel TARs were close to annotated genes, these constituted stress induced exons.

Reactive oxygen species are toxic by-products of aerobic metabolism that accumulate in cells during different stress conditions. They also act as important signal transduction molecules and there level in cells is tightly controlled by a vast network of genes termed the 'ROS gene network'. In a study conducted by Shi *et al.* (2014), it was found that, exogenous application of melatonin, a well-known animal hormone, which is also involved in plant development and abiotic stress responses conferred improved salt, drought and cold stress resistances in Bermuda grass. Exogenous melatonin treatment alleviated reactive oxygen species (ROS) which caused cell burst and damage induced by abiotic stress. Genome wide transcriptomic profiling identified 3933 transcripts (2361 up-regulated and 1572 down-regulated) that were differentially expressed in melatonin-treated plants versus controls. The genes involved in nitrogen metabolism, major carbohydrate metabolism, tricarboxylic acid (TCA)/org transformation, transport, hormone metabolism, metal handling, redox, and secondary metabolism were over-represented after melatonin pre-treatment. Further metabolic and transcriptomic analyses showed that the underlying mechanisms of melatonin could involve major reorientation of photorespiratory and carbohydrate and nitrogen metabolism. To identify barley genes involved in responses to low temperature and drought, Tommasini *et al*., (2008) studied the dehydrin gene expression. They found 4,153 genes responded to at least one component of these two stress regimes and about one fourth of all genes are present under any condition. About 82.4% of the responsive genes were similar to Arabidopsis genes.

Proteomic analysis

Proteomics is an important component of functional genomics. The term "proteomics" and "proteome" were coined by Mark Wilkins in 1997, when he was a Ph.D. student at Macquarie University, Australia. Proteome is the entire set of proteins produced or modified by an organism or system which varies with stresses that an organism undergoes. Plant adaptation to abiotic stress is directly linked with profound changes in proteome composition; thereby proteomic studies can predict possible relationship between protein abundance and plant's acclimation to abiotic stress. Qualitative and quantitative changes in proteome involving protein modifications, protein-protein interactions, stress dependent protein movements, de novo synthesis and controlled degradation result in changes in certain metabolic pathways. System Biology makes use of molecular parts (transcripts, proteins, metabolites) of an organism and attempts to fit them into functional networks and models designed to describe and predict the dynamic activities of that organism in different environments (Cramer *et al*., 2011). There are two major approaches to protein profiling. Traditional approach involves use of two dimensional gel electrophoresis and mass spectrometry to measure changes in protein quantity. Second approach focuses on specific modification of proteome like membrane protein phosphorylation (Bohnert *et al*., 2006).

Yan *et al*. (2005) have done proteomic analysis of salt-stress responsive proteins in rice roots. Using mass spectrometry analysis and database searching they have identified 12 spots representing 10 different proteins of which 4 were already known stress responsive proteins and 6 were novel ones, which includes UDP-glucose pyrophosphorylase, cytochrome c oxidase subunit 6b-1, glutamine synthetase root isozyme, putative nascent polypeptide associated complex alpha chain, putative splicing factor-like protein and putative actin-binding protein. These proteins are involved in regulation of carbohydrate, nitrogen and energy metabolism, reactive oxygen species scavenging, mRNA and protein processing and cytoskeleton stability. At the same time, Hajheidari *et al.* (2005) two dimensional gel electrophoresis used for study the changes induced in leaf proteins of sugerbeet under drought conditions. They detected around 500 spots of which 79 showed significant changes under drought conditions. Twenty protein spots were analyzed by liquid chromatography-tandem mass spectrometry (LC-MS/MS), which led them to identify Rubisco and 11 other proteins involved in redox regulation, oxidative stress, signal transduction, and chaperone activities. Comaprative proteomic analysis for chilling stress responses in rice was studied by Yan *et al.* (2005). Through mass spectrometry they identified 85 differentially expressed proteins. The identified proteins are involved in signal transduction, RNA processing, translation, protein processing, redox homeostasis, photosynthesis, photorespiration and metabolisms of carbon, nitrogen, sulfur, and energy.

To understand the salinity stress responses in Arabidopsis roots, a comparative proteomic analysis of roots that has been exposed to 150 mM NaCl was conducted by Jiang *et al.* (2007). They have used two dimensional gel electrophoresis used to detect more than 1000 spots where, 103 protein spots increased and 112 decreased in response to NaCl treatment. They identified many previously characterized stress-responsive proteins and others related to processes involving scavenging for reactive oxygen species; signal transduction; translation, cell wall biosynthesis, protein translation, processing and degradation; and metabolism of energy, amino acids, and hormones.

The effect of drought and salinity stress on the seedlings of the somatic hybrid wheat cv. Shanrong No. 3 (SR3) and its parent bread wheat cv. Jinan 177 (JN177) was investigated by Peng *et al.*, (2009). They found that of a set of 93 (root) and 65 (leaf) differentially expressed proteins, 34 (root) and 6 (leaf) DEPs were cultivar-specific. The remaining were salinity/drought stress-responsive but not cultivar specific. They also observed that many of the DEPs were expressed under both drought and salinity stresses.

Metabolomics analysis

Metabolome refers to complete set of small molecule metabolites like metabolic intermediates, hormones, signalling molecules and secondary metabolites etc. which are found within a living entity (Oliver *et al.*, 1998). Metabolomics is defined as the identification and quantification of low molecular weight metabolites in a given organism, at a given developmental stage and in a given organ, tissue or cell type (Arbona *et al.*, 2009). In addition to changes in transcriptome and proteome, environmental stresses alters plant's metabolism in various ways such as production of compatible solutes, which can help either in stabilizing proteins and cellular structures or in maintaining cell turgor pressure and cellular redox balance or by regulating all of these factors (Bartels and Sunkar, 2005; Valliyodan and Nguyen, 2006; Munns and Tester, 2008; Janska *et al.*; 2010). Therefore, profiling of stress related metabolites is crucial for targeted breeding programmes for abiotic stress tolerance. Metabolic profiling is useful for phenotyping and diagnostic analysis of plants and is an important tool for functional annotation of genes and understanding cellular responses. For metabolomics analysis, complex mixture of metabolomics sample first needs to be simplified by separating some analytes from others. This can be achieved by gas chromatography, high performance liquid chromatography or capillary electrophoresis. Mass spectrometry is used to identify and quantify metabolites after separation. For analysis by mass spectrometry, the analytes are imparted with a charge and transferred to the gas phase either by electron or chemical ionization or through electrospray ionization. Nuclear magnetic resonance spectroscopy does not rely on separation of analytes and has high analytical reproducibility.

Krasensky and Jonak (2012) has elaborately reviewed plant's metabolic adjustments in response to high salt stress, drought and extreme temperatures and concluded that metabolic networks are highly dynamic and metabolites can move between different cellular compartments, therefore metabolic analysis at subcellular level in specific tissues is a challenging need. Kaplan *et al.* (2004) has explored the temperature stress metabolome of Arabidopsis. They aimed metabolome profiling to determine metabolome temporal dynamics associated with the induction of acquired thermotolerance in response to heat shock and acquired freezing shock in response to cold shock. They used gas chromatography-mass spectrometry for metabolite analysis and were able to monitor 81 identified metabolites and 416 unidentified mass spectral tags. In their study, by comparing the heat and cold shock responses, they revealed a previously unknown relationship that majority of heat shock responses were shared with cold shock responses. To their further notice they found that many metabolites that showed increased response to both heat and cold shock were previously unlinked with temperature stress.

Widodo *et al.* (2009) has studied the metabolic responses of barley genotypes to salt stress. They found that salt sensitive plants had elevated levels of amino acids, including proline, GABA and polyamine putrescine, which are indicator of general cellular damage in plants, whereas in salt tolerant plants levels of hexose phosphatases, TCA cycle intermediates and metabolites involved in cellular protection increased in response to salt stress. To assess the relationship between stress tolerance, metabolome, water homeostasis and growth performance under osmotic stress, Lugan *et al.* (2010) has compared a salt tolerant *Brassicaceae* species, *Thellungiella salsuginea* with *Arabidopsis thaliana*. Through metabolic fingerprinting and profiling they found that apart from few notable differences in raffinose and secondary metabolites, the same metabolic pathways were regulated by salt stress in both species. Significant differences in two species were observed in terms of physicochemical properties of their metabolomes, like water solubility and polarity. Cook *et al*. (2004) has examined the changes that occur in Arabidopsis metabolome in response to low temperature and proposed that CBF cold responsive pathway has a prominent role in bringing about these modifications.

Conclusions

Biotechnology approaches have a great potential to enhance crop production under different stress conditions. On the one hand, abiotic stresses are complex in nature; but on the other hand, there are several challenges that have restricted the realization of the full potential of using biotechnology approaches in crop breeding and improvement. Although there are several reports on the identification or even validation of QTLs or markers for abiotic stress tolerance, but their successful deployment in the development of a superior cultivar has had only limited success in the past. This limited success of biotechnology for developing abiotic stress-tolerant cultivars indicates one or more of the following points (Steele *et al*., 2006): (i) the nature of abiotic stress is complex with variations in the timing, duration and intensity of stress affecting different stages of plant development. (ii) Abiotic stress tolerance is often measured using traits, such as yield under stress, that are integrators over time of many mechanisms. Therefore, approaches involving the introgression of one gene or QTL using GE or MB is usually not sufficient to develop drought- or heat-tolerant lines unless that gene or QTL has a large effect on a particular key process (e.g. disease resistance (Sundaram, *et al*., 2009). (iii) Our capacity to phenotyping is limited by our understanding of abiotic stress tolerance mechanisms, which ultimately limits all conventional or molecular plant breeding efforts. There is also a lack of appropriate large-scale phenotyping facilities in our research institutes, particularly in developing countries. (iv) the appropriate MB method (i.e. marker assisted recurrent selection or genome wide selection) needs to be used instead of MABC for achieving higher genetic gain for complex traits. (v) GE requires the identification of appropriate promoters, particularly for gene stacking. It is now time to use interdisciplinary approaches to tackle the serious challenges of complex abiotic stresses, and the scientific community should focus on multiple approaches (Varshney *et al*., 2011).

References

Abe, H., Urao, T., Ito, T., Seki, M., Shinozaki, K. and Yamaguchi-Shinozaki, K. 2003. Arabidopsis *AtMYC2* (bHLH) and *AtMYB2* (MYB) function as transcriptional activators in abscisic acid signaling. *Plant Cell.* 15:63–78.

Ahn, Y.J. and Zimmerman, L. 2006. Introduction of carrot HSP17.7 into potato (*Solanum tuberosum* L.) enhances cellular membrane stability and tuberization *in vitro. Plant Cell and Environment.* 29:95-104.

Allard, R.W. 1999 Principles of plant breeding, 2nd edn. Wiley, New York.

Almeida, A.M., Araujo, S.S., Cardoso, L.A., Fevereiro, M.P., Tarne, J.M., and dos Santos, D.M. 2004.Transformation of tobacco with an *Arabidopsis thaliana* gene involved in trehalose biosynthesis as a model to increase drought resistance in crop plants. Proceedings of the 17th EUCARPIA General Congress on Genetic Variation for Plant Breeding, Sep. 8-11, Tullen, Austria, pp:285-289.

Alscher, R.G., Erturk, N. and Heath, L.S. 2002. Role of superoxide dismutases (SODs) in controlling oxidative stress in plants. *Journal of Experimental Botany* 53:1331-1341.

Arbona, V., Iglesias, D.J., Talón, M. and Gómez-Cadenas, A. 2009.Plant phenotype demarcation using nontargeted LC-MS and GC-MS metabolite profiling. *Journal of Agricultural and Food Chemistry* 57:7338–7347.

Ashikari, M. & Matsuoka, M. 2006.Identification, isolation and pyramiding of quantitative trait loci for rice breeding. *Trends Pl. Sci.* 11:344–350.

Ashikari, M. *et al.*, 2005. Cytokinin oxidase regulates rice grain production. *Science*. 309:741–745.

Ashraf, M. 2009. Biotechnological approach of improving plant salt tolerance using antioxidants as markers. *Biotechnology Advances.* **27**:84-93.

Ashraf, M. and Foolad, M.R. 2007. Roles of glycine betaine and proline in improving plant abiotic stress resistance. *Environmental and Experimental Botany.* 59:206–16.

Bao-Yan, A.N., Yan, L., Jia-Rui, L., Wei-Hua, Q., Xian-Sheng, Z. and Xin-Qi, Z. 2008. Expression of a vacuolar Na+/H+ antiporter gene of alfalfa enhances salinity tolerance in transgenic Arabidopsis. *Acta Agronomica Sinica.* 34:557–64.

Barloy, D., Lemoine, J., Paulette, A., Tanguy, M., Roger, R., Joseph, J. 2006.Marker-assisted pyramiding of two cereal cyst nematode resistance genes from Aegilops variabilis in wheat. *Molecular Breeding.* 20:31-40.

Bartels, D. and Sunkar, R. 2005. Drought and salt tolerance in plants. *Critical Reviews in Plant Sciences.* 24:23-58.

Basavaraj, S.H., Vikas, K., Singh, A., Singh, A., Singh, A., Singh, D., Yadav, A.S., Gopala Krishnan, S., Nagarajan, M., Mohapatra, T., Prabhu, K.V. and Singh, A.K. 2010. Markerassisted improvement of bacterial blight resistance in 239 parental lines of Pusa RH10, a superfine grain aromatic rice hybrid. *Mol. Breeding*. 3:11032-010-9407.

Beckmann, J.S., and M. Soller, 1986. Restriction fragment length polymorphisms in plant genetic improvement. Oxford Surv. *Plant Molecular Cell Biology*. 3:196–250.

Benedict, C., Skinner, J.S., Meng, R., Chang, Y., Bhalerao, R., Huner, N.P.A., Finn, C.E., Chen, T.H.H. and Hurry, V. 2006. The CBF1-dependent low temperature signalling pathway, regulon and increase in freeze tolerance are conserved in Populus spp. *Plant Cell and Environment.*29:1259–1272.

Bohnert, H.J., Gong, Q., Li, P. and Ma, S. 2006. Unraveling abiotic stress tolerance mechanisms – getting genomics going. *Current Opinion in Plant Biology.* 9:180-188.

Bonnett, D.G., Rebetzke, G.J. and Spielmeyer, W. 2005. Strategies for efficient implementation of molecular markers in wheat breeding. *Molecular Breeding*. 15:75-85.

Brini, F., Hanin, M., Lumbreras, V., Amara, I., Khoudi, H., Hassairi, A., Pages, M. and Masmoudi, K. 2007a. Overexpression of wheat dehydrin DHN-5 enhances tolerance to salt and osmotic stress in *Arabidopsis thaliana. Plant Cell Reports.* 26:2017–2026.

Camargo, L.E.A., Osborn, T.C. 1996. Mapping loci controlling flowering time in Brassica oleracea. *Theoretical and Applied Genetics*. 92:610–616.

Capell, T. and Christou, P. 2004. Enhanced drought tolerance in transgenic rice. *Current Opinion in Biotechnology* 15:148-154.

Cardoza, V. & Stewart, C. N. 2004. Brassica biotechnology: progress in cellular and molecular biology. *In Vitro Cellular and Developmental Biology Plant*. 40:542–551.

Castiglioni, P., Warner, D., Bensen, R.J., Anstrom, D.C., Harrison, J., Stoecker, M., Abad, M., Kumar, G., Salvador, S., D'Ordine, R., Navarro, S., Back, S., Fernandes, M., Targolli, J., Dasgupta, S., Bonin, C., Luethy, M.H. and Heard, J.E. 2008. Bacterial RNA chaperones confer abiotic stress tolerance in plants and improved grain yield in maize under water-limited conditions. *Pl. Physiol.*. 147:446-455.

Cheema, K.K.., Navtej, S.B., Mangat, G.S., Das, A., Vikal, Y., Brar, D.S., Khush, G.S. and Singh, K. 2008. Development of high yielding IR64 × Oryza rufipogon (Griff.) introgression lines and identification of introgressed alien chromosome segments using SSR markers. *Euphytica*. 160:401–409.

Chen, J.Q., Meng, X.P., Zhang, Y., Xia, M. and Wang, X.P. 2008. Over-expression of *OsDREB* genes lead to enhanced drought tolerance in rice. *Biotechnology Letters*. 30:2191–2198.

Chen, S., Lin, X.H., Xu, C.G. and Zhang, Q.F. 2000.Improvement of bacterial blight resistance of 'Minghui 63', an elite restorer line of hybrid rice, by molecular marker-assisted selection. *Crop Sci.* 40:239-244.

Chen, S., Vaghchhipawala, Z., Li, W., Asard, H. and Dickman, M.B. 2004. Tomato phospholipid hydroperoxide glutathione peroxidase inhibits cell death induced by bax and oxidative stresses in yeast and plants. *Pl. Physiol.* 135:1630-1641.

Chen, W., Zhang, W-K., Liu X-G., Chen, B-N., Tu, J-G.;,and Tingdong, Fu. 2007. Detection of QTL for six yield-related traits in oilseed rape (Brassica napus) using DH and immortalized F2 populations. *Theoretical and Applied Genetics*. 115:849–858.

Choudhary, K., Choudhary, O.P. and Shekhawat, N.S. 2008. Marker Assisted Selection: A Novel Approach for Crop Improvement. *American-Eurasian Journal of Agronomy*. 1(2):26-30.

Collard, B.C.Y. and Mackill, D.J. 2007. Marker-assisted selection: an approach for precision plant breeding in the twenty-first century. *Philosophical Transactions of the Royal Society B*. 363:557572.

Collard, B.C.Y., Jahufer, M.Z.Z., Brouwer, J.B. and Pang, E.C.K. 2005. An introduction to markers, quantitative trait loci (QTL) mapping and marker-assisted selection for crop improvement: The basic concepts. *Euphytica*.142:169–196.

Cong, B., Liu, J. and Tanksley S.D. 2002. Natural alleles at a tomato fruit size quantitative trait locus differ by heterochronic regulatory mutations. *Proceedings of National Academy of Sciences, USA*. 99:13606–13611.

Cook, D., Fowler, S., Fiehn, O. and Thomashow, M.F. 2004. A prominent role for the CBF cold response pathway in configuring the low-temperature metabolome of Arabidopsis. *Proceedings of National Academy of Sciences, USA*. 101(42): 15243-15248.

Cramer, G.R., Urano, K., Delrot, S., Pezzotti, M. and Shinozaki, K. 2011. Effects of abiotic stress on plants: a systems biology perspective. BMC. *Plant Biology*. 11:163-176

Cuc, L.M., Huyen, L.T.N., Hien, P.T.M., Hang, V.T.T., Dam, N.Q., Mui, P.T., Quang, V.D., Ismail, A.M. and Ham, L.H. 2012. Application of marker assisted backcrossing to introgress the submergence tolerance QTL SUB1 into the Vietnam elite rice variety-AS996. *American Journal of Plant Sciences* 3:528-536.

De Ronde, J.A., Spreeth, M.H. and Cress, W.A. 2000. Effect of antisense pyroline-5-carboxylate reductase transgenic soybean plants subjected to osmotic and drought stress. *Plant Growth Regulation*. 32:13-26.

Diamant, S., Eliahu, N., Rosenthal, D. and Goloubinoff, P. 2001. Chemical chaperones regulate molecular chaperones *in vitro* and in cells under combined salt and heat stresses. *J. of boil. Chem.* 276:39586-39591.

Edmeades, G.O. 2009. Drought tolerance in maize: an emerging reality. In: Global Status of Commercial Biotech/GM Crops: 2008. ISAAA Brief No. 39, pp.197–217. Ithaca, NY: ISAAA.

Ergen, N.A., Thimmapuram, J., Bohnert, H.J. and Budak, H. 2009. Transcriptome pathways unique to dehydration tolerant relatives of modern wheat. *Functional & Integrative Genomics*. 9(3):377-396.

Flint-Garcia, S. A., Thornsberry, J. M. and Buckler, E. S. 2003.Structure of linkage disequilibrium in plants.Annual. *Rev. of Pl. Biol.* 54:357–374.

Friedt, W. and Ordon, F. 2007.Molecular markers for gene pyramiding and disease resistance breeding in barley. In Genomics assisted crop improvement, Vol. 2: Genomics applications in crops, Varshney, R. K., and Tuberosa, R. (eds.). Springer, Dordrecht, The Netherlands, pp.81-101.

Gaber, A., Yoshimura, K., Yamamoto, T., Yabuta, Y., Takeda, T., Miyasaka, H., Nakano, Y. and Shigeoka, S. 2006. Glutathione peroxidase-like protein of Synechocystis PCC 6803 confers tolerance to oxidative and environmental stresses in transgenic Arabidopsis. *Physiologiae Plantarum* 128:251–262.

Gahan, L.J., Ma, Y.T., Cobble, M.L.M., Gould, F., Moar, W.J., Heckel, D.G. 2005. Genetic basis of resistance to Cry 1Ac and Cry 2Aa in Heliothis virescens (Lepidoptera: Noctuidae). Journal Economics Entomology 98:1357-1368.

Gandhi, D. 2007. UAS scientist develops first drought tolerant rice. The Hindu.www.thehindu.com/2007/11/17/stories/2007111752560500.htm (varified 20 March 2009).

GMO Safety. 2008. Drought-tolerant wheat: "Promising results". Available online at: http://www.gmosafety.eu/en/news/.

Gopalkrishnan, R.K., Sharma, K. Anand Rajkumar, M. Josheph, V.P. Singh, A.K. Singh, K.V. Bhat, N.K. Singh and Mohapatra, T. 2008. Integrating marker assisted background analysis with foreground selection for identification of superior bacterial blight resistant recombinants in Basmati rice. Plant Breeding 127:131–139.

Gu, D., Liu, X., Wang, M., Zheng, J., Hou, W., Wang, G., *et al.* 2008. Overexpression of *ZmOPR1* in Arabidopsis enhanced the tolerance to osmotic and salt stress during seed germination. *Pl. Sci.* 174:124–30.

Guo, S.J., Zhou, H.Y., Zhang, X.S., Li, X.G. and Meng, Q.W. 2007. Overexpression of *CaHSP26* in transgenic tobacco alleviates photoinhibition of PSII and PSI during chilling stress under low irradiance. *J. of Pl. Physiol.* 164:126–136.

Gutterson, N. & Reuber, T. L. (2004).Regulation of disease resistance pathways by AP2/ERF transcription factors.Current. *Opinion in Biology.* 7:465–471.

Hajheidari, H., Abdollahian-Noghabi, M., Askari, H., Heidari, M., Sadeghian, S.Y., Ober, E.S. and Salekdeh, G.H. 2005. Proteome analysis of sugar beet leaves under drought stress. *Proteomics.* 5(4):950–960.

Hall, D. 2010. Using association mapping to dissect the genetic basis of complex traits in plants.Briefings in.Functional. *Genomics* 9:157–165.

Hash, C.T., Sharma, A., Kolesnikova-Allen, M.A., Singh, S.D., Thakur, R.P., Bhasker Raj, A.G., Ratnaji Rao, M.N.V., Nijhawan, D.C., Beniwal, C.R., Sagar, P., Yadav, H.P., Yadav, Y.P., Srikant Bhatnagar, S.K., Khairwal, I.S., Howarth, C.J., Cavan, G.P., Gale, M.D., Liu, C., Devos, K.M., Breese, W.A., Witcombe, J.R. 2006. Teamwork delivers biotechnology products to Indian small-holder crop-livestock producers: Pearl millet hybrid "HHB 67 Improved" enters seed delivery pipeline. *Journal of SAT Agricultural Research* 2(1): Hash, C.T., Yadav, R.S. Cavan, G.P., Howarth, C.J., Liu, H., Qi, X., Sharma, A., Allen,

M.K., Bidinger, F.R. and Witcombe, J.R. 2005. Marker-Assisted Backcrossing to Improve Terminal Drought Tolerance in Pearl Millet.

Hirschhorn, J. N. & Daly, M. J. 2005. Genome-wide association studies for common diseases and complex traits. *Nature Reviews Genetics.* 69:5–108.

Hittalmani, S., Shashidhar, H.E., Bagali, P.G., Huang, N., Sidhu, J.S., Singh, V.P., and Khush, G.S. 2002.Molecular mapping of quantitative trait loci for plant growth, yield and yield relatedtraits across three diverse locations in a doubled haploid rice population.*Euphytica.* 125:207-214.

Hmida-Sayari, A., Gargouri-Bouzid, R., Bidani, A., Jaoua, L., Savoure, A. and Jaoua, S. 2005. Overexpression of Ä[1]-pyrroline-5-carboxylate synthetase increases proline production and confers salt tolerance in transgenic potato plants. *Plant Science.* 169:746–752.

Hoewyk, D.V., Takahashi, H., Inoue, E., Hess, A., Tamaoki, M. and Pilon-Smits, E.A.H. 2008. Transcriptome analyses give insights into selenium-stress responses and selenium tolerance mechanisms in Arabidopsis. *Physiol. Plantarum.* 132(2):236–253.

Holland, J.B. 2004. Implementation of molecular markers for quantitative traits in breeding programs—challenges and opportunities. In Proc. 4th Int. Crop Sci. Congress., Brisbane, Australia, 26 September—1 October.

Hong, Z., Lakkineni, K., Zhang, Z., Pal, D. and Verma, S. Removal of feedback inhibition of Ä[1]-pyrroline-5-carboxylate synthetase results in increased proline accumulation and protection of plants from osmotic stress. *Pl. Physiol.* 122:1129–1136.

Hospital, F. 2003.Marker-assisted breeding. In: H.J. Newbury (ed.), Plant molecular breeding. Blackwell Publishing and CRC Press, Oxford and Boca Raton, pp. 30-59.

Hospital, F. 2005. Selection in backcross programmes. *Philosophical Transactions of the Royal Society B* 360:1503–1511.

Hospital, F. 2011. Size of donor chromosome segments around introgressed loci and reduction of linkage drag in marker assisted backcross programs. *Genetics.* 158(3):1363-1379.

Hospital, F. and Charcosset, A. 1997.Marker-assisted introgression of quantitative trait loci.*Genetics.* 147:1469–1485.

Houde, M., Sylvain, D., N'Dong, D. and Sarhan, F. 2004. Overexpression of the acidic dehydrin *WCOR410* improves freezing tolerance in transgenic strawberry leaves. *Pl. Biotech. J.* 2:381–387.

Hu, H., Dai, M., Yao, J., Xiao, B., Li, X., Zhang, Q. and Xiong, L. 2006.Overexpressing a NAM, ATAF and CUC (NAC) transcription factor enhancing drought resistance and salt tolerance in rice. *Proceedings of the National Academy of Sciences* 103:12987-12992.

Huang, N., Angeles, E.R., Domingo, J., Magpantay, G., Singh, S., Zhang, Q., Kumaravadivel N., Bennett J. and Khush G.S. 1997.Pyramiding of bacterial blight resistance genes in rice: marker-assisted selection using RFLP and PCR. *Theoretical and Applied Genetics.* 95:313-320.

Hussain, T.M., Chandrasekhar, T., Hazara, M., Sultan, Z., Saleh, B.K. and Gopal, G.R. 2008. Recent advances in salt stress biology-a review. *Biotechnology and Molecular Biology Reviews* 3:8-13.

Jaccoud, D., Peng, K., Feinstein, D. and Kilian, A. 2001. Diversity arrays: a solid state technology for sequence information independent genotyping. *Nucleic Acids Research.* 29(4):1-7.

Jackson, R.E., Bradley, J.R. and Van Duyn, J.W. 2003.Field performance of transgenic cottons expressing one or two Bacillus thuringiensis endotoxins against bollworm, Helicoverpa zea (Boddie). *J. of Cotton Sci.* 7:57-64.

James, R.A., Blake, C., Zwart, A.B., Hare, R.A., Rathjen, A.J. and Munns, R. 2012. Impact of ancestral wheat sodium exclusion genes *Nax1* and *Nax2* on grain yield of durum wheat on saline soils. *Functional Pl. Biol.* 39:609-618.

Janska, A., Marsik, P., Zelenkova, S. and Ovesna, J. 2010. Cold stress and acclimation: what is important for metabolic adjustment. *Pl. Biol.* 12:395-405.

Jean, M., Brown, G. G., Landry, B. S. 1997.Genetic mapping of nuclear fertility restorer genes for the 'polima' cytoplasmic male sterility in canola (Brassica napus L.) using DNA markers. *Theoretical and Applied Genetics*. 95:321–328.

Jiang, Y., Yang, B., Harris, N.S. and Deyholos, M.K. 2007. Understanding and Improving Salt Tolerance in Plants.Comparative proteomic analysis of NaCl stress-responsive proteins in Arabidopsis roots. *J. of Exp. Bot.* 58(13):3591-3607.

Joseph, M.S., Gopalakrishnan, R.K., Sharma, V.P., Singh, A.K., Singh, N.K., Singh and Mohapatra, T., 2004. Combining bacterial blight resistance and Basmati quality characteristics by phenotypic and molecular marker assisted selection in rice. *Mol. Breeding* 00:1–11.

Joshi, R.K. and Nayak, S. 2010. Gene pyramiding-A broad spectrum technique for developing durable stress resistance in crops. *Biotech. and Mol. Biol. Review* 5(3):51-60.

Kalita, M.C., Mohapatra, T., Dhandapani, A., Yadava, D.K., Srinivasan, K., Mukherjee, A.K. and Sharma, R.P. 2007. Comparative evaluation of RAPD, ISSR and Anchored-SSR markers in assessment of genetic diversity and fingerprinting of oilseed Brassica genotypes. *J. of Pl. Biochem. and Biotech.* 15(1):41–48

Kaplan, F., Kopka, J., Haskell, D.W., Zhao, W., Schiller, K.C., Gatzke, N., Sung, D.Y. and Guy, C.L. 2004. Exploring the Temperature-Stress Metabolome of Arabidopsis. *Pl. Physiol.* 136:4159-4168.

Kawakami, A., Sato, Y. and Yoshida, M. 2008. Genetic engineering of rice capable of synthesizing fructans and enhancing chilling tolerance. *J. of Exp. Bot.* 59:793–802.

Khairwal, I.S., Hash, C.T. 2007. : "HHB 67-Improved" – The first product of marker-assisted crop breeding in India. Asia-Pacific Consortium on Agricultural Biotechnology (APCoAB) e-News [http:// www.apcoab.org/special_news.html].

Kobayashi, F., Ishibashi, M. and Takumi, S. 2008.Transcriptional activation of *Cor/Lea* genes and increase in abiotic stress tolerance through expression of a wheat DREB2 homolog in transgenic tobacco.*Trans. Res.* 17:755–767.

Koundal, V., Parida, S.K., Yadava, D.K., Ali, A., Koundal, K.R. and, Mohapatra, T. 2008. Evaluation of microsattelite markers for genome mapping in Indian mustard (*Brassica juncea* L.). *J. of Pl. Biochem. and Biotech.* 17(1):69–72.

Krasensky, J. and Jonak, C. 2012. Drought, salt, and temperature stress-induced metabolic rearrangements and regulatory networks. *J. of Exp. Bot.* 63(4):1593-1608.

Kumar, A., Jain, A., Sahu, R.K., Srivastava, M.N., Nair, S., Mohan, M. 2005.Genetic analysis of resistance genes for the rice Gall Midge in two Rice Genotypes. *Crop Sci.* 5:1631-1635.

Kumar, R., Venuprasad, R. & Atlin, G.N. 2007. Genetic analysis of rainfed lowland rice drought tolerance under naturally occurring stress in eastern India: heritability and QTL effects. *Field Crops Res.* 103:42–52.

Kumari, M., Taylor,G.J. and Deyholos, M.K. 2008. Transcriptomic responses to aluminum stress in roots of Arabidopsis thaliana. *Mol. Gen. and Geno.* 279(4):339-357.

Kumari, S., Sabharwal, V., Kushwaha, H.R., Sopory, S.K., Singla-Pareek, S.L. and Pareek, A. 2009. Transcriptome map for seedling stage specific salinity stress response indicates a specific set of genes as candidate for saline tolerance in Oryza sativa L. *Functional & Integrative Genomics* 9(1):109-123.

Kumpatla, S.P., Buyyarapu R., Abdurakhmonov I.Y., and Mammadov J.A., 2012. Genomics-assisted plant breeding in the 21st century: technological advances and progress. In: I.Y. Abdurakhmonov (ed.), *Plant Breeding, InTech*, pp 131-184.

Kunert, A., Naz, A.A., Dedeck, O., Pillen, K. and Léon, J. 2007. AB-QTL analysis in winter wheat: I. Synthetic hexaploid wheat (T. turgidum ssp. dicoccoides × T. tauschii) as a source of favourable alleles for milling and baking quality traits. *Theoritically Applied Genetics*. 115:683–695.

Lal, S., Gulyani, V. and Khurana, P. 2008. Overexpression of *HVA1* gene from barley generates tolerance to salinity and water stress in transgenic mulberry (*Morus indica*). *Trans. Res.* 17:651–663.

Lang, N.T., Tao, N.V. and Buu, B.C. 2011. Marker-assisted backcrossing (mab) for rice submegence tolerance in Mekong delta.*Omonrice.* 18:11-21.

Lecomte, L., Duffe, P., Buret, M., Servin, B., Hospital, F. and Causse, M. 2004. Marker-assisted introgression of five QTLs controlling fruit quality traits into three tomato lines revealed interactions between QTLs and genetic backgrounds. *Theoretical and Applied Genetics* 109:658-668.

Lee, S.B., Kwon, H.B., Park, S.C., Jeong, M.J., Han, S.E., Byun, M.O. and Daniell, H. 2003. Accumulation of trehalose within transgenic chloroplast confer drought tolerance. *Mol. Breeding.* 11:1-13.

Li, H.Y., Chang, C.S., Lu, L.S., Liu, C.A., Chan, M.T. and Charng, Y.Y. 2003.Overexpression of *Arabidopsis thaliana* heat shock factor gene (*AtHsf1b*) enhance chilling tolerance in transgenic tomato. *Bot. Bull. of Academia Sinica.* 44:129-140.

Li, J.Z., Huang, XQ., Heinrichs, F., Ganal, M.W. and Röder, M,S. 2006. Analysis of QTLs for yield components, agronomic traits, and disease resistance in an advanced backcross population of spring barley. *Genome*. 49:454– 466.

Li, Z.C., Zhang, X.C., Zhang, L., Thuang, B.C., Zhang, C.L., Wang, G.Y. and Fu, Y.C. 2004. Expression of *MtD1* gene in transgenic rice leads to enhanced salt-tolerance. *China Agricultural University.* 9:38-43.

Lin, H.X., Zhu, M.Z., Yano, M., Gao, J.P., Liang, Z.W., Su, W.A., Hu, X.H., Ren, Z.H. and Chao, D.Y. 2004. QTLs for Na+ and K+ uptake of the shoots and roots controlling rice salt tolerance. *Theoretical and Applied Genetics* 108:253–260.

Lincoln, S., Daly M. and Lander E. 1993.Mapping genes controlling quantitative traits using MAPMAKER/QTL.Version 1.1.Whitehead Institute for Biomedical Research Technical Report, 2nd Edn.

Lindsay, M. P., Lagudah, E. S., Hare, R. A. & Munns, R. 2004.A locus for sodium exclusion (Nax1), a trait for salt tolerance, mapped in durum wheat. *Functional Pl. Biol.* 31:1105–1114.

Liu, J., Cong, B. & Tanksley, S.D. 2003. Generation and analysis of an artificial gene dosage series in tomato to study the mechanisms by which the cloned quantitative trait locus fw2.2 controls fruit size. *Pl. Physiol.* 132:292–299.

Lugan, R., Niogret, M.F., Leport, L., Guégan, J.P., Larher, F.R., Savouré, A., Kopka, J. and Bouchereau, A. 2010. Metabolome and water homeostasis analysis of Thellungiella salsuginea suggests that dehydration tolerance is a key response to osmotic stress in this halophyte. *The Pl. J.* 64:215-229.

Lynch, M. & Walsh, B. 1997. Genetics and the Analysis of Quantitative Traits (Sinauer Associates, Sunderland, Massachusetts,).

Mackay, T. F. 2001. The genetic architecture of quantitative traits. *Ann. Rev. of Genetics*. 35:303–339.

Mahmood, T., Ekuere, U.; Yeh, F., Good, A. G., Stringam, G. R. 2003.RFLP linkage analysis and mapping genes controlling the fatty acid profile of *Brassica juncea* using reciprocal DH populations. *Theoretical and Applied genetics*. 107:283–290.

Maiorino, F.M., Brigelius-Flohe, R., Aumann, K.D., Roveri, A., Schomburg, D. and Flohe, L. 1995.Diversity of glutathione peroxidases. *Methods in Enzymology.* 252:38–48.

Manavalan, L.P., Guttikonda, S.K., Tran L,S.P. and Nguyen, H.T. 2009. Physiological and molecular approaches to improve drought resistance in soybean. *Pl. and Cell Physiol.* 50:1260-76.

Mishra, S.K., Tripp, J., Winkelhaus, S., Tschiersch, B., Theres, K., Nover, L. and Scharf, K.D. 2002. In the complex family of heat stress transcription factors, HsfA1 has a unique role as master regulator of thermotolerance in tomato. *Genes & Development.* 16:1555–1567.

Mohapatra, T., Upadhyay, A., Sharma, A. and Sharma, R.P. 2002.Detection and mapping of duplicate loci in Brassica juncea. *J. of Pl. Biochem. and Biotech.* 11:37–42.

Morris, M., Dreher, K., Ribaut, J. M. and Khairallah, M. 2003. Money matters (II): Costs of maize inbred line conversion schemes at CIMMYT using conventional and marker-assisted selection. *Mol. Breeding.* 11:235-247.

Müller-Röber, B., Hucho, F., Van den Daele, W., Köchy, K., Reich, J., Rheinberger, H.J., Sperling, K., Wobus, A.M., Boysen, M. and Kölsch, M. 2007. Grüne Gentechnologie: aktuelle Entwicklungen in Wissenschaft und Wirtschaft – Supplement zum Gentechnologiebericht. - Forschungsberichte der Interdisziplinären Arbeitsgruppen der Berlin-Brandenburgischen Akademie der Wissenschaften.17; Berlin.

Munns, R. 2002. Comparative physiology of salt and water stress. *Plant Cell and Environment.* 25:239-250.

Munns, R. and Tester, M. 2008. Mechanisms of salinity tolerance. *Ann. Rev. of Pl. Biol.* 59:651-681.

Nair, S.K., Prasanna, B.M., Garg, A., Rathore, R.S., Setty, T.A.S. and Singh, N.N. 2005.Identification and validation of QTL conferring resistance to sorghum downy mildew (Peronoclerospora sorghi) and Rajasthan downy mildew (P. heteropogoni) in maize. *Theoretical and Applied genetics.* 110:184-1392.

Narayanan, N.N., Baisakh, N., Vera Cruz, C.M., Gnanamanickam, S.S., Datta, K., Datta, S.K. 2002.Molecular breeding for the development of Blast and Bacterial Blight resistance in Rice cv.IR50. *Crop Sci.* 42:2072- 2079.

Neeraja, C.N., Maghirang-Rodriguez, R., Pamplona, A., Heuer, S., Collard, B.C.Y., Septiningsih, E.M., Vergara, G., Sanchez, D., Xu, K., Ismail, A.M. and Mackill, D.J. 2007.A marker-assisted backcross approach for developing submergence-tolerant rice cultivars. *Theoretical and Applied Genetics.* 115:767-776.

Nelson, D.E. *et al.* 2007. Plant nuclear factor Y (NF -Y) B subunits confer drought tolerance and lead to improved corn yields on water-limited acres. *Proceedings of the National Academy of Sciences of the USA.*104:16450–16455.

Nelson, J.C., 1997. Qgene software for marker-based genomic analysis and breeding. *Molecular Breeding.* 3:239–245.

Niel, T.W. and Jonah, P.M. 2006. Biotechnology as a useful tool in wheat (*Triticum aestivum*) improvement. *J. of Res. in Agri.* 3:18-23.

Oh, S.J., Kwon, C.W., Choi, D.W., Song, S.I. and Kim, J.K. 2007. Expression of barley *HvCBF4* enhances tolerance to abiotic stress in transgenic rice. *Pl. Biotech. J.* 5:646-656.

Oliver, S.G., Winson, M.K., Kell, D.B. and Baganz, F. 1998. Systematic functional analysis of the yeast genome. *Trends in Biotech.* 16 (9):373-378.

Padan, E., Venturi, M., Gerchman, Y. and Dover, N. 2001. Review: Na+/H+ antiporters. *Biophysica Acta* 1505:144–57.

Padmaja, K., Arumugam, N., Gupta V., Mukhopadhyay , A., Sodhi, Y. S., Pental, D. and Pradhan, A. K. 2005). Mapping and tagging of seed coat colour and the identification of microsatellite markers for marker-assisted manipulation of the trait in Brassica juncea. *Theoretical and Applied genetics.* 111:8–14.

Parker, G.D., Langridge, P. 2000. Development of a STS marker linked to a major locus controlling flour colour in wheat (Triticum aestivum L.). *Molecular Breeding.* 6:169-174.

Parvanova, D., Povova, A., Zaharivena, I., Lambrev, P. and Kostantinova, T. *et al.* 2004. Low temperature tolerance of tobacco plants transferred to accumulate proline, fructans, or glycine betaine. Variable chlophyll fluorescence evidence. *Phytosynthetica.* 42:179-185.

Pasquali, G., Biricolti, S., Locatelli, F., Baldoni, E. and Mattana, M. 2008.*Osmyb4* expression improves adaptive responses to drought and cold stress in transgenic apples. *Plant Cell Reports.* 27:1677–1686.

Pellegrineschi, A., Reynolds, M., Pacheco, M., Bitro, R.M., Almeraya, R., Yamaguchi-Shinazaki, K. and Hoisington, D. 2004. Stress-induced expression in wheat of the *Arabidopsis thalianaDREB1A* gene delays water stress symptoms under greenhouse condition. *Genome.* 47:493-500.

Pellegrineschi, A., Ribaut, J.M., Tretowan, R., Yamaguchi-Shinozaki, K., Hoisington, D. and Vasil, I.K. 2003.Preliminary characterization of DREB genes in transgenic wheat. Proceedings of the 10th IAPTC and B Congress on Plant Biotechnolgy, June 23-28, Orlando, Florida, USA, pp:1836-1837.

Peng, Z., Wang, M., Li, F., Lv, H., Li, C. and Xia, G. 2009.A Proteomic Study of the Response to Salinity and Drought Stress in an Introgression Strain of Bread Wheat. *Molecular & Cellular Proteomics.* 8(12):2676-2686.

Perera, I.Y., Hung, C.Y., Moore, C.D., Stevenson-Paulik, J. and Boss, W.F. 2008. Transgenic Arabidopsis plants expressing the type 1 inositol 5-phosphatase exhibit increased drought tolerance and altered abscisic acid signaling. *Pl. Cell.*20:2876-2893.

Prashanth, S.R., Sadhasivam, V. and Parida, A. 2008. Over expression of cytosolic copper/zinc superoxide dismutase from a mangrove plant *Avicennia marina* in indica Rice var Pusa Basmati-1 confers abiotic stress tolerance. *Transgenic Research.*17:281–291.

Priyamedha, Singh, B. K., Kaur, G., Sangha, M. K. and Banga, S. S. 2012. RAPD, ISSR and SSR Based Integrated Linkage Map From an F2 Hybrid Population of Resynthesized and Natural *Brassica carinata* National Academy Science Letters 35, (4):303-308.

Puhakainen, T., Hess, M.W., Ma¨kela,¨ P., Svensson, J., Heino, P. and Palva, E.T. 2004. Overexpression of multiple dehydrin genes enhances tolerance to freezing stress in Arabidopsis. *Pl. Mol. Biol.* 54:743-753.

Qin, Q.L., Liu, J.G., Zhang, Z., Peng, R.H., Xiong, A.S., Yao, Q.H. and Chen, J.M. 2007. Isolation, optimization, and functional analysis of the cDNA encoding transcription factor OsDREB1B in *Oryza Sativa* L. *Mol. Breeding.* 19:329–340.

Qiu, Q., Ma, T., Hu, Q., Liu, B., Wu, Y., Zhou, H., Wang, Q., Wang, J. and Liu, J. 2011.Genome-scale transcriptome analysis of the desert poplar, Populuseuphratica. *Tree Physiol.* 31(4):452-61.

Rajgopal, D., Agarwal, P., Tyagi, W., Singla-Pareek, S.L., Reddy, M.K. and Sopory, S.K. 2007.*Pennisetum glauca* Na+/H+ antiporter confers high level of salinity tolerance in transgenic *Brassica juncea. Mol. Breeding.* 19:137-151.

Ren, Z.H. *et al.*, 2005. A rice quantitative trait locus for salt tolerance encodes a sodium transporter. *Nature Genetics.* 371:141–1146.

Rhodes, D. and Hanson, A.D. 1993. Quaternary ammonium and tertiary sulfonium compounds in higher-plants. *Ann. Rev. of Pl. Physiol. and Pl. Mol. Biol.* 44:357–84.

Ribaut, J. M. and Betran, J. 1999. Single large-scale marker-assisted selection (SLS-MAS). *Mol. Breeding.* 5:531-541.

Ribaut, J.M., Jiang, C. and Hoisington, D. 2002. Simulation experiments on efficiencies of gene introgression by backcrossing. *Crop Sci.* 42:557–565.

Ribaut, J.M., X. Hu, D. Hoisington & D. Gonzalez-De-Leon, 1997.Use of STSs and SSRs as rapid and reliable preselection tools in marker-assisted selection backcross scheme. *Pl. Mol. Biol. Reports.* 15:156–164.

Robert, V.J.M. 2001. Marker assisted introgression of black mold resistance QTL alleles from wild Lycopersicon cheesmanii to cultivated tomato (L. esculentum) and evaluation of QTL phenotypic effects. *Molecular Breeding.* 8:217-223.

Rohila, J.S., Jain, R.K. and Wu, R. 2002. Genetic improvement of Basmati rice for salt and drought tolerance by regulated expression of a barley Hva1 cDNA. *Pl. Sci.* 163:525-532.

Roy, B. and Basu, A.K. 2009. Abiotic stresses in crop plants-breeding and biotechnology. New Delhi, ISBN 10: 81-89422-94-4, pp:1-544.

Sahi, C., Singh, A., Kumar, K. Blumwald, E. & Grover, A. 2006. Salt stress response in rice: genetics, molecular biology, and comparative genomics. *Functional and Integrated Genomics.* 6:263–284.

Sakuma, Y., Maruyama, K., Qin, F., Osakabe, Y., Shinozaki, K. and Yamaguchi-Shinozaki, K. 2006a.Dual function of an Arabidopsis transcription factor DREB2A in in water-stress-responsive and heat-stress-responsive gene expression. *Proceedings of the National Academy of Sciences.* 103:18822–18827.

Sakuma, Y., Maruyama, K., Qin, F., Osakabe, Y., Shinozaki, K. and Yamaguchi-Shinozaki, K. 2006b.Functional analysis of an Arabidopsis transcription factor DREB2A drought-responsive gene expression. *Pl. Cell.* 18:1292–1309.

Salina, E., Dobrovolskaya, O., Efremova, T., Leonova, I. & Roder, M.S. 2003.Microsatellite monitoring of recombination around the Vrn-B1 locus of wheat during early backcross breeding. *Pl. Breeding.* 122:116–119.

Salvi, S., Tuberosa R., Chiapparino E., Maccaferri M., Veillet S., Van Beuningen L., Issac P., Edards K., and Phillip R.L. 2002. Towards positional cloning of vgt1, a QTL controlling the transition from the vegetative to the reproductive phase in Maize. *Pl. Mol. Biol.* 48:601-613.

Sanchez, A. C., Subudhi, P. K., Rosenow, D. T. & Nguyen, H. T. 2002. Mapping QTLs associated with drought resistance in sorghum (Sorghum bicolor L. Moench). *Pl. Mol. Biol.* 48:713–726.

Santra, D., DeMacon, V.K., Garland-Campbell, K., Kidwell, K. 2006. Markerassisted backcross breeding for simultaneous introgression of stripe rust resistance genes yr5 and yr15 into spring wheat (triticum aestivum l.). In 2006 international meeting of ASA-CSSA-SSSA. pp74-75.

Sawahel, W.A. and Hassan, A.H. 2002. Generation of transgenic wheat plants producing high levels of the osmoprotectant proline. *Biotechnology Letters.* 24:721–5.

Schwartz, S.H., Qin, X. and Zeevaart, J.A.D. 2003.Elucidation of the indirect pathway of abscisic acid biosynthesis by mutants, genes, and enzymes. *Pl. Physiol.*131:1591–1601.

Septiningsih, E.M. *et al.* 2009. Development of submergence-tolerant rice cultivars: the Sub1 locus and beyond. *Annals of Bot.* 103:151–160.

Serrano, R., Culianz-Macia, F.A. and Moreno, V. 1998.Genetic engineering of slat and drought tolerance with yeast regulatory genes. *Scientia Horticulturae* 78:261-269.

Shaomin, G., Jun- Zou, Ruiyan, L., Long, Y,N., Chen, S.G. and Meng, J.G. 2012. A genetic linkage map of Brassica carinata constructed with a doubled haploid population. *Theoretical and Applied Genetics* 125:1113-1124.

Shi, H., Jiang, C., Ye, T., Tan, D., Reiter, R.J., Zhang, H., Liu, R. and Chan, Z. 2014. Comparative physiological, metabolomic, and transcriptomic analyses reveal mechanisms of improved abiotic stress resistance in bermudagrass [Cynodondactylon(L). Pers.]by exogenous melatonin. *J. of Exp. Bot.* doi:10.1093/jxb/eru373

Shinozaki, K. and Yamaguchi-Shinozaki, K. 1996.Molecular responses to drought and cold stress.*Current Opinion in Biotech.* 7:161–167.

Siangliwa, J.L., Jongdeeb, B., Pantuwanc, G. and Toojinda, T. 2007. Developing KDML105 backcross introgression lines using marker-assisted selection for QTLs associated with drought tolerance in rice. *ScienceAsia.* 33:207-214.

Song, X. J., Huang, W., Shi, M., Zhu, M. Z. & Lin, H. X. 2007. A QTL for rice grain width and weight encodes a previously unknown RING-type E3 ubiquitin ligase. *Nature Genetics.* 39:623–630.

Steele, K.A. *et al.*, 2006. Field evaluation of upland rice lines selected for QTLs controlling root traits. *Field Crops Res.* 101:180–186.

Stoskopf, N.C., Tomes, D.T. and Christie, B.R. 1993 Plant breeding: theory and practice. San Francisco, CA; Oxford: Westview Press Inc.

Strand, A., Foyer, C.H., Gustafsson, P., Gardestrom, P. and Hurry, V. 2003. Altering flux through the sucrose biosynthesis pathway in transgenic *Arabidopsis thaliana* modifies photosynthetic acclimation at low temperatures and the development of freezing tolerance. *Pl. Cell and Environment.* 26:523–535.

Su, J. and Wu, R. 2004. Stress-inducible synthesis of proline in transgenic rice confers faster growth under stress conditions than that with constitutive synthesis. *Pl. Sci.* 166:941–8.

Suji, K.K., Prince, K.S.J., Mankhar. P.S., Kanagaraj, P., Poornima, R., Amutha, K., Kavitha, S., Biji, K.R., Gomez, M. and Babu, R.C. 2012. Evaluation of rice (*Oryza sativa* L.) near isogenic lines with root QTLs for plant production and root traits in rainfed target populations of environment. *Field Crops Res.* 137:89-96.

Sundaram, R.M. *et al.* 2009. Introduction of bacterial blight resistance into Triguna, a high yielding, mid-early duration rice variety. *J. of Biotech.* 4:400–407.

Sundaram, R.M., Priya, M.R.V., Laha, G.S., Shobha Rani, N., Rao, S. P., Balachandran, S.M., Ashok Reddy, G.,Sharma, N.P. and Sonti, R.V., 2009. Introduction of bacterial blight resistance into Triguna, a high yielding, mid-early duration rice variety by molecular marker assisted breeding. *Biotech. J.* 4:400-407.

Suresh S. and Malathi D. 2013. Gene Pyramiding For Biotic Stress Tolerance In Crop Plants. *Weekly Sci. Res. J.* 23:2321-7871.

Swarnendu, R. and Chakraborty, U. 2014. Salt tolerance mechanisms in Salt Tolerant Grasses (STGs) and their prospects in cereal crop improvement. *Botanical Studies* 55:31.

Takeuchi, N., Ebitani, T., Yamamato, T., Sato, H., Ohta, H., Nemoto, H., Imbe, T. and Yano, M. 2006.Development of isogenics of rice cultivar Koshihikari with early and late heading by marker-assisted selection. *Breeding Sci.* 56:405-413.

Tang, L., Kwon, S.Y., Kim, S.H., Kim, J.S., Choi, J.S., Cho, K.Y., Sung, C.K., Kwak, S.S. and Lee, H.S. 2006. Enhanced tolerance of transgenic potato plants expressing both superoxide dismutase and ascorbate peroxidase in chloroplasts against oxidative stress and high temperature. *Pl. Cell Reports.* 25:1380-1386.

Tanksley, S.D. 1983. Molecular markers in plant breeding. *Pl. Mol. Biol. Reports* 1:1–3.

Then, C. 2005. Gen-Pflanzen: Alles unter Kontrolle? - *GRÜN de - Das Magazin der GAL für Hamburg.* 12:13.

Tommasini, L., Svensson, J.T., Rodriguez, E.M., Wahid, A., Malatrasi, M., Kato, K., Wanamaker, S., Resnik, J. and Close, T.J. 2008. Dehydrin gene expression provides an indicator of low temperature and drought stress: transcriptome-based analysis of Barley (Hordeumvulgare L.). *Functional & Integrative Genomics.* 8(4):387-405.

Toojinda, T., Tragoonrung, S., Vanavichit, A., Siangliw, J.L., Pa-In, N., Jantaboon, J., Siangliw, M. and Fukai, S. 2005. Molecular breeding for rainfed lowland rice in the Mekong region. *Plant Production Sci.* 8:330-333.

Torres, A.M., Avila, C.M., Gutierrez, N., Palomino, C., Moreno, M.T. and Cubero, J.I. 2010. Marker assisted selection in fababean (*Vicia faba* L.). *Field Crops Res.* 115:243-25.

Tuberosa, R. & Salvi, S. 2006. Genomics-based approaches to improve drought tolerance of crops. *Trends in Pl. Sci.* 11:405–412.

Turhan, H. 2005. Salinity response of transgenic potato genotypes expressing the oxalate oxidase gene. *Turkish J. of Agri. and Forestry.* 29:187-195.

Umezawa, T., Fujita, M., Fujita, Y., Yamaguchi-Shinozaki, K. and Shinozaki, K. 2006. Engineering drought tolerance in plants: discovering and tailoring genes to unlock the future. *Current Opinion in Biotech.* 17:113-22.

Valliyodan, B. and Nguyen, H.T. 2006. Understanding regulatory networks and engineering for enhanced drought tolerance in plants. *Curent Opinion in Pl. Biol.* 9:189–195.

Varshney, R.K., Bansal, K.C., Aggarwal, P.K., Datta. S.K., and Craufurd, P.Q. 2011. Agricultural biotechnology for crop improvement in a variable climate: hope or hype? *Trends in Pl. Sci.* 16(7):1360–1385.

Veeresha, B. A., Rudra Naik, V., Chetti, M. B., Desai, S. A. and Suma S. Biradar 2015. QTL mapping in crop plants: principles and applications. *International J. of Develop. Res.* 5:2961-2965.

Vijn, I. and Smeekens, S. 1999. Fructan: more than a reserve carbohydrate? *Pl. Physiol.* 120:351–359.

Vinocur, B. and Altman, A. 2005. Recent advances in engineering plant tolerance to abiotic stress: achievements and limitations. *Current Opinion in Biotech.* 16:123-132.

Vu, H.T.T., Le, D.D., Ismail, A.M. and Le, H.H. 2012.Marker-assisted backcrossing (MABC) for improved salinity tolerance in rice (*Oryza sativa* L.) to cope with climate change in Vietnam. *Aus. J. of Crop Sci.* 6(12):1649-1654.

Wang, G.L., Mackill, D.J., Bonman, J.M., McCouch, S.R., Champoux, M.C. and Nelson, R.J. 1994. RFLP mapping of genes conferring complete and partial resistance to blast in a durably resistance rice cultivar. *Genetics*. 136:1421-1434.

Wang, W., Vinocur, B. and Altman, A. 2003. Plant responses to drought, salinity and extreme temperatures: towards genetic engineering for stress tolerance. *Planta* 218:1-14.

Wang, Y. *et al*. 2009. Shoot-specific down-regulation of protein farnesyltransferase (a subunit) for yield protection against drought in canola. *Mol. Pl.* 2:191–200.

Wang, Y., Ying, J., Kuzma, M., Chalifoux, M. and Sample, A. *et al*. 2005.Molecular tailoring of farnesylation for plant drought tolerance and yield protection. *The Pl. J.* 43:413-424.

Widodo, Patterson, J.H., Newbigin, E., Tester, M., Bacic, A. and Roessner, U. 2009. Metabolic responses to salt stress of barley (*Hordeumvulgare* L.) cultivars, Sahara and Clipper, which differ in salinity tolerance. *J. of Exp. Bot.* 60(14):4089-4103.

Witcombe, J.R. & Virk, D.S. 2001. Number of crosses and population size for participatory and classical plant breeding. *Euphytica.* 122:451–462.

Witcombe, J.R., Hollington, P.A., Howarth, C.J., Reader, S. and Steele, K.A. 2008.Breeding for abiotic stresses for sustainable agriculture. *Philosophical Transactions of the Royal Society B*. 363:703–716.

Wu, L., Fan, Z., Guo, L., Li, Y., Chen, Z.L. and Qu, L.J. 2005. Over-expression of the bacterial *nhaA* gene in rice enhances salt and drought tolerance. *Pl. Sci.* 168:297-302.

Xiao, B.Z. *et al*. 2008. Evaluation of seven function-known candidate genes for their effects on improving drought resistance of transgenic rice under field conditions. *Mol. Pl.* 2:1–11.

Xu, K., Xu, X., Fukao, T., Canlas, P., Maghirang-Rodriguez, R., Heuer, S., Ismail, A. M., Bailey-Serres, J., Ronald, P. C. and Mackill, D. J. 2006. Sub1A is an Ethylene-Response Factor-Like Gene That Confers Submergence Tolerance to Rice. *Nature* 442:705-708.

Xu, K., Xu, X., Ronald, P.C. & Mackill, D.J. 2000. A high-resolution linkage map of the vicinity of the rice submergence tolerance locus Sub1. *Molecular and General genetics.* 263:681–689.

Xue, Z.Y., Zhi, D.V., Xue, G.P., Zhang, H., Zhao, Y.X. and Xia, G.M. 2004. Enhance salt tolerance of transgenic wheat (*Triticum aestivuns* L.) expressing a vacuolar Na+/H+ antiporter gone with improved grain yields in saline soils in the field and a reduced level of leaf Na+. *Pl. Sci.* 167:849-859.

Yan, S., Tang, Z., Su, W. and Sun, W. 2005.Proteomic analysis of salt stress-responsive proteins in rice root. *Proteomics.* 5:235–244.

Yan, S., Zhang, Q.Y., Tang, Z.C., Su, W.A. and Sun, W.N. 2005. Comparative Proteomic Analysis Provides New Insights into Chilling Stress Responses in Rice. *Molecular & Cellular Proteomics.* 5(3):484-496.

Yang, W.J., Rich, P.J., Axtell, J.D.,Wood, K.V., Bonham, C.C., Ejeta, G., *et al.* 2003. Genotypic variation for glycine betaine in sorghum. *Crop Sci.* 43:162–9.

Yang, X., Liang, Z. and Lu, C. 2005. Genetic engineering of the biosynthesis of glycinebetaine enhances photosynthesis against high temperature stress in transgenic tobacco plants. *Pl. Physiol.* 138:2299- 2309.

Yoshimura, S., Yoshimura, A., Iwata, N., McCouch, S.R., Abenes, M.N., Baraoidan, M.R., Mew, T.W., and Nelson, R.J. 1995. Tagging and combining bacterial blight resistance genes using RAPD and RFLP markers. *Mol. Breeding* 1:375-387.

Yu, J. and Buckler, E.S. 2006.Genetic association mapping and genome organization of maize.Current. *Opinion in Biotech.* 17:155–160.

Zeller, G., Henz, S.R., Widmer, C.K., Sachsenberg, T., Ratsch, G., Weigel, D. and Laubinger, S. 2009. Stress-induced changes in the Arabidopsis thaliana transcriptome analyzed using whole-genome tiling arrays. *The Pl. J.* 58:1068-1082.

Zeng, Z.B., 1994. Precision mapping of quantitative trait loci. *Genetics*. 136:1457–1468.

Zhang, S.Z., Yang, B.P., Feng, C.L. and Tang, H.L. 2005. Genetic transformation of tobacco with the trehalose synthase gene from *Grifola forndosa* Fr. enhances the resistance to drought and salt in tobacco. *J. of Pl. Biol. Res.* 47: 579-587.

Zhao, J.; Meng, J. 2003. Genetic analysis of loci associated with partial resistance to Sclerotinia sclerotiorum in rapeseed (*Brassica napus* L.). *Theoretical and Applied genetics*. 106:759–764.

Zhao, K. *et al.*, 2007. An Arabidopsis example of association mapping in structured samples. *PLoS Genetics*. 3:71–82.

Zheng, J., Fu, J., Gou, M., Huai, J., Liu, Y., Jian, M., Huang, Q., Guo, X., Dong, Z. and Wang, H. 2010. Genome-wide transcriptome analysis of two maize inbred lines under drought stress. *Pl. Mol. Biol.* 72(4-5):407-421.

Zhu, J.K. 2001.Plant salt tolerance.*Trends in Pl. Sci.* 6:66-71.

6

Biotechnological Approaches to Improve Abiotic Stress Tolerance-II

Naveen Kumar and Renu Munjal

Environmental constraints that include abiotic stress factors such as salt, drought, cold, extreme temperatures and heavy metalsalready a major limitingfactor for crop productivity and will soon become even more severe due to climate change conditions.Together, these stresses constitute the primary causes of crop losses worldwide, reducing average yields of most major crop plants by more than 50% (Boyer, 1982; Bray *et al.*, 2000; Wang *et al.*, 2003). Current climate change scenarios predict an increase in mean surface temperatures and drought that will drastically affect global agriculture in the near future (Le Treut *et al.*, 2007).

Abiotic stresses trigger many biochemical, molecular and physiological changes and responses that influence various cellular and wholeplant processes (Wang *et al.*, 2001, 2003). For example, drought, salinityand low temperature stress lead to reduced availability of water (alsoknown as dehydration/osmotic stress) characterized by a decreasedturgor pressure and water loss (Dhariwal *et al.*, 1998; Boudsocq and Lauriere, 2005). Osmotic stress promotes the synthesis of the phytohormoneabscisic acid (ABA)which then triggers a major change in gene expressionand adaptive physiological responses (Seki *et al.*, 2002; Yamaguchi and Shinozaki, 2006; Shinozaki andYamaguchi-Shinozaki, 2007).

Mechanisms of abiotic stress tolerance that operate signal perception, transduction and downstream regulatory factors are now being examined and an understanding of cellular pathways involved in abiotic stress responses provide valuable information on such responses.Of the various general types of plant response to abiotic stress, avoidance mechanisms mainly result from morphological and physiological changes at the whole-plant level. By contrast, tolerance mechanisms are caused by cellular, molecular and biochemical modifications that lend themselves to biotechnological manipulation. All types of abiotic stress evoke cascades of physiological and molecular events and some of these can result in similar responses; for example, drought, high temperature, salinity and freezing can all be manifested at the cellular level as physiological dehydration.Improvement of crop plants with traits that confer tolerance to these stresses was practiced using conventional and modern breeding methods. Molecular breeding and genetic engineering contributed substantially to our understanding of the complexity of stress response.

This chapter presents the biotechnological approaches for abiotic stress tolerance which includes the molecular breeding and genomic-assisted breeding methods, An integrative genomic and breeding approach to reveal developmental programs that enhance yield stability under unfavourable environmental conditions of abiotic stresses is discussed. Omics for crop tolerance, microRNA-based biotechnology, double haploid technique,and transgenic approach that have helped to reveal complex regulatory networks controlling abiotic stress tolerance mechanisms by high-throughput expression profiling and gene inactivation techniques. Further, an account of stress-inducible regulatory genes which have been transferred into crop plants to enhance stress tolerance is discussed as possible modes of integrating information gained from functional genomics into knowledge-based breeding programs.

Molecular breeding for abiotic stress tolerance

Traditional approaches are limited by the complexity of stress tolerance traits, low genetic variance of yield components under stress conditions and the lack of efficient selection techniques (Ribaut *et al.,* 1996, 1997; Frova *et al.,* 1999a, 1999b). Many crop traits including abiotic stress tolerance are quantitative, complex, and controlled by multiple interacting genes. The improvement of crop abiotic stress tolerance by traditional breeding is filled with difficulties because of the multigenic nature of this trait. Furthermore, quantitative trait loci (QTLs) that are linked to tolerance at one stage in development can differ from those linked to tolerance at other stages.

Recent progress in molecular biology provides the tools to study the genetic make-up of plants, which allows us to unravel the inheritance of all traits whether they are controlled by single genes or many genes acting together, known as the quantitative trait loci (QTL). The molecular marker technologies available since the 1980s, allows dissecting the variation in traits. With the progress of QTL mapping, new breeding approaches such as marker assisted selection and breeding by design have emerged (Peleman *et al.,* 2003).

The application of quantitative trait loci (QTL) mapping is one approach to dissect the complex issue of plant stress tolerance. When fully developed, this approach will be of great significance to breeding for abiotic stress tolerance in plants (Flower, 2004). Once identified, desirable QTLs can require extensive breeding to restore desirable traits along with the introgressed tolerance trait. QTLs associated with abiotic stress tolerance have been identified in many important crop species (e.g. salt stress in rice (Lin *et al.*,2004), drought stress in cotton (Saranga *et al.*, 2004), and cold stress in the woody plant Salix (Tsarouhas *et al.*,2004). A high salt-tolerant *indica* rice variety was crossbred with a susceptible *japonica* rice variety (Lin *et al.*,2004) and QTLs were detected for seedling survival under stress, which correlated well with the degree of leaf damage and Na^{+} accumulation in the shoots.

From an evolutionary point of view, all plant responses to stress and all tolerance mechanisms are programmed and genotype-specific. However, although adaptation to stress has some ecological advantages, its metabolic energy costs can result in yield penalties, consequently limiting their benefit for agricultural plants. Therefore, efficient plant breeding for abiotic stress tolerance can be achieved only by combining traditional and molecular breeding.

Marker assisted selection (MAS)

Marker assisted selection (MAS) is a combined product of traditional genetics and molecular biology. Marker assisted selection is the process of using the results of DNA testing in the selection of individuals to become parents for the next generations. The information from the DNA testing, combined with the observed performance records for individuals, is intended to improve the accuracy of selection and increase the possibility of identifying organisms carrying desirable and undesirable traits at an earlier stage of development. MAS allows for the selection of genes that control traits of interest.

It is important to combine DNA results with performance and phenotype information to maximize the effectiveness of selection for traits of interest. Combining information from performance records and genetic tests into the selection process will be better than using performance, phenotype, and markers separately. With the development and availability of an array of molecular markers and dense molecular genetic maps in crop plants, MAS has become possible for traits both governed by major genes as well as quantitative trait loci (QTLs). Most of the physiological traits associated with abiotic stress tolerance are quantitative in nature. Thus, with the advances in MAS approaches, it is now possible to evaluate trait-based approaches for addressing abiotic stress adaptation in crops on a much wider scale than was previously possible.

The submergence tolerant rice cultivars was developed through Marker-assisted back cross (MABC), which has improved yields in >15 million hectares of rain-fed lowland rice in South and Southeast Asia (Septiningsih *et al.*,2009). A recent *ex ante* economic study of molecular breeding of rice for tolerance to salty and low-phosphorus soils in selected Asian countries has estimated that the method saves a minimum of 2–3 years, which results in significant incremental benefits in the range of $300–800 million USD (Alpuerto *et al.*,2009).

Large-scale genomic resources and specialized genetic stocks that have become available in tier 2 and 3 crops, which have been less-studied crops until recently, are expected to enhance molecular breeding such as MABC and marker-assisted recurrent selection (MARS) (Ribaut *et al.*, 2010). This will lead to enhanced crop productivity and, in turn, increased food security in developing countries.

Nonetheless, marker-assisted selection of specific secondary traits that are indirectly related to yield (e.g. the interval between anthesis and silking (Ribaut *et al.*, 1996, 1997), osmotic adjustment (Zhang *et al.*, 1999), membrane stability (Frova *et al.*, 1999b) or physiological tolerance indices (Frova *et al.*, 1999a), might prove increasingly useful as the resolution of the genetic and physical chromosome maps of the major crops improves. This strategy could be used in combination with 'pyramiding' strategies or consecutive selection for, and accumulation of, physiological yield-component traits (Flowers and Yeo, 1995).

Genomics-assisted breeding

Most of the key developing country crops now find themselves in a situation in which genomic resources are sufficient to support meaningful genetic studies and molecular breeding (Ribaut *et al.*, 2010). Several successful genomics-assisted breeding programmes have been built through collaborations between CGIAR (Consortium of International Agricultural Research) institutes and NARS (National Agricultural Research System) partners (Okogbenin *et al.*, 2007; Hash *et al.*, 2006; Thomson *et al.*,2010). Various bottlenecks stillimpede adoptioninthese countries; limited human resources, inadequate field infrastructure and limited capacity in information management remain major challenges. The magnitude of these challenges is exacerbated where it is important to breed for biotic (pests or diseases) and abiotic (drought, heat, cold or salinity) stresses, thus making accurate phenotyping challenging. The task remains challenging as the Platformis intended to serve a broad range of users who areworking on different crops for different environments (Ribaut *et al.*, 2010).

Genome sequencing

Recently, high-throughput sequencing has brought powerful and efficient research tools that can lead to a better understanding of the molecular mechanisms behind stress in plants. Over the last ten years with the rapid development of genomic and next generation sequencing, many plant genomes have been sequenced. Many molecular markers, functional and regulatory genes have been discovered based on the genome sequencing. The new technologies, such as transcriptome analysis, digital gene expression, deep sequencing of small RNAs, proteomics, metabolomics, etc. should pave new avenues for studying stress tolerance in crop plants.

The first genomic sequence of the model plant, *A. thaliana*, was published (Arabidopsis Genome Initiative, 2000), followed by rice (*Oryza sativa* L.) (Goff *et al.,* 2002; Yu *et al.,* 2002), and then poplar (*Populus trichocarpa* L.) (Tuskan *et al.,* 2006). In 2007, the first fruit genome, grapevine (*Vitis vinifera* L.) genome was sequenced, analyzed and published independently by two groups (Jaillon *et al.,* 2007; Velasco *et al.,* 2007), and then subsequently in 2009, the first important vegetable genome, cucumber (*Cucumis sativus* L.), was sequenced and published (Huang *et al.,* 2009).

Identification of molecular markers across whole genome in crop plants

The availability of genome sequence data has provided new potential resources for vegetable crop improvement. Almost unlimited molecular markers, such as SSRs (simple sequence repeats) and SNPs (single-nucleotide polymorphisms) can greatly increase the breeding efficiency through marker-assisted selection (MAS), genetic mapping and trait identification (Gao *et al.,* 2012). In fact, a large number of genetic markers have been identified at the genome-wide levels for crop plants that provided new approaches and strategies for germplasm characterization and for genetic improvement, including breeding for resistance to abiotic stresses (Dubey *et al.,* 2011).

A total of 309,052 SSR markers were identified and analyzed from the Pigeonpea genome (Varshney *et al.,* 2012). Varshney's group also identified 28,104 novel SNP markers across 12 genotypes in the Pigeonpea (Dubey *et al.,* 2011; Varshney *et al.,* 2012). Over 3.67 million SNPs have also been identified in the potato genomes (Potato Genome Sequencing Consortium, 2011). Nearly 5.4 million SNPs were distributed along the chromosomes of tomato genome (Tomato Genome Consortium, 2012). Transcriptome sequencing technology also provides rapid and cost-effective strategy for producing a large number of diverse expressed sequence tags (ESTs) that could support computational identification of molecular markers such as SNPs and SSRs in vegetable crops (Barbazuk *et al.,* 2007; Blanca *et al.,* 2011; 2012; Nicolai *et al.,* 2012). This massive amount of markers is valuable resources for facilitating breeding vegetable crop lines with abiotic stress tolerance. Meantime, it will also aid in gene discovery and the marker-assisted breeding to improve vegetable crops varieties (Lee and Choi, 2013). Based on 1.696 million transcriptome sequence reads, around 3,700 SSR loci were identified in Pigeonpea (Dutta *et al.,* 2011).

Over 5.4 million SNPs were found to be distributed along the tomato chromosomes (Tomato Genome Consortium, 2012).

Omics for abiotic stress tolerance

The emergence of the novel "omics" technologies, such as genomics, proteomics, and metabolomics, is now allowing researchers to identify the genetic behind plant stress responses. "Omics" studies are primarily aimed at the detection of biological components such as genes, mRNA, proteins and metabolites being described by terms genomics, transcriptomics, proteomics and metabolomics respectively. These approaches are means for the deduction of different pathways which are altered in response to environmental stresses. These omics technologies enable a direct and unbiased monitoring of the factors affecting plant growth and development and provide the data that can be directly used to investigate the complex interplay between the plant, its metabolism, and also the stress caused by the environmental constraints(abiotic stress factors such as salt, drought, cold, extreme temperatures and heavy metals) or the biological threats (insects, fungi, or other pathogens). Omics investigations are increasingly being used by various groups to study plant responses to abiotic stress *via* profound changes in gene expression which result in changes in composition of plant transcriptome, proteome, and metabolome (Perez-Alfocea *et al.*,2011).

Genomics

A gene by gene approach has been typically used to understand its function, while the functional genomics allows large-scale gene function analysis with high throughput technology and incorporates interaction of gene products at cellular and organism level. The information coming from sequencing programs is providing enormous input about genes to be analyzed. The availability of many plant genomes nowadays (Chain *et al.,* 2009; Feuillet *et al.,* 2010) facilitates studying the function of genes on a genomewide scale. The lack of information from other plant genomes will also be compensated in part by the availability of large collection of expressed sequence tags (ESTs) and cDNA sequences (Marques *et al.,* 2009). The basic interest behind these EST projects is to identify genes responsible for critical functions. ESTs, cDNA libraries, microarray, and serial analysis of gene expression (SAGE) are used to analyze global gene expression profiles in a functional genomics program.

Advances in plant genomics research have opened up new perspectives and opportunities for improving crop plants and their productivity. The genomics technologies have been found useful in deciphering the multigenicity of biotic and abiotic plant stress responses through genome sequences, stress-specific cell and tissue transcript collections, protein and metabolite profiles and their dynamic changes, protein interactions, and mutant screens. Moreover, the understanding of the complexity of stress signaling and plant adaptive processes would require the analysis of the function of numerous genes involved in stress response. Elucidating the molecular mechanism that mediates the complex stress responses in plants system is an important step to develop improved variety of stress tolerant crops.

Functional genomics for abiotic stress tolerance

The yield and quality of crop plants are affected by various abiotic stresses, such as drought, salinity and low and high temperatures. Higher plants have evolved a series of complex responses in order to adapt to a single or multiple stresses which include biochemical, physiological, cellular and molecular processes (Bartels and Sunkar, 2005; Yamaguchi and Shinozaki, 2006; Negraoa *et al.,* 2011; Sasidharana *et al.,* 2011).A series of complex responses have evolved in order to adapt and cope with abiotic stresses that are critical for the survival of all plants (Thomashow, 1999; Jamila *et al.,* 2011; Shahbaz *et al.,* 2012).

When plants are exposed to various abiotic stresses, a large number of stress-induced genes, which are involved in stress tolerance, transcription regulation or signal transduction, are activated and many proteins are produced to activate and adjust the physiological and biochemical pathways (Shinozaki *et al.,* 2003; Nakashima *et al.,*2009; Lee *et al.,* 2010). Significant progress has been made in dissecting the biochemical and physiological pathways involved in higher plants responses to abiotic stresses. Numerous reports have documented the molecular and cellular mechanisms underlying abiotic stress adaptation in higher plants, especially in model plants, such as *Arabidopsis thaliana* (Walley and Dehesh, 2010; Daszkowska- Golec, 2011), rice (Santos *et al.,* 2011; Wang *et al.,* 2012; Lei *et al.,* 2013), and poplar (Popko *et al.,* 2010; Ye *et al.,*2011; Janz *et al.,*2012).

Full-length cDNAs (fl-cDNAs) are important resources for the characterization of gene function, since they contain all the information required for the production of functional RNAs and proteins. Large sets of fl-cDNA clones have been collected from several plant species and have become available for functional genomic analysis. Higuchi *et al.,* 2011 developed a system for the identification of gene function by screening for transgenic plants ectopically expressing fl-cDNAs and named it the FOX (fl-cDNA overexpressor gene) hunting system. This system can be applied to almost all plant species without prior knowledge of their genome sequences because only fl-cDNAs are required. For utilization of the FOX hunting system, *Agrobacterium* libraries and Arabidopsis seeds carrying rice and Arabidopsis fl-cDNAs are available. Higuchi *et al.,* 2011 described the procedure followed in the FOX hunting system from the generation of expression vectors carrying fl-cDNAs to the confirmation of phenotype in retransformed plants.

Responses to environmental stresses in higher plants are controlled by a complex web of abscisic acid (ABA)-dependent and independent signaling pathways. Papdi *et al.,* 2008 performed genetic screens for identification of novel Arabidopsis (*Arabidopsis thaliana*) loci involved in the control of abiotic stress responses, a complementary DNA (cDNA) expression library was created in a Gateway version of estradiol-inducible XVE binary vector (controlled cDNA overexpression system [COS]). The COS system was tested in three genetic screens by selecting for ABA insensitivity, salt tolerance, and activation of a stress-responsive *ADH1-LUC* (alcohol dehydrogenase-luciferase) reporter gene. Twenty-seven cDNAs conferring dominant, estradiol-dependent stress tolerance phenotype, were identified by polymerase chain reaction amplification and sequence analysis. Several cDNAs were recloned into the XVE vector and transformed recurrently into Arabidopsis, to confirm that the observed conditional phenotypes were due to their estradiol-dependent expression. Characterization of a cDNA conferring insensitivity to ABA in germination assays has identified the coding region of heat shock protein HSP17.6A suggesting its implication in ABA signal transduction. Screening for enhanced salt tolerance in germination and seedling growth assays revealed that estradiol-controlled overexpression of a 2-alkenal reductase cDNA confers considerable level of salt insensitivity. Screening for transcriptional activation of stress- and ABA-inducible ADH1-LUC reporter gene has identified the ERF/AP2-type transcription factor RAP2.12, which sustained high level ADH1-LUC bioluminescence, enhanced ADH1 transcription rate, and increased ADH enzyme activity in the presence of estradiol. These data illustrate that application of the COS cDNA expression library provides an efficient strategy for genetic identification and characterization of novel regulators of abiotic stress responses.

Micro RNA-based biotechnology approach for abiotic stress tolerance

MicroRNAs (miRNAs) are an extensive class of newly discovered endogenous small RNAs, which negatively regulate gene expression at the post-transcription levels. As the application of next-generation deep sequencing and advanced bioinformatics, the miRNA-related study has been expended to non-model plant species and the number of identified miRNAs has dramatically increased in the past years. The miRNAs play a critical role in almost all biological and metabolic processes, and provide a unique strategy for plant improvement. Plant tolerance to abiotic stress was significantly enhanced by regulating the expression of an individual miRNA. Both endogenous and artificial miRNAs may serve as important tools for plant improvement.

Abiotic stresses are the major factors limiting plant growth, yield and sustainable agriculture worldwide. The common abiotic stress include drought, high salinity, high and low temperature, nutrient deficiency, hypoxia, and pollutants. Thus, breeding new plant cultivars with improving plant tolerance to environmental stresses is necessary for maximizing plant biomass and crop yield. In the past decades, although several protein-coding genes have been identified for controlling plant response to environmental abiotic stress (Zhu, 2001; Singh *et al.*, 2002; Wang *et al.*, 2003; Fujita *et al.*, 2006), the knowledge on the regulatory mechanism of plant response to abiotic stress is still limited and transformative tools are needed to adapt crops to harsh environments.

Recently discovered miRNAs may play an important role in plant adoption to abiotic stresses. It is well known that multiple signaling pathways and their key gene players, such as transcription factors, are differentially expressed during plant response to abiotic stress of interest (Castro *et al.*, 2012; Mizoi *et al.*,2012; Nakashima *et al.*, 2012; Obata and Fernie, 2012; Suzuki *et al.*, 2012), which are targeted by specific miRNAs (Khraiwesh *et al.*, 2012). For example, NAC transcription factors are plant-specific transcription factors, which are critical important not only in plant development but also in abiotic stress responses (Nakashima *et al.*, 2012); overexpression stress-responsive NAC (SNAC) genes have exhibited improved drought tolerance in Arabidopsis and rice plants; a study has shown that NAC transcriptional factors are regulated by miRNAs (Guo *et al.*, 2005). Another well studies drought/salinity-induced transcript factor AP2 (Kizis *et al.*, 2001; Mizoi *et al.*, 2012) is also targeted by another well studies miRNA miR172 (Aukerman and Sakai, 2003; Chen, 2004; Lauter *et al.*, 2005). Our recent studies also show that miR 172 was altered by drought and salinity stress in cotton (Wang *et al.*, 2013) and tobacco (Frazier *et al.*, 2011). These suggest that miRNAs may play a key role during plant response to abiotic stress of interest and regulating miRNA expression may improve plant tolerance to environmental stresses.

Role of miRNAs for improving drought and salinity tolerance

Drought and high salinity stress are two major constraints to plant growth and development and further to agricultural productivity around the world. Although, many genes are induced by these two classes of stresses and some of them were employed to improve plant tolerance to drought and salinity through overexpressing their expression levels using transgenic technology; however, the majority of these transgenic plants exhibited very small increase and even no increase on the improvement of plant tolerance to drought and salinity stresses (Bartels and Sunkar, 2005; Sunkar *et al.*, 2012). One of the major reasons is that it is a complicated mechanism for plant response to environmental stresses, in which many genes, including protein-coding genes, transcription factors aswell small RNAs are involved in this gene network. Thus, the recently identified miRNAs may play an important cross-link role during this process.

Currently, numerous drought- and/ or- salinity stress-responsive miRNAs have been identified in a wide range of plant species, including not only model plant species, such as Arabidopsis (Sunkar and Zhu, 2004) and rice (Zhao *et al.*, 2007), but also no-model agriculturally important crops, such as soybean (Kulcheski *et al.*,2011; Dong *et al.*, 2013), cotton (Yin *et al.*, 2012; Wang *et al.*, 2013), barley (Kantar *et al.*, 2010), peach (Eldem *et al.*, 2012), and switchgrass (Xie *et al.*, 2014). Although the reported results varied from research to research among different tissues at different plant species, many miRNAs exhibited the same response to drought or salinity stress in different plant species; such as miR 160, miR 167, miR 393, and miR 394 were commonly induced under drought and/or salinity in different plant species.

Several studies show that regulating the expression of a single miRNA can enhance or decrease plant tolerance to drought and/or salinity stress. Transgenic Arabidopsis plants overexpressing miR169a exhibited more sensitive to drought stress than its wild-type plants due to enhanced leaf water loss (Li *et al.*, 2008). This is because mR169 target nuclear factor YA5 (NFYA5) that is important for drought tolerance in an ABA-dependent manner (Li *et al.*, 2008). However, another experiment demonstrates a different role for miR169 in tomato, in which constitutive over-expression of miR169 enhanced drought tolerance in tomato potentially by reducing stomatal opening, decreased transpiration rate,and lowered leaf water loss (Zhang *et al.*, 2011). Constitutive expression of miR 319 significantly enhanced transgenic plant tolerance to drought and salinity stress in creep bentgrass (Zhou *et al.*, 2013).

There are several reasons for enhancing abiotic stress in transgenic plants: (1) Transgenic plants with overexpressing miR 319 accumulated less Na^+ in the cytoplasm then their wild-type plant under salinity stress condition when treated with 200mM NaCl; interestingly, transgenic plants accumulated more K^+, phosphate, and other nutrients than its wild-type control, which may contribute to better growth and development of transgenic plants; it has been reported that enhanced K^+/Na^+ ratio may enhance plant tolerance to salinity stress (Asch *et al.*, 2000; Moller *et al.*, 2009); (2) transgenic plants with overexpressing miR 319 exhibited enhanced water retention and cell membrane integrity, including higher stomatal conductance; (3) the transgenic plants also increased their photosynthesis capacity; and (4) miR 319 transgenic plants increased leaf wax content which significantly inhibited water loss (Zhou *et al.*, 2013). Other miRNAs also regulate plant tolerance to drought and salinity stresses. Overexpression of miR394 enhanced plant tolerance to drought in Arabidopsis majorly due to lowered leaf water loss (Ni *et al.*, 2012) in an ABA-dependent manner (Song *et al.*, 2013). However, the same transgenic plants with overexpression of miR 394 show sensitive to salinity (Song *et al.*, 2013). Overexpression of osamiR 396c decreased salt and alkali stress tolerance in both Arabidopsis and rice (Gao *et al.*, 2010).

Transgenic approaches for abiotic stress tolerance in crop plants

Use of modern molecular biology tools for elucidating the control mechanisms of stress tolerance and for engineering stress tolerant plants is based on the expression of specific stress-related genes. The successes in genetic improvement of environmental stress resistance have involved manipulation of a single or a few genes involved in signalling/ regulatory pathways or that encode enzymes involved in these pathways (Jewell *et al.*, 2010).

The plant hormone abscisic acid (ABA) regulates the adaptive response of plants to environmental stresses such as drought, salinity, and chilling via diverse physiological and developmental processes (Arbona *et al.*,2008). The ABA biosynthetic pathway has been deeply studied, and many of the key enzymes involved in ABA synthesis have been used in transgenic plants to improve abiotic stress tolerance (Ji *et al.*,2011). Transgenic plants overexpressing the genes involved in ABA synthesis showed increased tolerance to drought and salinity stress (Ji *et al.*, 2011). Similarly, many studies have illustrated the potential of manipulating *CBF/DREB* genes to confer improved drought tolerance (Xiong *et al.*, 2006).

Another mechanism involved in plant protection to osmotic stress associated to many abiotic stresses such as drought and salinity implies the accumulation of compatible solutes involved in avoiding oxidative damage and chaperoning through direct stabilization of membranes and/or proteins (Zhang *et al.*, 2008). Many genes involved in the synthesis of these osmoprotectants have been explored for their potential in engineering plant abiotic stress tolerance (Zhang *et al.*, 2008). The amino acid proline is known to occur widely in higher plants and normally accumulates in large quantities in response to environmental stresses (Ashraf *et al.*, 2007). The osmoprotectant role of proline has been verified in some crops by overexpressing genes involved in proline synthesis (Hmida-Sayari *et al.*, 2005). The results of transgenic modifications of biosynthetic and metabolic pathways in most of the previously mentioned cases indicate that higher stress tolerance and the accumulation of compatible solutes may also protect plants against damage by scavenging of reactive oxygen species (ROS) and by their chaperone-like activities in maintaining protein structures and functions (Umezawa *et al.*, 2006).

Polyamines, being polycationic compounds of low molecular weight, are involved in many cellular processes, such as replication, transcription, translation, membrane stabilization, enzyme activity modulation, plant growth, and development (Liu and Moriguchi, 2007). It has been reported that stress results in an accumulation of free or conjugated polyamines, indicating that polyamine biosynthesis might serve as an integral component of plant response to stress (Wang*et al.*,2011;Liu*et al.*, 2011).Polyamines metabolic pathways are regulated by a limited number of key enzymes, among them ornithine decarboxylase (*ODC*) and arginine decarboxylase (*ADC*). Transgenic plants overexpressing *ADC* gene showed increase in biomass and better performance under salt stress conditions. It has also been described that genetic transformation with genes encoding *ADC* improved environmental stress tolerance in various plant species (Wang*et al.*,2011).

A common factor among most stresses is the active production of reactive oxygen species (Hirayama and Shinozaki, 2010). ROS are not only toxic to cells but also play an important role as signalling molecules. Under normal growth conditions, there is equilibrium between the production and the scavenging of ROS, but abiotic stress factors may disturb this equilibrium, leading to a sudden increase in intracellular levels of ROS. In order to control the level of ROS and protect the cells from oxidative injury, plants have developed a complex antioxidant defense system to scavenge these ROS (Hossain*et al.*, 2009). These antioxidant systems include various enzymes and nonenzymatic metabolites that may also play a significant role in ROS signalling in plants. A number of transgenic improvements for abiotic stress tolerance have been achieved through detoxification strategy (Amuda and Balasubramani, 2011). These include transgenic plants overexpressing enzymes involved in oxidative protection, such as glutathione peroxidase, superoxide dismutase, ascorbate peroxidases, and glutathione reductases (Tang *et al.*, 2006).

LEA proteins, including several groups of high molecular weight, accumulate in response to different environmental stresses. It has been reported that constitutive overexpression of the *HVA1*, a group 3 LEA protein from barley, conferred tolerance to soil water deficit and salt stress in transgenic rice plants (Rohila*et al.*, 2002). It has also been reported that plants expressing a wheat LEA group 2 protein (*PMA80*) gene or the wheat LEA group 1protein (*PMA1959*) gene resulted in increased tolerance to dehydration and salt stresses (Amuda and Balasubramani, 2011).

An important strategy for achieving greater tolerance to abiotic stress is to help plants to re-establish homeostasis under stressful environments, restoring both ionic and osmotic homeostasis. This is a major approach to improve salt tolerance in plants through genetic engineering, where the target is to achieve Na^+ excretion out of the root, or their storage in the vacuole (Wu *et al.*, 2005).

Transgenic approaches also aim to improve photosynthesis under abiotic stress conditions through changes in the lipid biochemistry of the membranes. Genetically engineered plants overexpressing chloroplast glycerol-3-phosphate acyltransferase gene (involved in phosphatidyl glycerol fatty acid desaturation) showed an increase in the number of unsaturated fatty acids and a corresponding decrease in the chilling sensitivity (Sui *et al.,* 2007).

The heat shock response is a highly conserved biological response, occurring in all organisms in response to heat or other toxic agent exposures (Miller and Mittler, 2006). Genetic engineering for increased thermotolerance by enhancing heat shock protein synthesis in plants has been achieved in a number of plant species. Some authors have reported the positive correlation between the levels of heat shock proteins and stress tolerance (Bhatnagar-Mathur *et al.,* 2008).

A special case of study is the heavy metal contamination.In spite of the natural occurrence of heavy metals as rare elements, diverse anthropogenic practices have contributed to spread them in the environment. Plants have developed mechanisms that can protect cells from heavy metal cytotoxicity,as the cytosolic detoxification by binding to the metal binding molecules as phytochelatins, and metallothioneins which play an important role in heavy metal detoxification and homeostasis of intracellular metal ions in plant tissues. Overexpression of phytochelatin synthase in Arabidopsis leads to enhanced arsenic tolerance but surprisingly to cadmium hypersensitivity (Li *et al.,* 2004). Therefore, new approaches could contribute to uncovering the complexity of plant tolerance to heavy metal stress (Chaffai and Koyama, 2011).

The transcription factors activate cascades of genes that act together in enhancing tolerance towards multiple stresses as indicated before. On the other hand, some stress responsive genes may share the same transcription factors, as indicated by the significant overlap of the gene expression profiles that are induced in response to different stresses (Qui *et al.,* 2009). Transcriptional activation of stress-induced genes has been possible in transgenic plants over expressing one or more transcriptionfactors that recognize promoter regulatory elements of these genes (Munns, 2005; Bhatnagar-Mathur *et al.,* 2008). Two families, *bZIP* and *MYB*, are involved in ABA signaling and its gene activation. Introduction of transcription factors in the ABA signalling pathway can also be a mechanism of genetic improvement of plantstress tolerance. Constitutive expression of *ABF3* or *ABF4*demonstrated enhanced drought tolerance in Arabidopsis,with altered expression of ABA/stress-responsive genes, for example, *rd29B*, *rab18*, *ABI1*, and *ABI2* (Yamaguchi and Shinozaki, 2006).

It is important to point that genetic modification of higherplants by introducing DNA into their cells is a highly complex process. Practically any plant transformation experiment relies at some point on cell and tissue culture. Although the development transformation methods that avoid plant tissueculture have been described for Arabidopsis and have been extended to a few crops, the ability to regenerate plants from isolated cells or tissues *in vitro* is needed for most plant transformation systems. Not all plant tissue is suited to everyplant transformation method, and not all plant species can be regenerated by every method (Tzfira and Citovsky, 2012). Therefore, need to find both the suitable plant tissue culture/regeneration regime and a compatible plant transformation methodology (Perez-Clemente and Gomez-Cadenas, 2012).

Chloroplast transformation

Chloroplasts of higher plants are semi-autonomous organelles with a small, highly polyploid genome and their own transcription-translation machinery (Maliga, 2004). If transgenes are integrated into the chloroplast genome of species that inherit chloroplasts from the female parent, then pollen produced from the resulting transgenic plants should be transgene free.

Chloroplast transformation offers several advantages (Bock, 2001; Maliga, 2004; Grevich and Daniell, 2005; Sharma *et al.,* 2005; Maliga and Bock, 2011). First, each plant cell contains a large number of chloroplasts, with some having up to 100 chloroplasts per cell (Martin and Borst, 2003). Also, each chloroplast contains 50–100 or more copies of the chloroplast genome (Franklin and Mayfield, 2004). As a consequence, if a trait gene is incorporated into the chloroplast genome, the gene can be enriched up to 100,000 copies per each cell. High copy number of transgenes per cell leads to high expression levels. Second, transgene expression via chloroplast transformation is more stable because gene silencing that can affect expression of nuclear transgenes appears to have little effect on transgenes that are introduced into the chloroplast genome (Van Bel *et al.,* 2001). Third, for plant species in which chloroplasts are inherited exclusively through the female line, chloroplast transformation can provide an effective means of preventing transgenes from moving to non-transgenic crops or wild relatives through pollen (Rigano *et al.,* 2012)

A plastid transformation procedure for generating plants with an enhanced ability to tolerate abiotic stresses has also been built in tobacco (Bansal *et al.,* 2012). The stress tolerance gene was designed to be inserted into the plastid transformation vector with an *aadA* gene that encodes resistance to spectinomycin as a selectable marker. Genetic transformation of chloroplasts should provide an effective tool to reduce pollen-mediated transgene flow in crops that inherit chloroplasts through the female parent. On the other hand, problems associated with seed-mediated transgene escape will not be addressed with a chloroplast transformation strategy, so additional methods must be applied if complete transgene containment is required. Another major problem is that chloroplast transformation has been achieved only in a small number of plant species so far due to the fact that it is technically difficult to engineer chloroplast genome and to regenerate plants from transformed chloroplasts (Rigano *et al.*, 2012).

Double Haploid technique

This technique has twin advance of speed and efficiency and due to this it has become most effective tools for the plant breeders to attain homozygosity of recombinants in shortest possible time. This technique has been used in one of the Indo Australian Collaboration Project on water logging tolerance under sodic soils and has yielded double haploids tolerant than both of the parents. These double haploids can be produced by fertilizing wheat ears of crosses between two parents with maize pollens. The maize pollens are subsequently eliminated during development leaving haploid wheat embryo which is rescued after 14-21 days and transferred into nutrient culture medium. Plantlets later treated with colchicines for the doubling of the chromosomes to get 100% homozygous lines (Singh and Tyagi, 1997).

Conclusions

In a world where population growth is outstripping food supply, agricultural and especially plant-biotechnology, needs to be swiftly implemented in all walks of life. Achievements today in plant biotechnology have already surpassed all previous expectations, and the future is even more promising. Biotechnology should be fully integrated with classical physiology and breeding as an aid to classical breeding, and for generation of engineered plants for abiotic stress tolerance into agricultural production systems. Biotechnology is nowadays changing the agricultural and plant scene in protecting plants against the ever-increasing threats of abiotic stresses.

Strategies for the manipulation of abiotic stress tolerance in plants might include: expression of osmoprotectants and compatible solutes, ion and water transport and channels, expression of water-binding and membrane-associated dehydrins and other proteins, transcription factors and DNA-binding proteins, etc. Also of specific interest are the intervening stages of stress perception, signal transduction (ABA and others), and protein modification. The discovery of new stress-related genes and the design of stress-specific promoters are equally important. Transgenic plant technology offers a highly efficient and powerful tool to create abiotic stress tolerant crop plants. Even though a large number of field evolution tests for stress tolerant transgenic plants have been conducted in the past decades, stress tolerant transgenic plants have not been commercialized, partially due to perceived environmental impact of undesirable escape of stress tolerance genes into native plant populations with a resulting increase in invasive or weedy potential. The full realisation of the agricultural biotechnology revolution depends on both continued successful and innovative research and development activities and on a favourable regulatory climate and public acceptance.

References

Alpuerto, V.E.B., Norton, G.W., Alwang, J. and Ismail, A.M. 2009. Economic impact analysis of marker assisted breeding for tolerance to salinity and phosphorous deficiency in rice. *Rev. Agr. Econ.* 31(4): 779-792.

Amudha, J. and Balasubramani, G. 2011. Recent molecular advances to combat abiotic stress tolerance in crop plants. *Biotechnology and Molecular Biology Review.* 6(2): 31-58.

Arabidopsis Genome Initiative. 2000. Analysis of the genome sequence of the flowering plant, *Arabidopsis thaliana. Nature.* 408: 796-815.

Arbona, V. and Gomez-Cadenas, A. 2008. Hormonal modulation of citrus responses to flooding. *Journal of Plant Growth Regulation.* 27(3): 241-250.

Asch, F., Dingkuhn, M., Dorffling, K. and Miezan, K. 2000. Leaf K/Na ratio predicts salinity induced yield loss in irrigated rice. *Euphytica.* 113: 109-118.

Ashraf, M. and Foolad, M.R. 2007. Roles of glycine betaine and proline in improving plant abiotic stress resistance. *Environmental and Experimental Botany.* 59(2): 206-216.

Aukerman, M.J. and Sakai, H. 2003. Regulation of flowering time and floral organ identity by a microRNA and its *APETALA2*-like target genes. *Plant Cell.* 15: 2730-2741.

Bansal, K.C., Singh, A.K. and Wani, S.H. 2012. Plastid transformation for abiotic stress tolerance in plants. *Methods Mol. Biol.* 913: 351-358.

Barbazuk, W.B., Emrich, S.J., Chen, H.D., Li, L. and Schnable, P.S. 2007. SNP discovery via 454 transcriptome sequencing. *Plant J.* 51: 910-918.

Bartels, D. and Sunkar, R. 2005. Drought and salt tolerance in plants. *Crit. Rev. Plant Sci.* 24: 25-38.

Bhatnagar-Mathur, P., Vadez, V. and Sharma, K.K. 2008. Transgenic approaches for abiotic stress tolerance in plants: retrospect and prospects. *Plant Cell Reports.* 27(3): 411-424.

Blanca, J., Canizares, J., Roig, C., Ziarsolo, P., Nuez, F. and Pico, B. 2011. Transcriptome characterization and high throughput SSRs and SNPs discovery in *Cucurbita pepo* (Cucurbitaceae). *BMC Genomics.* 12: 104.

Blanca, J., Esteras, C., Ziarsolo, P., Perez, D., Ferna Ndez-Pedrosa, V., Collado, C., Rodra Guez de Pablos, R., Ballester, A., Roig, C., Canizares, J. and Pico, B. 2012. Transcriptome sequencing for SNP discovery across *Cucumis melo. BMC Genomics.* 13: 280.

Bock, R. 2001. Transgenic plastids in basic research and plant biotechnology. *J. Mol. Biol.* 312: 425-438.

Boyer, J.S. 1982. Plant productivity and environment. *Science.* 218: 443-448.

Boudsocq, M. and Lauriere, C. 2005. Osmotic Signaling in Plants. Multiple pathways by emerging kinase families. *Plant Physiol.* 138(3): 1185-1194.

Bray, E.A., Bailey-Serres, J. and Weretilnyk, E. 2000. Responses to abiotic stresses. In: Buchannan, B.B., Gruissem, W. and Jones, R.L. (ed). Biochemistry and molecular biology of plants. Rockville, MD: ASPB. pp. 1158-1249.

Castro, P.H., Tavares, R.M., Bejarano, E.R. and Azevedo, H. 2012. SUMO, a heavyweight player in plant abiotic stress responses. *Cell Mol. Life Sci.* 69: 3269-3283.

Chaffai, R. and Koyama, H. 2011. Heavy metal tolerance in *Arabidopsis thaliana. Advances in Botanical Research.* 6: 1-49.

Chain, P.S.G., Grafham, D.V. and Fulton, R.S. *et al.* 2009. Genome project standards in a new era of sequencing. *Science.* 326(5950): 236-237.

Chen, X.M. 2004. A microRNA as a translational repressor of *APETALA2* in Arabidopsis flower development. *Science.* 303: 2022-2025.

Daszkowska-Golec, A. 2011. Arabidopsis seed germination under abiotic stress as a concert of action of phytohormones. *OMICS* 15: 763-774.

Dhariwal, H.S., Kwai, M. and Uchimiya, H. 1998. Genetic engineering for abiotic stress tolerance in plants. *Plant Biotechnol.* 15: 1-10.

Dong, Z.H., Shi, L., Wang, Y.W., Chen,, L., Cai, Z.M., Wang, Y.N., Jin, J.B. and Li, X. 2013. Identification and dynamic regulation of microRNAs involved in salt stress responses in functional soybean nodules by high-throughput sequencing. *Int. J. Mol. Sci.* 14: 2717-2738.

Dubey, A., Farmer, A., Schlueter, J., Cannon, S.B., Abernathy, B., Tuteja, R. and Woodward, J. *et al.* 2011. Defining the transcriptome assembly and its use for genome dynamics and transcriptome profiling studies in Pigeonpea (*Cajanus cajan* L.). *DNA Res.* 18: 153-164.

Dutta, S., Kumawat, G., Singh, B. P., Gupta, D. K., Singh, S., Dogra, V. and Gaikwad, K. *et al.* 2011. Development of genic-SSR markers by deep transcriptome sequencing in Pigeonpea [*Cajanus cajan* (L.) Millspaugh]. *BMC Plant Biol.* 11: 17.

Eldem, V., Akcay, U.C., Ozhuner, E., Bakir, Y., Uranbey, S. and Unver, T. 2012. Genome-wide identification of miRNAs responsive to drought in Peach (*Prunus persica*) by high throughput deep sequencing. *PLoS ONE.* 7: 14.

Feuillet, C., Leach, J.E., Rogers, J., Schnable, P.S. and Eversole, K.2010. Crop genome sequencing: lessons and rationales. *Trends in Plant Science.* 16(2): 77-88.

Flowers, T.J. 2004. Improving crop salt tolerance. *J. Exp. Bot.* 55: 307-319.

Flowers, T.J. and Yeo, A.R. 1995. Breeding for salinity resistance in crop plants: where next? *Aust. J. Plant Physiol.* 22: 875-884.

Franklin, S.E. and Mayfield, S.P. 2004. Prospects for molecular farming in the green alga *Chlamydomonas*. *Curr. Opin. Plant Biol.* 7: 159-165.

Frazier, T.P., Sun, G.L., Burklew, C.E. and Zhang, B.H. 2011. Salt and drought stresses induce the aberrant expression of microRNA genes in tobacco. *Mol. Biotechnol.* 49: 159-165.

Frova, C., Caffulli, A. and Pallavera, E. 1999b. Mapping quantitative trait loci for tolerance to abiotic stresses in maize. *J. Exp. Zool.* 282: 164-170.

Frova, C., Krajewski, P., Di-Fonzo, N., Villa, M. and Sari-Gorla, M. 1999a.Genetic analysis of drought tolerance in maize by molecular markers. 1. Yield components. *Theor. Appl. Genet.* 99: 280-288.

Fujita, M., Fujita, Y., Noutoshi, Y., Takahashi, F., Narusaka, Y., Yamaguchi-Shinozaki, K. and Shinozaki, K. 2006. Crosstalk between abiotic and biotic stress responses: A current view from the points of convergence in the stress signaling networks. *Curr. Opin. Plant Biol.* 9: 436-442.

Gao, P., Bai, X., Yang, L., Lv, D.K., Li, Y., Cai, H., Ji, W., Guo, D.J. and Zhu, Y.M. 2010. Over-expression of *osa-MIR396c* decreases salt and alkali stress tolerance. *Planta.* 231: 991-1001.

Gao, Q., Yue, G., Li, W., Wang, J., Xu, J. and Yin, Y. 2012. Recent progress using high-throughput sequencing technologies in plant molecular breeding. *J. Integr. Plant Biol.* 54: 215-227.

Goff, S.A., Ricke, D., Lan, T.H., Presting, G., Wang, R., Dunn, M. and Glazebrook, J. *et al.* 2002. A draft sequence of the rice genome (*Oryza sativa* L. ssp. *japonica*). *Science.* 296: 92-100.

Grevich, J.J. and Daniell, H. 2005. Chloroplast genetic engineering: recent advances and future perspectives. *Crit. Rev. Plant Sci.*24: 83-107.

Guo, H.S., Xie, Q., Fei, J.F. and Chua, N.H. 2005. MicroRNA directs mRNA cleavage of the transcription factor *NAC1* to downregulate auxin signals for Arabidopsis lateral root development. *Plant Cell.* 17: 1376-1386.

Hash, C.T., Sharma, A., Kolesnikova- Allan, M.A., Singh, S.D. and Thakur, R.P. *et al.*2006. Teamwork delivers biotechnology products to Indian small-holder crop-livestock producers, pearl millet hybrid "HHB 67 improved" enters seed delivery pipeline. *SAT eJournal, ejournal.icrisat.org* 2(1): 3.

Higuchi, M., Konduo, Y., Ichikawa, T. and Matsui, M. 2011. Full length cDNA overexpressor gene hunting system (FOX Hunting System). In: Pereira, A. (ed). Plant Reverse Genetics: Methods and Protocols, Methods in Molecular Biology, Springer. 678: 77.

Hirayama, T. and Shinozaki, K. 2010. Research on plant abiotic stress responses in the post-genome era: past, present and future. *Plant Journal.* 61(6): 1041-1052.

Hmida-Sayari, A., Gargouri-Bouzid, R., Bidani, A., Jaoua, L., Savoure, A. and Jaoua, S. 2005. Overexpression of D1-pyrroline-5- carboxylate synthetase increases proline production and confers salt tolerance in transgenic potato plants. *Plant Science.* 169(4): 746-752.

Hossain, Z., Lopez-Climent, M.F., Arbona, V., Perez- Clemente, R.M. and Gomez-Cadenas, A. 2009. Modulation of the antioxidant system in citrus under waterlogging and subsequent drainage. *Journal of Plant Physiology.* 166(13): 1391-1404.

Huang, S., Li, R., Zhang, Z., Li, L., Gu, X., Fan, W. and Lucas, W.J. *et al.* 2009. The genome of the cucumber, *Cucumis sativus* L. *Nat. Genet.* 41: 1275-1281.

Jaillon, O., Aury, J.M., Noel, B., Policriti, A., Clepet, C., Casagrande, A. and Choisne, N. *et al.* 2007. The grapevine genome sequence suggests ancestral hexaploidization in major angiosperm phyla. *Nature.* 449: 463-467.

Janz, D., Lautner, S.,Wildhagen, H., Behnke, K., Schnitzler, J.P., Rennenberg, H., Fromm, J. and Polle, A. 2012. Salt stress induces the formation of a novel type of 'pressure wood' in two *Populus* species. *New Phytol.* 194: 129-141.

Jamila, A., Riaza, S., Ashrafbc, M. and Fooladd, M.R. 2011. Gene expression profiling of plants under salt stress. *Crit. Rev. Plant Sci.* 30: 435-458.

Jewell, M.C., Campbell, B.C. and Godwin, I.D. 2010. Transgenic plants for abiotic stress resistance. In: Kole, C. *et al.* (ed).Transgenic Crop Plants. Springer, Heidelberg, Germany. pp. 67-132.

Ji, X., Dong, B. and Shiran, B., Talbot, M.J., Edlington, J.E., Hughes, T., White, R.G., Gubler, F. and Dolferus, R. 2011. Control of abscisic acid catabolism and abscisic acid homeostasis is important for reproductive stage stress tolerance in cereals.*Plant Physiology.* 156(2): 647-662.

Kantar, M., Unver, T. and Budak, H. 2010. Regulation of barley miRNAs upon dehydration stress correlated with target gene expression. *Funct Integr Genomics.* 10: 493-507.

Khraiwesh, B., Zhu, J.K. and Zhu, J.H. 2012. Role of miRNAs and siRNAs in biotic and abiotic stress responses of plants. *Biochim Biophys Acta.* 1819: 137-148.

Kizis, D., Lumbreras, V. and Pages, M. 2001. Role of *AP2/EREBP* transcription factors in gene regulation during abiotic stress. *FEBS Lett.*498: 187-189.

Kulcheski, F.R., De-Oliveira, L.F.V., Molina, L.G., Almerao, M.P., Rodrigues, F.A., Marcolino, J., Barbosa, J.F., Stolf-Moreira, R., Nepomuceno, A.L., Marcelino-Guimaraes, F.C., Abdelnoor, R.V., Nascimento, L.C., Carazzolle, M.F., Pereira, G.A.G. and Margis, R. 2011. Identification of novel soybean microRNAs involved in abiotic and biotic stresses. *BMC Genomics.* 12: 307.

Lauter, N., Kampani, A., Carlson, S., Goebel, M. and Moose, S.P. 2005. MicroRNA172 down-regulates glossy 15 to promote vegetative phase change in maize. *Proc. Natl. Acad. Sci. USA* 102: 9412-9417.

Le Treut, H., Somerville, R., Cubasch, U., Ding, Y., Mauritzen, C. andMokssit, A.2007.Historical overviewof climate change. In: Solomon, S., Qin, D.,Manning, M., Chen, Z.,Marquis, M., Averyt, K.B., Tignor, M. and Miller, H.L.(ed). Climate Change 2007: The physical science basis.Contribution ofworking group I to the fourth assessment report of the

IntergovernmentalPanel on Climate Change. Cambridge and New York: Cambridge UniversityPress. pp. 96-127.

Lee, S. and Choi, D. 2013. Comparative transcriptome analysis of pepper (*Capsicum annuum*) revealed common regulons in multiple stress conditions and hormone treatments. *Plant Cell Rep.*32: 1351-1359.

Lee, S.J., Kang, J.Y., Park, H.J., Kim, M.D., Bae, M.S., Choi, H.I. and Kim, S.Y. 2010.*DREB2C* interacts with *ABF2*, a *bZIP* protein regulating abscisic acid-responsive gene expression, and its overexpression affects abscisic acid sensitivity. *Plant Physiol.* 153: 716-727.

Lei, D., Tan, L., Liu, F., Chen, L. and Sun, C. 2013. Identification of heat sensitive QTL derived from common wild rice (*Oryza rufipogon* Griff.). *Plant Sci.* 201–202: 121-127.

Li, W.X., Oono, Y., Zhu, J.H., He, X.J., Wu, J.M., Iida, K., Lu, X.Y., Cui, X.P., Jin, H.L. and Zhu, J.K. 2008. The Arabidopsis NFYA5 transcription factor is regulated transcriptionally and post transcriptionally to promote drought resistance. *Plant Cell.* 20: 2238-2251.

Li, Y., Dhankher, O.P. and Carreira, L. *et al.* 2004. Overexpression of phytochelatin synthase in Arabidopsisleads to enhanced arsenic tolerance and cadmium hypersensitivity. *Plant and Cell Physiology.* 45(12): 1787-1797.

Lin, H.X., Zhu, M.Z., Yano, M., Gao, J.P., Liang, Z.W., Su, W.A., Hu, X.H., Ren, Z.H. and Chao, D.Y. 2004. Mapping of quantitative trait loci controlling low-temperature germinability in rice (*Oryza sativa* L.). *Theor. Appl. Genet.* 108: 794-799.

Liu, J.H. and Moriguchi, T. 2007. Changes in free polyamines and gene expression during peach flower development. *Biologia Plantarum.* 51(3): 530-532.

Liu,J.H., Nakajima, I. and Moriguchi, T. 2011. Effects of salt and osmotic stresses on free polyamine content and expression of polyamine biosynthetic genes in *Vitis vinifera. Biologia Plantarum* 55(2): 340-344.

Maliga, P. 2004. Plastid transformation in higher plants. *Annu. Rev. Plant Biol.* 55: 289-313.

Maliga, P. and Bock, R. 2011. Plastid biotechnology: food, fuel, and medicine for the 21st century. *Plant Physiol.* 155: 1501-1510.

Marques, M.C., Alonso-Cantabrana, H. and Forment, J. *et al.* 2009. A new set of ESTs and cDNA clones from full-length and normalized libraries for gene discovery and functional characterization in citrus. *BMC Genomics.* 10: 428.

Martin, W. and Borst, P. 2003. Secondary loss of chloroplasts in trypanosomes. *Proc. Natl. Acad. Sci. USA.* 100: 765-767.

Miller, G. and Mittler, R. 2006. Could heat shock transcription factors function as hydrogen peroxide sensors in plants? *Annals of Botany.* 98(2): 279-288.

Mizoi, J., Shinozaki, K. and Yamaguchi-Shinozaki, K. 2012. AP2/ERF family transcription factors in plant abiotic stress responses. *Biochim Biophys Acta.* 1819: 86-96.

Moller, I.S., Gilliham,, M., Jha, D., Mayo, G.M., Roy, S.J., Coates, J.C., Haseloff, J. and Tester, M. 2009. Shoot Na^+ exclusion and increased salinity tolerance engineered by cell type-specific alteration of Na^+ transport in Arabidopsis. *Plant Cell.* 21: 2163-2178.

Munns, R. 2005. Genes and salt tolerance: bringing them together. *New Phytologist.* 167(3): 645-663.

Nakashima, K., Ito, Y. and Yamaguchi-Shinozaki, K. 2009. Transcriptional regulatory networks in response to abiotic stresses in Arabidopsis and grasses. *Plant Physiol.* 149: 88-95.

Nakashima, K., Takasaki, H., Mizoi, J., Shinozaki, K. and Yamaguchi-Shinozaki, K. 2012. NAC transcription factors in plant abiotic stress responses. *Biochim. Biophys. Acta.* 1819: 97-103.

Negraoa, S., Courtoisb, B., Ahmadib, N., Abreuac, I., Saiboa, N. and Oliveiraa, M.M. 2011. Recent updates on salinity stress in rice: from physiological to molecular responses. *Crit. Rev. Plant Sci.* 30: 329-377.

Ni, Z.Y., Hu, Z., Jiang, Q.Y. and Zhang, H. 2012. Overexpression of gma-MIR394a confers tolerance to drought in transgenic *Arabidopsis thaliana*. *Biochem Biophys Res. Commun.* 427: 330-335.

Nicolai, M., Pisani, C., Bouchet, J.P., Vuylsteke, M. and Palloix, A. 2012. Discovery of a large set of SNP and SSR genetic markers by high-throughput sequencing of pepper (*Capsicum annuum*). *Genet. Mol. Res.* 11: 2295-2300.

Obata, T. and Fernie, A.R. 2012. The use of metabolomics to dissect plant responses to abiotic stresses. *Cell. Mol. Life Sci.* 69: 3225-3243.

Okogbenin, E., Porto, M.C.M., Egesi, C., Mba, C., Espinosa, E., Santos, L.G., Ospina, C., Marin, J., Barrera, E., Gutierrez, J., Ekanayake, I., Iglesias, C. and Fregene, M.A. 2007. Marker-assisted introgression of resistance to cassava mosaic disease into latin American germplasm for the genetic improvement of cassava in Africa. *Crop. Sci.* 47(5): 1895-1904.

Papdi, C., Abraham, E., Joseph, M.P., Popescu, C., Koncz, C. and Szabados, L. 2008. Functional identification of Arabidopsis stress regulatory genes using the controlled cDNA overexpression system. *Plant Physiol.* 147(2): 528-542.

Peleman, J.D. and van der Voort, J.R.2003. Breeding by design. *Trends in Plant Science.* 8(7): 330-334.

Perez-Alfocea, F., Ghanem, M.E., Gomez-Cadenas, A. and Dodd, I.2011. Omics of root-to-shoot signaling under salt stress and water deficit. *OMICs A Journal of Integrative Biology.* 15(12): 893-901.

Perez-Clemente, R.M. and Gomez-Cadenas, A. 2012.*In vitro* tissue culture, a tool for the study and breeding of plants subjected to abiotic stress conditions. In: Leva, A. and Rinaldi, L.M.R. (ed). Agricultural and Biological Sciences: *Recent Advances in Plant In VitroCulture.* pp. 91-108.

Popko, J., Hansch, R., Mendel, R.R., Polle, A. and Teichmann, T. 2010. The role of abscisic acid and auxin in the response of poplar to abiotic stress. *Plant Biol (Stuttg).* 12: 242-258.

Potato Genome Sequencing Consortium. 2011. Genome sequence and analysis of the tuber crop potato. *Nature.* 475: 189-195.

Qiu, D., Xiao, J., Xie, W., Cheng, H., Li, X. and Wang, S. 2009. Exploring transcriptional signalling mediated by *OsWRKY13*, a potential regulator of multiple physiological processes in rice. *BMC Plant Biology.* 9: 74.

Ribaut, J,M., Hosington, D.A., Deitsch, J.A., Jiang, C. and Gonzalez-de-Leon, D. 1996. Identification of quantitative trait loci under drought conditions in tropical maize. 1. Flowering parameters and the anthesis–silking interval. *Theor. Appl. Genet.* 92: 905-914.

Ribaut, J.M., Jiang, C., Gonzalez-de-Leon, D., Edmeades, G.O. and Hosington, D.A. 1997.Identification of quantitative trait loci under drought conditions in tropical maize. 2. Yield components and marker-assisted selection strategies. *Theor. Appl. Genet.* 94: 887-896.

Ribaut, J.M. Vicente, M.C. and Delannay, X.2010. Molecular breeding in developing countries, challenges and perspectives. *Curr. Opin. Plant Biol.* 13(2): 213-218.

Rigano, M.M., Scotti, N. and Cardi, T. 2012. Unsolved problems in plastid transformation. *Bioengineered.* 3: 329-333.

Rohila, J.S., Jain, R.K. and Wu, R. 2002. Genetic improvement of Basmati rice for salt and drought tolerance by regulated expression of a barley *Hva1* cDNA. *Plant Science.* 163(3): 525-532.

Santos, A.P., Serra, T., Figueiredo, D.D., Barros, P., Lourenco, T., Chander, S., Oliveira, M.M. and Saibo, N.J. 2011. Transcription regulation of abiotic stress responses in rice: a combined action of transcription factors and epigenetic mechanisms. *OMICS.* 15: 839-857.

Saranga, Y., Jiang, C.X., Wright, R.J., Yakir, D. and Paterson, A.H. 2004. Genetic dissection of cotton physiological responses to arid conditions and their inter-relationships with productivity. *Plant Cell Environ.* 27: 263-277.

Sasidharana, R., Voeseneka, L.A.C.J. and Pierika, R. 2011. Cell wall modifying proteins mediate plant acclimatization to biotic and abiotic stresses. *Crit. Rev. Plant Sci.* 30: 548-562.

Seki, M., Ishida, J., Narusaka, M., Fujita, M., Nanjo, T. and Umezawa, T. *et al.* 2002. Monitoring the expression pattern of around 7,000 Arabidopsis genes under ABA treatments using a full-length cDNA microarray. *Funct Integr Genomics.* 2: 282-291.

Septiningsih, E.M. Pamplona, A.M., Sanchez, D.L., Neeraja, C.N., Vergara, G.V., Heuer, S., Ismail, A.M. and Mackill, D.J.2009. Development of submergence-tolerant rice cultivars: the *Sub1* locus and beyond. *Ann. Bot.* 103: 151-160.

Shahbaz, M., Ashraf, M., Al-Qurainy, F. and Harrisc, P.J.C. 2012. Salt tolerance in selected vegetable crops. *Crit. Rev. Plant Sci.* 31: 303-320.

Sharma, K.K., Bhatnagar, P. and Thorpe, T.A. 2005. Genetic transformation technology: status and problems. *In Vitro Cell Dev. Biol. Plant.* 41: 102-112.

Shinozaki, K. and Yamaguchi-Shinozaki, K. 2007. Gene networks involved in drought stress response and tolerance. *J. Exp. Bot.* 58: 221-227.

Shinozaki, K., Yamaguchi-Shinozaki, K. and Seki, M. 2003. Regulatory network of gene expression in the drought and cold stress responses. *Curr. Opin. Plant Biol.* 6: 410-417.

Singh, K.B., Foley, R.C. and Onate-Sanchez, L. 2002. Transcription factors in plant defense and stress responses. *Curr. Opin. Plant Biol.* 5: 430-436.

Singh, K.N. and Tyagi, N.K. 1997. Genetic improvement for suppressive/salt affected soils. In: Nagarajan, S. Gyanendra, S. and Tyagi B.S. (ed). Proceedings of the international group Meeting on "Wheat research needs beyond 2000 AD" held at Directorate of Wheat Research, Karnal, Narosa Publishing House. pp. 199-207.

Song, J.B., Gao, S., Sun, D., Li, H., Shu, X.X. and Yang, Z.M. 2013. MiR394 and LCR are involved in Arabidopsis salt and drought stress responses in an abscisic acid-dependent manner. *BMC Plant Biol.* 13: 210.

Sui, N., Li, M., Zhao, S.J., Li, F., Liang, H. and Meng, Q.W. 2007. Overexpression of glycerol-3-phosphate acyltransferase gene improves chilling tolerance in tomato. *Planta.* 226(5): 1097-1108.

Sunkar, R. and Zhu, J.K. 2004. Novel and stress-regulated microRNAs and other small RNAs from Arabidopsis. *Plant Cell.* 16: 2001-2019.

Sunkar, R., Li, Y.F. and Jagadeeswaran, G. 2012. Functions of microRNAs in plant stress responses. *Trends Plant Sci.* 17: 196-203.

Suzuki, N., Koussevitzky, S., Mittler, R. and Miller, G. 2012. ROS and redox signalling in the response of plants to abiotic stress. *Plant Cell Environ.* 35: 259-270.

Tang, L., Kwon, S.Y. and Kim S.H. *et al.* 2006. Enhanced tolerance of transgenic potato plants expressing both superoxide dismutase and ascorbate peroxidase in chloroplasts against oxidative stress and high temperature. *Plant Cell Reports.* 25(12): 1380-1386.

Thomashow, M.F. 1999. Plant cold acclimation: Freezing tolerance genes and regulatory mechanisms. *Annu. Rev. Plant Physiol. Plant Mol. Biol.* 50: 571-599.

Thomson, M.J. *et al.* 2010. Marker assisted breeding. In: Pareek, A. *et al.* (ed). Abiotic Stress Adaptation in Plants: Physiological, Molecular and Genomic Foundation, Springer. pp. 451-469.

Tomato Genome Consortium. 2012. The tomato genome sequence provides insights into fleshy fruit evolution. *Nature.* 485: 635-641.

Tsarouhas, V., Gullberg, U. and Lagercrantz, U. 2004. Mapping of quantitative trait loci (QTLs) affecting autumn freezing resistance and phenology in Salix. *Theor Appl Genet.* 108: 1335-1342.

Tuskan, G.A., Di-Fazio, S., Jansson, S. and Bohlmann, J.*et al.* 2006. The genome of black cottonwood, *Populus trichocarpa* (Torr. & Gray). *Science.* 313(5793): 1596-1604.

Tzfira, T. and Citovsky, V. 2006. Agrobacterium-mediated genetic transformation of plants: biology and biotechnology. *Current Opinion in Biotechnology.* 17(2): 147-154.

Umezawa, T., Fujita, M., Fujita, Y., Yamaguchi-Shinozaki, K. and Shinozaki, K. 2006. Engineering drought tolerance in plants: discovering and tailoring genes to unlock the future. *Current Opinion in Biotechnology.* 17(2): 113-122.

Van Bel, A.J., Hibberd, J., Prufer, D. and Knoblauch, M. 2001. Novel approach in plastid transformation. *Curr. Opin. Biotechnol.* 12: 144-149.

Varshney, R.K., Chen, W., Li, Y., Bharti, A.K., Saxena, R.K., Schlueter, J.A. and Donoghue, M.T. 2012. Draft genome sequence of Pigeonpea (*Cajanus cajan*), an orphan legume crop of resource-poor farmers. *Nat. Biotechnol.* 30: 83-89.

Velasco, R., Zharkikh, A., Troggio, M., Cartwright, D.A., Cestaro, A., Pruss, D. and Pindo, M. 2007. A high quality draft consensus sequence of the genome of a heterozygous grapevine variety. *PLoS ONE.* 2(12): e1326.

Walley, J.W. and Dehesh, K. 2010. Molecular mechanisms regulating rapid stress signaling networks in Arabidopsis. *J. Integr. Plant. Biol.* 52: 354-359.

Wang, J., Sun, P.P., Chen, C.L., Wang, Y., Fu, X.Z. and Liu, J.H. 2011. An arginine decarboxylase gene *PtADC* from *Poncirus trifoliata* confers abiotic stress tolerance and promotes primary root growth in Arabidopsis. *J. of Exp. Bot.* 62(8): 2899-2914.

Wang, M., Wang, Q.L. and Zhang, B.H. 2013. Response of miRNAs and their targets to salt and drought stresses in cotton (*Gossypium hirsutum* L.). *Gene.* 530: 26-32.

Wang, W., Vinocur, B. and Altman, A. 2003. Plant responses to drought, salinity and extreme temperatures: towards genetic engineering for stress tolerance. *Planta.* 218: 1-14.

Wang, W.X., Vinocur, B., Shoseyov, O. and Altman, A. 2001. Biotechnology of plant osmotic stress tolerance: physiological and molecular considerations. *Acta Horticult.* 560: 285-292.

Wang, Z., Chen, Z., Cheng, J., Lai, Y., Wang, J., Bao, Y., Huang, J. and Zhang, H. 2012. QTL analysis of Na^+ and K^+ concentrations in roots and shoots under different levels of NaCl stress in Rice (*Oryza sativa* L.). *PLoS One.* 7(12): e51202.

Wu, Y.Y., Chen, Q.J., Chen, M., Chen, J. and Wang, X.C. 2005. Salt-tolerant transgenic perennial ryegrass (*Lolium perenne* L.) obtained by *Agrobacterium tumefaciens*-mediated transformation of the vacuolar Na^+/H^+ antiporter gene. *Plant Science.* 169(1): 65-73.

Xie, F., Stewart, C.N., Taki, F.A., He, Q., Liu, H. and Zhang, B. 2014. High-throughput deep sequencing shows that microRNAs play important roles in switchgrass responses to drought and salinity stress. *Plant Biotechnol. J.* 12: 354-366.

Xiong, Y. and Fei, S.Z. 2006. Functional and phylogenetic analysis of a DREB/CBF-like gene in perennial ryegrass (*Lolium perenne* L.). *Planta.* 224(4): 878-888.

Yamaguchi, S.K. and Shinozaki, K. 2006. Transcriptional regulatory networks in cellular responses and tolerance to dehydration and cold stresses. *Annu. Rev. Plant Biol.* 57: 781-803.

Ye, X., Busov, V., Zhao, N., Meilan, R., McDonnell, L.M., Coleman, H.D., Mansfield, S.D., Chen, F., Li, Y. and Cheng, Z.M. 2011. Transgenic *Populus* trees for forest products, bioenergy, and functional genomics. *Crit. Rev. Plant Sci.* 30: 415-434.

Yin, Z.J., Li, Y., Yu, J.W., Liu, Y.D., Li, C.H., Han, X.L. and Shen, F.F. 2012. Difference in miRNA expression profiles between two cotton cultivars with distinct salt sensitivity. *Mol. Biol. Rep.* 39: 4961-4970.

Yu, J., Hu, S., Wang, J., Wong, G.K., Li, S., Liu, B. and Deng, Y. 2002. A draft sequence of the rice genome (*Oryza sativa* L. ssp. *indica*). *Science.* 296: 79-92.

Zhang, J.X., Nguyen, H.T. and Blum, A. 1999.Genetic analysis of osmotic adjustment in crop plants. *J. Exp. Bot.* 50: 291-302.

Zhang, X.H., Zou, Z., Gong, P.J., Zhang, J.H., Ziaf, K., Li, H.X., Xiao, F.M. and Ye, Z.B. 2011. Overexpression of microRNA169 confers enhanced drought tolerance to tomato. *Biotechnol Lett.*33: 403-409.

Zhang, Y.Y., Li, Y., Gao, T., Zhu, H., Wang, D.J., Zhang, H.W., Ning, Y.S., Liu, L.J., Wu, Y.R., Chu, C.C., Guo, H.S. and Xie, Q. 2008. Arabidopsis*SDIR1* enhances drought tolerance in crop plants. *Bioscience, Biotechnology and Biochemistry.* 72(8): 2251-2254.

Zhao, B.T., Liang, R.Q., Ge, L.F., Li, W., Xiao, H.S., Lin, H.X., Ruan, K.C. and Jin, Y.X. 2007. Identification of drought-induced microRNAs in rice. *Biochem Biophys Res Commun.* 354: 585-590.

Zhou, M., Lim D.Y., Li, Z.G., Hu,Q., Yang, C.H., Zhu, L.H. and Luo, H. 2013. Constitutive expression of a miR319 gene alters plant development and enhances salt and drought tolerance in transgenic Creeping bentgrass. *Plant Physiol/* 161: 1375-1391.

Zhu, J.K. 2001. Plant salt tolerance. *Trends Plant Sci.* 6: 66-71.

7

Nutritional Poverty in Wheat Under Abiotic Stress Scenario

Naveen Kumar and Renu Munjal

Agricultural research has been primarily focused to assure food self-sufficiency and security, and little concern on nutritional value "nutritional security" or health promoting qualities of food being produced. Food security has three dimensions, namely: (1) endemic hunger caused by poverty-induced under-nutrition and malnutrition; (2) hidden hunger caused by the deficiency of micronutrients like iron, iodine, zinc and vitamin A in the diet; and (3) transient hunger caused by natural calamities or civilian conflicts. Micronutrient malnutrition is recognized as a massive and rapidly growing public health issue especially among poor people living on an unbalanced diet.

Although great progress has been made in reducing the prevalence of hunger, over 805 million people are undernourished in 2012-14, still unable to meet their daily calorie needs for living healthy lives. This number has fallen by 100 million over the last decade (Table: 1 & Table: 2; Source: FAO, 2014). Undernourishment refers to food intake that is insufficient to meet dietary energy requirements for an active and healthy life. About one in nine people go to bed daily on an empty stomach. In cases where food is available, often the quality of the food does not meet micronutrient (vitamin and mineral) needs. More than two billion people continue to suffer from nutritional deficiencies such as vitamin A, iron, zinc and iodine. Despite progress, the number is still high, and marked differences across regions persist.

Table 1: Nutritional deficiency and dietary indicator status in world (FAO, 2014)

	World		
	1992	2002	2014
Total population (mln)	5494.8	6280.8	7243.7
Prevalence of undernourishment (%)	18.7	14.9	11.3
Number of people undernourished (mln)	1014.5	929.9	805.3
Depth of food deficit (kcal/cap/day)	139	111	84
Dietary energy supply (kcal/cap/day)	2595	2719	2881
Average dietary energy supply adequacy (%)	113	116	122
Average protein supply (g/cap/day)	69	75	79
Share of cereals (excluding beer) dietary energy supply (%)	50.6	48.1	34.8

Table 2: Nutritional deficiency and dietary indicator status in India (FAO, 2014)

	India		
	1992	2002	2014
Total population (mln)	903.8	1076.7	1267.4
Anemia, women (pregnant/non-pregnant, %)	52.5/54.2	55.2/54.5	54.2/50.2
Anemia, children under-5 (%)	74.0	66.7	60.9
Vitamin A deficiency, total pop. (%)	20.0	62.0	-
Iodine deficiency, children (%)	81.0	11.3	50.9
Prevalence of undernourishment (%)	23.8	17.6	15.2
Number of people undernourished (mln)	210.8	186.2	190.7
Depth of food deficit (kcal/cap/day)	166	123	109
Dietary energy supply (kcal/cap/day)	2278	2330	2455
Average dietary energy supply adequacy (%)	105	105	108
Average protein supply (g/cap/day)	55	56	58
Share of cereals (excluding beer) dietary energy supply (%)	64.3	61.1	58.0

Wheat (*Triticum aestivum* L.) is the most important crop in the world as it is the staple food to one third of the world population. The native Indian wheat had been globally recognized for its quality and was always in demand from overseas clients. Economically, this is one of the major food crops both in terms of area and production. Wheat is grown in diverse environments, from cool rain-fed to hot dry-land areas around the world. This wide-spread cultivation of the crop around the globe is largely due to high versatility of its genome.

Wheat is also a main source of energy in the average Indian diet, accounting for 38 percent of calories from cereals for rural households in India (NSSO 2011, 2012). It is the cheapest source of carbohydrates and proteins. But, in changing climate scenario, abiotic stresses are serious threat for future crop production and cause the nutritional poverty in crop plants. Considering the above facts and the severity of widespread micronutrient malnutrition, improving nutritional quality would have a significant impact on global human health. This calls for new, abiotic stress tolerant, high-yielding and high-quality wheat varieties containing higher levels of bioavailable proteins, vitamins, minerals and essential amino acids for nutrition.

Abiotic stress to plants is arising from an excess or deficit in the physical or chemical environment. The unfavourable climatic and soil conditions resulting in salt stress, drought stress, low and high temperature stress, flooding stress, chemical stress, oxidative stress and other related stress types are the major impediments to increased crop production. There is hardly a landmass in India, which is not influenced by one or the other of these stress factors. In fact, most of these factors co-occur resulting in a compound effect. The drought stress is mostly accompanied by high temperature stress, salt stress is often associated with water stress and low temperature stress is associated with drought stress. The contribution due to osmotic stress is a common denominator in water stress, salt stress and low temperature stress. Likewise, the contribution due to oxidative damage is a common factor in stresses caused by excess light, excess or shortage of water, and low and high temperatures.

These abiotic stresses cause an array of morpho-anatomical, physiological and biochemical changes in plants, which affect plant growth and development, and may lead to a drastic reduction in economic yield and nutritional state of the plants. The adverse effects of nutritional poverty due to abiotic stress can be mitigated by developing crop plants with improved tolerance using various genetic approaches. For this purpose, however, a thorough understanding of physiological responses of plants to abiotic stress, mechanisms of stress tolerance and possible strategies for improving nutritional poverty is crucial. Based on a complete understanding of such mechanisms, potential genetic strategies to improve plant nutritional poverty and abiotic stress tolerance include the traditional and modern molecular breeding protocols and transgenic approaches. The major stresses to wheat crop are heat and drought stress, which are discussed in this chapter.

Effect of heat and drought stress on wheat

Wheat (*Triticum aestivum* L.) is very sensitive to high temperature. High temperature stress causes some physiological, biochemical and molecular changes in plant metabolism such as protein denaturation, lipid peroxidation, membrane injury, inhibition of photosynthesis, enzyme inactivation etc. Wheat experiences heat stress to varying degrees at different phenological stages, but heat stress during the reproductive phase is more harmful than during the vegetative phase due to the direct effect on grain number and dry weight (Wollenweber *et al.*, 2003). Wheat is very sensitive crop with respect to terminal temperature or post anthesis heat stress (Barnabas *et al.*, 2008). The optimum temperature for wheat anthesis and grain filling ranges from 12 - 22 °C. Exposure to temperatures above this can significantly reduce grain yield (Tewolde *et al.*, 2006). Heat stress accelerates the rate of grain filling whereas grain-filling duration is shortened (Dias and Lidon, 2009). For instance, 5°C increases in temperature above 20°C increased the rate of grain filling and reduced the grain filling duration by 12 days in wheat (Yin *et al.*, 2009). Under these conditions, the supply of photoassimilates may be limited. Elevated temperatures reduce the duration between anthesis and physiological maturity (Warrington *et al.*, 1977), which is associated with a reduction in grain weight (Warrington *et al.*, 1977; Shpiler and Blum, 1986). Reduced grain weight (H=1.5 mg per day) can occur for every 1°C above 15–20°C.

At moderately high temperatures, injuries or death may occur only after long-term exposure. Direct injuries due to high temperatures include protein denaturation and aggregation, and increased fluidity of membrane lipids (Dhanda and Munjal, 2012). Indirect or slower heat injuries include inactivation of enzymes in chloroplast and mitochondria, inhibition of protein synthesis, protein degradation and loss of membrane integrity (Howarth, 2005). These injuries eventually lead to starvation, inhibition of growth, reduced ion flux, production of toxic compounds and reactive oxygen species (ROS) (Schoffl *et al.*, 1999; Howarth, 2005).

Heat stress, singly or in combination with drought, is a common constraint during anthesis and grain filling stages in many cereal crops of temperate regions. For example, heat stress lengthened the duration of grain filling with reduction in kernel growth leading to losses in kernel density and weight by up to 7% in spring wheat (Guilioni *et al.*, 2003). Studies of heat stress on wheat have been focusing on the period of grain filling (Borghi *et al.*, 1995; Corbellini *et al.*, 1997, 1998; Stone and Nicolas, 1995), and have shown that two typical heat stresses are common during wheat grain filling. "Heat shock" is characterized by sudden, extreme high temperatures (>32 °C) for a short duration (3-5 days), while "chronic heat stress" consists of moderately high maximum temperatures (20-30 °C) for a longer duration. Heat shock takes different forms, which are characterized by timing (days after anthesis) and by duration, which may also gave rise to different effects on the durum wheat quality (Corbellini *et al.*, 1997). In wheat, both grain weight and grain number appeared to be sensitive to heat stress, as the number of grains per ear at maturity declined with increasing temperature (Ferris *et al.*, 1998).

Furthermore, high temperatures during grain filling can modify flour and bread quality and other physio-chemical properties of grain crops such as wheat (Perrotta *et al.*, 1998), including changes in protein content of the flour (Wardlaw *et al.*, 2002). Thus, for crop production under high temperatures, it is important to know the developmental stages and plant processes that are most sensitive to heat stress, as well as whether high day or high night temperatures are more injurious. Such insights are important in determining heat-tolerance potential of crop plants. Moisture stress caused an increase in protein content and a reduction in thousand kernel weight (Rharrabtia *et al.,* 2003a). Limited water input during grain filling decreased grain quality by reducing test weight and SDS sedimentation volume, and by increasing ash content (Rharrabtia *et al.,* 2003b).

High temperatures during seed development may adversely affect seed quality. The sensitivity of seeds to environmental stress depends on the stage of development (Monjardino *et al.,* 2005). Seed germination and vigor rapidly decreased seeds exposed to adverse environmental conditions (Hasan *et al.,* 2013). An increase in 1°C can decrease grain weight up to 4 mg (Ishag and Mohamed, 1996). Parental plants exposed to high temperature influenced seed quality (Hasan *et al.,* 2013) because soluble starch synthase involved in synthesis and deposition of starch is extremely sensitive to high temperature and decreases its activity when temperature touches the level beyond 20°C (Keeling *et al.,* 1994; Bansal *et al.,* 2012). As a consequence, smaller and shrivelled grains of low vigor and viability and if used as seed for next crop, do not perform well.

Besides heat stress, drought or continuous water deficit is also one of the most important factors affecting plant growth, development, survival and crop productivity. Physiological responses to drought include stomatal closure, decreased photosynthetic activity, altered cell wall elasticity, and even generation of toxic metabolites causing plant death. Water stress at grain filling stage of wheat crop is detrimental to crop mineral nutrition because it reduces soil moisture and inhibits the optimal crop growth and development by decreasing the availability of nutrients uptake. In wheat, grain filling stage has been observed most sensitive to drought stress and highly responsive to exogenous application of potassium or glycine betaine (Raza *et al.*, 2013, 2015).

In order to cope with heat and drought stress, plants implement various mechanisms, including maintenance of membrane stability, scavenging of ROS, production of antioxidants, accumulation and adjustment of compatible solutes, induction of mitogen-activated protein kinase (MAPK) and calcium-dependent protein kinase (CDPK) cascades, and, most importantly, chaperone signaling and transcriptional activation. All these mechanisms, which are regulated at the molecular level, enable plants to thrive under heat and drought stress. Based on a complete understanding of such mechanisms, potential genetic strategies to improve plant heat and drought stress tolerance include traditional and modern molecular breeding protocols and transgenic approaches. In addition to genetic approaches, crop heat and drought tolerance can be enhanced by preconditioning of plants under different environmental stresses or exogenous application of osmoprotectants such as glycinebetaine and proline. Acquiring thermotolerance is an active process by which considerable amounts of plant resources are diverted to structural and functional maintenance to escape damages caused by heat stress. The studies combined with genetic approaches to identify and map genes (or QTLs) conferring thermotolerance and drought tolerance will not only facilitate marker-assisted breeding for heat and drought tolerance but also pave the way for cloning and characterization of underlying genetic factors which could be useful for engineering plants with improved heat and drought tolerance.

So, the research on the genetic mechanism of heat tolerance is getting more and more important to the utilization of heat tolerant gene and the development of new wheat varieties with heat tolerance. However, estimating inheritance of heat tolerance from their overall response to high temperature was difficult because of the low precision due to effects of plant development status and interactions of environmental factors with genes for the trait (Ottaviano *et al.*, 1991; Dhanda and Munjal, 2006).

Nutritional poverty

Wheat provides 500 kcal of food energy capita^{-1} day^{-1} in the two most populous countries in the world, China and India (FAO, 2007). Overall across in the developing world, 16 % of total dietary calories come from wheat (cf. 26 % in developed countries) second only to rice in importance. As the most traded food crop internationally, wheat is the single largest food import into developing countries and, also, a major portion of emergency food aid.

Elements that generally occur at relatively low concentrations in living tissues are designated 'trace elements' or 'microminerals'. Seventeen elements essential for higher plants include carbon (C), hydrogen (H), oxygen (O), nitrogen (N), phosphorus (P), potassium (K), sulphur (S), magnesium (Mg), calcium (Ca), iron (Fe), manganese (Mn), zinc (Zn), copper (Cu), boron (B), molybdenum (Mo), chlorine (Cl) and nickel (Ni). Out of these Fe, Mn, Zn, Cu, B, Mo, Cl and Ni have been classified as the micronutrients - nutritionally important elements absolutely essential for completion of life cycle of the organisms (plants or humans) needed in relatively smaller quantities. Out of seven elements (Fe, Zn, Cu, Mn, Ni, B and Mo) common to plants and humans, malnutrition of Zn and Fe afflicts over two billion people, especially resource-poor women and children in the developing world. Problem of micronutrient malnutrition was categorically emphasized because more than half of the humanity - mostly the poor in developing countries - suffers from the devastating consequences of micronutrient malnutrition. No other problem of this magnitude is afflicting such a huge portion of the world population. According to WHO (2002), deficiencies of zinc and iron occupy 5th and 6th place, respectively among top ten leading causes of illness and diseases in low income countries.

In a way, like the concerns about abiotic stress, the new concerns of nutritional poverty have been brought into sharp focus by the data and statistics of the nutrition community and World Health Organization in the last few years. The food supply, while it has been sufficient, is simply not nutritious. No previous disease or deficiency has ever affected over half of the world population than the micronutrient deficiency. In reality most of the women and children, together with a surprisingly large number in developed countries are micronutrient deficient. Some blame the green revolution in that the new highly productive cereals did not provide nutrient balance, but we believe it is more rational to attribute any blame to population increase. As the basis of the effort to increase food production in poor countries, highly productive cereals have displaced other crops that are higher in iron. For example, in India where cereal production has increased more than four times in the two decades since 1970 (while the population has less than doubled), pulse production actually declined. During the green revolution push towards food security, precious little thought was given to nutritional value, and certainly almost none to the iron content of the new cereal varieties being bred, let alone to the iron content of the changing diets. To be fair, it must be said that the nutritional poverty in the crop plants is far worse than the problems we must now tackle. The challenge now is to support a new paradigm for agriculture - an agriculture which aims not only for productivity and sustainability, but also, for balanced nutrition, what we have called the productive, sustainable, nutritious food systems paradigm.

A protein change in response to abiotic stress is a complex mechanism, in most cases species- and genotype-dependent. For instance, in wheat, the tolerant genotype 'Khazar-1' showed an increase in accumulation of thioredoxin (Trx h) under drought field conditions while, in contrast, the sensitive genotypes 'Afghani' and 'Arvand' showed a decrease (Hajheidari *et al.,* 2007).

Abiotic stress impact on nutritional quality and safety

Improving plant survival or performance under abiotic stress will certainly be based on proteins known to have defensive or protection functions. Since, it is already known that proteins belonging to the plant defence system are often also allergens (Breiteneder and Radauer, 2004); one may speculate that alterations in their expression may increase plants potential allergenicity. Moreover, since abiotic stimuli can alter global protein expression and differences in the environmental growth conditions of a given food plant may also have consequences on its protein composition. These hypotheses suggest a broad impact that climate change may have on the safety (Beggs and Walczyk, 2008) and quality (DaMatta *et al.,* 2009) of plant food products.

There is a great lack of studies correlating the effects of altered environmental conditions on plant food protein composition and the quality and safety of derived products. This may be a problem in a world with a rising population and constant demand of protein-rich diets. Also, in some cases a reduction in protein concentration may also affect food products performance. For example for adequate bread-making quality, wheat grain protein concentrations higher than 11.5% are required. Under CO_2 enrichment, protein concentrations in wheat grains may decrease to values below the minimum quality standard for bread-making (Hogy and Fangmeier, 2008). There are some recent studies that clearly relate increases in atmospheric temperatures and carbon dioxide concentrations, with the rise of pollen concentration and consequent contribution to an increased incidence and prevalence of allergic diseases (Beggs and Walczyk, 2008; Reid and Gamble, 2009; Ziello *et al.,* 2012).

In order to prevent and control potential allergy-related health problems associated with the urgent need of getting higher yields of more nutritious food in abiotic stress conditions, it will be crucial to develop plants able tolerate these abiotic stresses, while maintaining or increasing their nutritional quality, and maintaining or reducing their allergenic potential.

Shinde and Deokule, 2015 estimated the protein content in *Triticum aestivum* wheat under control and water stress condition and observed that the protein content was decreased under stress condition. Inhibition of protein synthesis was induced by water stress (Badiani *et al.,* 1990; Price and Hendry, 1991). Contribution of cysteine proteases to total proteolytic activity increases drastically in response to water deficit in wheat (Zagdanska and Winievski, 1996). It is established that water deficit stress induces the expression of many genes among which are some genes coding proteases (Bray, 2002; Cruz de Carvalho *et al.,* 2001). Intracellular proteases have an important role in the degradation of damaged or unnecessary proteins, metabolism reorganisation and nutrient remobilization under stress (Grudkowska and Zagdanska, 2004). Some experimental evidence suggests that drought sensitive species and varieties have higher proteolytic activity compared to resistant ones (Roy-Macauley *et al.,* 1992; Zagdanska and Winievski, 1996; Hieng *et al.*, 2004).

Seed priming is a low risk technology (Harris *et al.,* 1999) and low cost solution for poor stand establishment (Farooq *et al.,* 2006). It improves seed germination, emergence and seedling growth by altering seed vigor and/or the physiological state of the seed (Black and Peter, 2006). Seed priming improves germination by repairing damaged proteins, RNA and DNA of low vigor and aged seeds that may accumulate during seed development (McDonald, 2000; Netondo *et al.,* 2004). Certain proteins like globulins and cruciferin are identified only during priming and not during seed germination. Similarly, low molecular weight heat shock proteins (HSPs) are specifically synthesized during osmopriming and not during imbibition in water. These proteins function as molecular chaperones and protect the cell from moisture stress during osmopriming and also protect those proteins which are damaged naturally (Varier *et al.,* 2010). Number of productive tillers, biological yield, grain yield and harvest index was higher in osmoprimed wheat under late condition (Farooq *et al.,* 2008; Hussain *et al.,* 2013). Seed priming improves the performance of poor quality wheat seed under drought stress (Hussain *et al.,* 2014).

Improvement of wheat grain quality for bread making

For most traditional uses, wheat quality derives mainly from two interrelated characteristics: grain hardness and protein content. Grain hardness is a heritable trait but it can be strongly affected by abnormal weather conditions such as excessive rainfall during the harvest period. Protein content is weakly heritable and strongly dependent on environmental factors such as available soil nitrogen and moisture during the growing season (Belderok *et al.*, 2000). In addition, each end-use requires a specific 'quality' in the protein. The quality requirements of wheat for various products like chapati, bread, biscuit and pasta are different. Like hard wheat with strong gluten (> 60 ml sedimentation value), > 12.0 % protein, 5+10 Molecular Weight Gluten Subunit with 9 or 10 Glu-1 score are required for making good bread. For biscuit, the quality requirements are weak and soft wheat with < 10.0 % protein, < 30 ml sedimentation value and ~ 50 % Alkaline Water Retention Capacity (AWRC). For chapati, the quality requirements are in between to those required for bread and biscuit (Gupta, 2008). Durum wheat cultivars have the hardest grain texture and are usually high in protein content. Recent research has shown that the presence of *ã*-gliadin 45 is a reliable marker of good cooking quality. This marker is now used for screening early generation material in many durum wheat breeding programs. Bread (also common or hexaploid) wheat cover a wide range of grain hardness and protein content.

Among all the cereals, only the flour of hexaploid wheat (*Triticum aestivum*) is able to form dough that exhibits the rheological properties required for the production of leavened bread. This property results from the ability of wheat storage proteins, gliadins and glutenins, to form special protein complex known as gluten. Biochemical and genetic evidence has demonstrated that high molecular weight glutenin subunit (HMW-GS) plays a major role in determining the viscoelastic properties thereby determining bread making qualities. Considerable progress has been achieved in research of the molecular properties of flour proteins that are required for highest bread quality. Segregating breeding populations can be screened by electrophoresis or high performance liquid chromatography for the presence of desirable glutenin subunits. However, because of the tight linkage of the HMW-GS alleles, it is quite difficult to manipulate them by traditional breeding methods. In near future, molecular approaches including genetic transformation and marker assisted selection will provide an opportunity for improving further the wheat processing qualities by introduction of genes associated with good bread-making qualities into a cultivar which is agronomically desirable but which has poor bread making qualities. This would avoid or minimize the necessity of blending flour from different cultivars in milling operations.

Biofortification

Biofortification, which refers to the breeding of plants/crops with high bioavailable micronutrient content using conventional breeding and genetic engineering approaches, is being used to improve the nutritional quality of major crops (Bouis *et al*., 2003; Welch and Graham, 2004; White and Broadely, 2005). Biofortification has the potential to provide coverage for remote rural population, and it inherently targets the poor who consume high levels of staple foods and little else. This can complement current efforts of nutritional supplementation and food fortification to address micronutrient deficiencies. Moreover, biofortification is beneficial for individuals who find it difficult to change their dietary habits owing to financial, cultural, regional or religious restrictions. Wheat is a particularly suitable target for biofortification because it is a major staple crop.

Kumar, 2014 worked on F_3, F_4, BC_1F_2 and BC_1F_3 populations derived from the cross between high-yielding (PAU201) and iron-rich (Palman 579) *indica* rice varieties which displayed large variation for various physio-morphological traits including grain yield per plant and mineral (iron and zinc) contents. Iron and zinc content varied from 0.9- 149.9 and 0-143.1 μg/g respectively in all the four populations (F_3, F_4, BC_1F_2 and BC_1F_3). Transgressive segregation for grain iron content was noticed in F_3 population with one of the plants having exceptionally higher iron (746.8 μg/g) content. Although rice is not considered a major mineral source in the diet, any increase in its mineral concentration could significantly help to reduce iron and zinc deficiency in humans because of the high levels of rice consumption among the poor in Asia.

Kanpur based Chandra Shekhar Azad University of Agriculture and Technology (CSAUAT), a state university, along with Directorate of Wheat Research, Karnal (now named as Indian Institute of wheat and Barley Research, Karnal), had developed "K-1006" wheat variety with the biofortification of iron and zinc (Sharma, 2013). The Directorate of Wheat Research, Karnal, the nodal agency to undertake wheat research in the country and has coordinated successful identification and testing exercise of the "K-1006" variety. Normally wheat contains 40-45 ppm of zinc and 30-35 ppm of iron, while 'K-1006" has 49.2 ppm zinc and 45.4 ppm iron content. It is a major step which can offer nutritious option of wheat to the end users.

A QTL (*Gpc-B1*) has been reported associated with 12 % higher concentration of Zn, 18 % higher concentration of Fe, 29 % higher concentration of Mn and 38 % higher concentration of protein in the grain as compared to RSLs carrying alleles from cultivated wheat (*Triticum durum*). Distelfeld *et al.* (2007) suggested that the *Gpc-B1* locus is involved in more efficient remobilization of protein, zinc, iron and manganese from leaves to grains, in addition to its effect on earlier senescence of the green tissues.

Escalating of nutritious components

The nutritional value of wheat is extremely important as it takes an important place among the few crop species being extensively grown as staple food sources. The importance of wheat is mainly due to the fact that it presents the main source of nutrients such as proteins, carbohydrates, lipids, fibre and vitamins, to the most of the world population. Although not much work of this nature has been carried out in wheat, there is no reason why the success which has been achieved in crops such as rice, maize and soybean, or demonstrated in model species such as *Arabidopsis*, cannot be extended to wheat (Vasil, 2007).

Protein quality is based on their amino acid composition (particularly their relative content of essential amino acids) and their digestibility. Therefore, high quality proteins are those that are easily digested and contain the essential amino acids in quantities that correspond to human requirements. Deficiency in certain amino acids reduces the availability of others present in abundance. In general, cereal proteins are low in lysine (1.5–4.5% vs. 5.5 % of WHO recommendation), tryptophan (Trp, 0.8– 2.0% vs. 1.0%), and threonine (Thr, 2.7–3.9% vs. 4.0%). Because of this deficiency, these essential amino acids (EAAs) become the limiting amino acids in cereals. It is thus of economic and nutritional significance to enhance the EAAs in plant proteins.

The correlation between nutritional quality and yield has been a serious issue over the years, since the two factors appear to be negatively correlated. This problem appears to have been overcome since the introduction of modifier genes (*mo2* genes) that changed the opaque-2 phenotype of the maize seed, thus allowing wild-type-like seed characteristics to be maintained, resulting in normal yield but conserving the high lysine and high tryptophan concentrations (Gaziola *et al.,* 1999). These new maize lines have been designated QPM (Quality Protein Maize) and several hybrids were produced and introduced into the market.

For a long time there has been much interest in developing high-amylose wheat as a source of resistant starch (RS), which is one of the major sources of dietary fiber and its many related benefits (e.g. prevention of coronary heart disease, cancers of the colon and rectum, diabetes) to humans. Suppression of starch branching enzyme II (SBEIIa and SBEIIb) expression by RNA interference was used to produce high amylose wheat which was shown to be healthful for rats (Regina *et al*., 2006). The $(1 \rightarrow 3;1 \rightarrow 4)$-$\beta$-D-glucans, found exclusively in the cell walls of cereal and grass species, are important components of dietary fiber and highly beneficial in the prevention and treatment of serious human health conditions, including colorectal cancer, high serum cholesterol and cardiovascular diseases, obesity and non-insulin dependent diabetes. β-D-Glucans were shown to have immune stimulating activity (Dalmo and Bogwald, 2008). Genes responsible for $(1 \rightarrow 3;1 \rightarrow 4)$-$\beta$-D-glucan synthesis in grasses have been identified and provide an excellent opportunity to enhance the dietary fiber content of cereal and other food crops through transformation (Burton *et al*., 2006).

A well-known success of genetic transformation in cereals represents the development of the golden rice. It was first engineered with the insertion of the *PSY* gene from daffodil (*Narcissus pseudonarcissus*) and the bacterial phytoene desaturase (*CrtI*) gene from *Erwinia uredovora* (Ye *et al*., 2000). Bacterial *CrtI* can catalyze three enzymatic steps from phytoene to all-trans-lycopene. The *PSY* gene is under the control of an endosperm-specific glutelin promoter. To localize the product in plastids, *CrtI* was designed as a fusion with the ribulose-1,5-bisphosphate carboxylase/oxygenase (Rubisco) small subunit. An alternative construct was made by co-transformation with constructs carrying the *PSY/CrtI* gene as described above and the *LCY* gene under the control of a glutelin promoter. By the latter approach, the carotenoid content of edible rice endosperm was 1.6 μg/g dry weight (Ye *et al*., 2000). However, in 2005, Golden Rice-2 was developed and the carotenoid content was increased up to 23- fold (37 μg/g of dry weight) compared to the original Golden Rice. This content is close to a realistic level for palliating VAD (vitamin A deficiency) in children (Paine *et al.,* 2005). Expression of carotenoid biosynthetic genes in other cereals, such as wheat, requires further scientific investigation.

The International Maize and Wheat Improvement Center, along with its many partners, has identified several maize and wheat varieties with 25% to 30% higher grain iron and zinc concentrations. Wild relatives of wheat have been found to contain some of the highest iron and zinc concentrations in the grains. Backcrossing to bread wheat could result in highly nutritious cultivars (Ozturk *et al.,* 2006; Peleg *et al.,* 2008). Uauy *et al.* (2006) have characterized and cloned *Gpc-B1*, a quantitative trait locus from wild emmer wheat that is associated with increased levels of grain protein, zinc and iron as a consequence of accelerated senescence and increased nutrient mobilization from leaves to the developing grains. In ancestral wild wheat the allele encodes a NAC transcription factor (*NAM-B1*), while only a non-functional *NAM-B1* allele is present in modern cultivated wheat varieties. Silencing of the multiple *NAM* homologues by RNA interference in transgenic plants caused delayed maturation and reduced grain protein, iron and zinc content by more than 30%. The cloning of *Gpc-B1* provides a direct link between the regulation of senescence and nutrient remobilization and an entry point to characterize the genes regulating these two processes. This may contribute to their more efficient manipulation in crops and translate into food with enhanced nutritional value.

Phosphate, which is stored in the form of phytic acid in plant seeds (including wheat), is indigestible in monogastric animals including humans due to the fact that they lack phytases which degrade phytic acid in the digestive tract. Transgenic wheat plants expressing the *Aspergillus niger* phytase encoding gene *phyA* accumulate phytase in their seed (Brinch-Pedersen *et al.,* 2003; Brinch-Pedersen *et al.,* 2006). Further improvement in the expression and thermostability of phytases in transgenic wheat plants has the potential to increase the bioavailability of Zn^{2+}, Ca^{2+} and Fe^{2+} by breaking down their otherwise indigestible complexes with phytic acid.

Guttieri *et al.* (2007) identified a wheat mutant (*Lpa1-1*) with reduced phytic acid phosphorus and increased inorganic phosphorus (Pi). During germination there is a large decrease in phytine, which can make up to 80% of the total phosphate found in seeds (Reddy *et al.,* 1982) and a concomitant increase of Pi suggesting that phytine acts as a storage pool of phosphate during germination. Thus, it can be disputable whether the phytine decrease could result in growth disorders of the wheat plant and negatively affect all the metabolic reactions where phosphorus is a limiting factor (biosynthesis of nucleic acids, phospholipids proteins and other energy-generating processes).

Large numbers of alleles of puroindoline genes have been identified for influencing grain hardness. All of the known High Molecular Weight glutenin genes have been characterised and the sequence information has been used the development of transgenics to enhance gluten strength and extensibility (Sewa Ram, 2008). Molecular markers have been identified for some of allele of low molecular weight glutenins and gliadins. Characterisation of starch biosynthetic enzymes and their corresponding genes along with improved understanding of structure and properties of starch will help in manipulating starch functionality for different end-use products and nutritional quality. The possibility of engineering of metabolic pathway for improving lysine content in seeds has increased with the understanding of the genes involved in its biosynthesis. Molecular markers for higher protein content and micronutrient density have been identified. Current understanding of the molecular basis of processing and nutritional quality will speed up wheat improvement.

Conclusions

The world's agricultural community should adopt plant breeding and other genetic technologies to improve human health, and the world's nutrition and health communities should support these efforts. Plant foods can be improved as sources of essential trace elements either by increasing the concentrations of the microminerals in the food, increasing the bioavailability of the micronutrients in food, or increasing both of these. Trace element concentrations in edible portions of forage and food crops are influenced by a myriad of complex, dynamic and interacting factors, including plant genotype, soil properties, environmental conditions and nutrient interactions. Also, many dietary and host factors interact to affect the bioavailability of mineral nutrients to people in plant foods. Molecular tools have been used to understand factors regulating micronutrient contents/ bioavailability, rapid discovery of genes involved in nutritional elements uptake and storage in target tissues. The application of molecular genetic markers is efficient for the backcrossing of single gene traits and will become increasingly important for more complex traits. The use of biotechnological tools, such as molecular marker-assisted selection, will significantly increase the pace and prospects of success for breeding to improve the nutritional value of staple food crops.

Agricultural approaches are required to counteract for sustainable solutions to the enormous global problem of 'hidden hunger'. Biofortification has the potential to contribute to increased micronutrient intakes and improved micronutrient status. The success of this strategy will require the collaboration between health and agriculture sectors. However, because of the critical nutritional status of human population, there is an urgent need for development of such wheat varieties that would be more nutritious (with improved protein, zinc, iron, etc. value), meeting our health demands and resilent to climate change.

References

Badiani, M., Debiasi, M.G., Colognola, M. and Artemi, F. 1990. Catalase, peroxidase and superoxide dismutase activities in seedlings submitted to increasing water deficit. *Agrochimica.* 34: 90-102.

Bansal, K., Munjal, R., Madan, S. and Arora, V. 2012. Influence of high temperature stress on starch metabolism in two durum wheat varieties differing in heat tolerance. *J. Wheat Res.* 4(1): 43-48.

Barnabas, B.K., Jager and Feher, A. 2008. The effect of drought and heat stress on reproductive processes in cereals. *Plant, Cell Env.* 31: 11-38.

Beggs, P.J. and Walczyk, N.E. 2008. Impacts of climate change on plant food allergens: a previously unrecognized threat to human health. *Air Qual Atmos Health.* 1: 119-123.

Belderok, B., Mesdag, H. and Donner, D.A. 2000. Bread-making quality of wheat. *Springer* 3.

Black, M.H. and Peter, H. 2006. The encyclopedia of seeds: science, technology and uses. CAB Int. Wallingford, UK. pp. 224.

Bouis, H.E., Chassy, B. and Ochanda, J.O. 2003. Genetically modified food crops and their contribution to human nutrition and food quality. *Trends Food Sci. Technol.* 14: 191–209.

Bray, E.A. 2002. Classification of genes differentially expressed during water-deficient stress in *Arabidopsis thaliana*: an analysis using microarray and differential expression data. *Annals of Botany.* 89: 803-811.

Breiteneder, H. and Radauer, C. 2004. A classification of plant food allergens. *Am Acad Allergy Asthma Immunol.* 113: 821-830.

Brinch-Pedersen, H., Hatzack, F., Sorensen, L.D. and Holm, P.B. 2003. Concerted action of endogenous and heterologous phytase on phytic acid degradation in seed of transgenic wheat (*Triticum aestivum* L.). *Transgen Res.* 12: 649-659.

Brinch-Pedersen, H., Hatzack, F., Stoger, E., Arcalis, E., Pontopidan, K. and Holm, P.B. 2006. Heat-stable phytases in transgenic wheat (*Triticum aestivum* L.): deposition pattern, thermostability, and phytate hydrolysis. *J. Agric. Food Chem.* 54: 4624-4632.

Borghi, B., Corbellini, M., Ciaffi, M., Lafiandra, D., De Stefanis, E., Sgrulletta, D., Boggini, G. and Di Fonzo, N. 1995. Effect of heat shock during grain filling on grain quality of bread and *durum* wheats. *Australian Journal of Agricultural Research* 46: 1365-1380.

Burton, R.A., Wilson, S.M., Hrmova, M., Harvey, A.J., Shirley, N.J., Medhurst, A., Stone, B.A., Newbigin, N.J., Bacic, A., Fischer, G.B. 2006. Cellulose synthase-like *CslF* genes mediate the synthesis of cell wall (1, 3; 1, 4)-*â*-D glucans. *Science.* 311: 1940-1942.

Corbellini, M., Canevar, M.G., Mazza, L., Ciaffi, M., Lafiandra, D. and Borghi, B. 1997. Effect of the duration and intensity of heat shock during grain filling on dry matter and protein accumulation, technological quality and protein composition in bread and durum wheat. *Functional Plant Biology.* 24: 245-260.

Corbellini, M., Mazza, L., Ciaffi, M., Lafiandra, D. and Borghi, B. 1998. Effect of heat shock during grain filling on protein composition and technological quality of wheats. *Euphytica.* 100: 147-154.

Cruz de Carvalho, M.H., Arcy-Lameta, A.D., Roy-Macauley, H., Gareil, M., Maarouf, H.E., Pham-Thi, A.T. and Zuily-Fodil, Y. 2001. Aspartic protease in leaves of common bean (*Phaseolus vulgaris* L.) and cowpea (*Vigna unguiculata* L. Walp): enzymatic activity, gene expression and relation to drought susceptibility. *FEBS Letters.* 492: 242-246.

Dalmo, R.A. and Bogwald, J. 2008. ß-glucans as conductors of immune symphonies. *Fish Shellfish Immunol.* 25: 384-396.

DaMatta, F.M., Grandis, A., Arenque, B.C. and Buckeridge, M.S. 2009. Impacts of climate changes on crop physiology and food quality. *Food Res. Int.* 43: 1814-1823.

Dhanda, S.S. and Munjal, R. 2006. Inheritance of cellular thermotolerance in bread wheat. *Plant Breeding, Germany* 125: 557-564.

Dhanda, S.S. and Munjal, R. 2012. Heat tolerance in relation to acquired thermotolerance for membrane lipids in bread wheat. *Field Crops Research.* 135: 30-37.

Dias, A.S. and Lidon, F.C. 2009. Evaluation of grain filling rate and duration in bread and *durum* wheat, under heat stress after anthesis. *Journal of Agronomy and Crop Science.* 195(2): 137-147.

Distelfeld, A., Cakmak, I., Peleg, Z., Ozturk, L., Yazici, A.M., Budak, H., Saranga, Y. and Fahima, T. 2007. Multiple QTL effects of wheat *Gpc-B1* locus on grain protein and micronutrient concentrations. *Physiologia Plantarum.*129(3): 635-643.

FAO. 2007. Commission on genetic resources for food and agriculture. Food and Agriculture Organization, Rome. http://www.fao.org/ag/cgrfa/itpgr.htm (January 2007).

FAO. 2014. Food and nutrition in numbers. Food and Agriculture Organisation of United Nations, Rome. (*www.fao.org/publications*).

Farooq, M., Basra, S.M.A. and Cheema, M.A. 2006. Integration of pre-sowing soaking, chilling and heating treatments for vigor enhancement in rice (*Oryza sativa* L.). *Seed Sci. Technol.* 34: 521-528.

Farooq, M., Basra, S.M.A., Rehman, H. and Saleem, B.A. 2008. Seed priming enhances the performance of late sown wheat (*Triticum aestivum* L.) by improving chilling tolerance. *J. Agron. Crop Sci.* 194: 55-60.

Ferris, R., Ellis, R.H., Wheeler, T.R. and Hadley, P. 1998. Effect of high temperature stress at anthesis on grain yield and biomass of field-grown crops of wheat. *Annals of Botany.* 82: 631-639.

Gaziola, S.A., Alessi, E.S., Guimarães, P.E.O., Damerval, C. and Azevedo, R.A. 1999. Quality protein maize: a biochemical study of enzymes involved in lysine metabolism. *J. Agric. Food Chem.* 47: 1268-1275.

Grudkowska, M. and Zagdanska, B. 2004. Multifunctional role of plant cysteine proteinases. *Acta Biochimica Polonica.* 51(3): 609-624.

Guilioni, L., Wery, J. and Lecoeur, J. 2003. High temperature and water deficit may reduce seed number in field pea purely by decreasing plant growth rate. *Funct. Plant Biol.* 30: 1151–1164.

Guttieri, M.J., Peterson, K.M. and Souza, E.J. 2007. Nutritional and baking quality of low phytic acid wheat. In: Buck, H.T. (ed). Wheat Production in Stressed Environments. pp. 487-493.

Gupta, R.K. 2008. Quality status of Indian wheat. In: Mishra, B., Chatrath, R. and Singh, S.K. (ed). Advances in genetic enhancement and resource conservation technologies for enhanced productivity, sustainability and profitability in rice wheat cropping system. Compendium of ICAR sponsored winter school organised by DWR, Karnal. pp. 222-224.

Hajheidari, M., Eivazi, A., Buchanan, B.B., Wong, J.H., Majidi, I. and Salekdeh, G.H. 2007. Proteomics uncovers a role for redox in drought tolerance in wheat. *J Proteome Res.* 6: 1451-1460.

Harris, D.A., Joshi, P.A., Khan, P., Gothkar, P. and Sodhi, S. 1999. On-farm seed priming in semi-arid agriculture: Development and evaluation in maize, rice and chickpea in India using participatory methods. *Exp. Agri.* 35: 15-29.

Hasan, M.A., Ahmed, J.U., Hossain, T., Mian, M.A.K. and Haque, M.M. 2013. Evaluation of the physiological quality of wheat seed as influenced by high parent plant growth temperature. *J. Crop Sci. Biotech.* 16: 69-74.

Hieng, B., Ugrinovic, K., Sustar-Vozlic, J. and Kidric, M. 2004. Different classes of proteases are involved in the response to drought of *Phaseolus vulgaris* L. cultivars differing in sensitivity. *J. Plant Physiology.* 161: 519-530.

Hogy, P. and Fangmeier, A. 2008. Effects of elevated atmospheric CO_2 on grain quality of wheat. *J Cereal Sci.* 48: 580-591.

Howarth, C.J. 2005. Genetic improvements of tolerance to high temperature. In: Ashraf, M. and Harris, P.J.C. (ed). Abiotic Stresses: Plant resistance through breeding and molecular approaches. Howarth Press Inc., New York.

Hussain, I., Ahmad, R., Farooq, M. and Wahid, A. 2013. Seed priming improves the performance of poor quality wheat seed. *Int. J. Agric. Biol.* 15: 1343-1348.

Hussain, I., Ahmad, R., Farooq, M., Rehman, A.U. and Amin, M. 2014. Seed priming improves the performance of poor quality wheat seed under drought stress. *App. Sci. Report.* 7(1): 12-18.

Sewa Ram. 2008. Molecular basis of wheat grain quality improvement. In: Mishra, B., Chatrath, R. and Singh, S.K. (ed). Advances in genetic enhancement and resource conservation technologies for enhanced productivity, sustainability and profitability in rice wheat cropping system. Compendium of ICAR sponsored winter school organised by DWR, Karnal. pp. 228-232.

Sharma, I. 2013. North east to get wheat fortified with iron, zinc. In: Hindustan times, newspaper. (dated: 27 September, 2013).

Ishag, H.M. and Mohamed, B.A. 1996. Phasic development of spring wheat and stability of yield and its components in hot environments. *Field Crops Res.* 46: 169-176.

Keeling, P.L., Banisadr, R., Barone, L., Wasserman, B.P. and Singletary, G.W. 1994. Effect of temperature on enzymes in the pathway of starch biosynthesis in developing wheat and maize grain. *Aust. J. Plant Physiol.* 21: 807-827.

Kumar, N. 2014. Selection of high-yielding iron-rich PAU201/Palman579 segregating rice (*Oryza sativa* L.) lines using conventional and molecular marker techniques. Thesis, Ph.D. CCS HAU- Hisar, Haryana (India).

McDonald, M.B. 2000. Seed priming. Black, M. and Bewley, J.D. (ed). Seed technology and its biological basis. Sheffield Academic Press Ltd, Sheffield, UK. pp. 287-325.

Monjardino, P., Smith, A.G. and Jones, R.J. 2005. Heat stress effects on protein accumulation of maize endosperm. *Crop Sci.* 45: 1203-1210.

Netondo, G.W., Onyango, J.C. and Beck, E. 2004. Sorghum and salinity: II. Gas exchange and chlorophyll fluorescence of sorghum under salt stress. *Crop Sci.* 44: 806-811.

NSSO (National Sample Survey Office, Government of India). 2011. Key Indicators of Household Consumer Expenditure in India 2009–2010. National Sample Survey 66th Round Report NSS KI (66/1.0). New Delhi.

NSSO (National Sample Survey Office, Government of India). 2012. Nutritional Intake in India. National Sample Survey 66th Round Report 540. New Delhi.

Ottaviano, E., Gorla, M.S., Pe, E. and Frova, C. 1991. Molecular markers (RFLPs and HSPs) for the genetic dissection of thermotolrance in maize. *Theor. Appl. Genet.* 81: 713-719.

Ozturk, L., Yazici, M.A., Yucel, C., Torun, A., Cekic, C., Bagci, A., Ozkan, H., Braun, H.J., Sayers, Z. and Cakmak, I. 2006. Concentration and localization of zinc during seed development and germination in wheat. *Physiol. Plant.* 128: 144-152.

Paine, J.A., Shipton, C.A., Chaggar, S., Howells, R.M., Kennedy, M.J., Vernon, G., Wright, S.Y., Hinchliffe, E., Adams, J.L., Silverstone, A. and Drake, R. 2005. Improving the nutritional value of Golden Rice through increased pro-vitamin A content. *Nat. Biotechnol.* 23: 482-487.

Peleg, Z., Saranga, Y., Yazici, A., Fahima, T., Ozturk, L. and Cakmak, I. 2008. Grain zinc, iron and protein concentrations and zinc-efficiency in wild emmer wheat under contrasting irrigation regimes. *Plant Soil.* 306: 57-67.

Perrotta, C., Treglia, A.S., Mita, G., Giangrande, E., Rampino, P., Ronga, G., Spano, G. and Marmiroli, N. 1998. Analysis of mRNAs from ripening wheat seeds: the effect of high temperature. *J. Cereal Sci.* 27: 127-132.

Price, A.H. and Hendry, G.A.F. 1991. Iron-catalysed oxygen radical formation and its possible contribution to drought damage in nine native grasses and three cereals. *Plant Cell Environ.* 14: 477-484.

Regina, A., Bird, A., Topping, D., Bowden, S., Freeman, J., Barsby, T., Kosar- Hashemi, B., Li, Z., Rahman, S. and Morell, M.K. 2006. High-amylose wheat generated by RNA interference improves indices of largebowel health in rats. *Proc. Natl. Acad. Sci. USA.* 103: 3546-3551.

Rharrabtia, Y., Royo, C., Villegas, D., Aparicio, N. and García del Moral, L.F. 2003a. *Durum* wheat quality in Mediterranean environments I. Quality expression under different zones, latitudes and water regimes across Spain. *Field Crops Research.*80: 123-131.

Rharrabtia, Y., Villegas, D., Royo, C., Martos-Nunez, V. and García del Moral, L.F. 2003b. *Durum* wheat quality in Mediterranean environments II. Influence of climatic variables and relationships between quality parameters. *Field Crops Research.* 80: 133-140.

Raza, M.A.S., Saleem, M.F., Shah, G.M., Jamil, M. and Khan, I.H. 2013. Potassium applied under drought improves physiological and nutrient uptake performances of wheat (*Triticum aestivum* L.). *Journal of Soil Science and Plant Nutrition.* 13: 175-185.

Raza, M.A.S., Saleem, M.F. and Khan, I.H. 2015. Combined application of glycine betaine and potassium on the nutrient uptake performance of wheat under drought stress. *Pak. J. Agri. Sci.* 52(1): 19-26.

Reid, C.E. and Gamble, J.L. 2009. Aeroallergens, allergic disease, and climate change: impacts and adaptation. *Ecohealth.* 6: 458-470.

Roy-Macauley, H., Zuilfy-Fodil, Y., Kidric, M, Pham-Thi, A.T. and Vieira-de-Silva, J. 1992. Effect of drought stress on proteolytic activities in *Phaseolus* and *Vigna* leaves from sensitive and resistant plants. *Physiologia Plantarum.* 85(1): 90-96.

Schoffl, F., Prandl, R. and Reindl, A. 1999. Molecular responses to heat stress. In: Shinozaki, K. and Yamaguchi-Shinozaki, K. (ed). Molecular responses to cold, drought, heat and salt stress in higher plants. R.G. Landes Co., Austin, Texas. pp. 81-98.

Sharkova, V.E. 2001. The effect of heat shock on the capacity of wheat plants to restore their photosynthetic electron transport after photo inhibition or repeated heating. *Russ. J. Plant Physiol* .48: 793-797.

Shinde, S.S. and Deokule, S.S. 2015. Studies on different physiological parameters under water Stress condition in different wheat cultivars. *International Journal of Science and Research.* 4(5): 640-644.

Shpiler, L. and Blum, A. 1986. Differential reaction of wheat cultivars to hot environments. *Euphytica.* 35: 483-492.

Smertenko, A., Draber, P., Viklicky, V. and Opatrny, Z. 1997. Heat stress affects the organization of microtubules and cell division in *Nicotiana tabacum* cells. *Plant Cell Environ.* 20: 1534-1542.

Stone, P.J. and Nicolas, M.E. 1995. A survey of the effects of high temperature during grain filling on yield and quality of 75 wheat cultivars. *Australian Journal of Agricultural Research.* 46: 475-492.

Tewolde, H., Fernandez, C.J. and Erickson, C.A. 2006. Wheat cultivars adapted to post-heading high temperature stress. *J. Agron. Crop Sci.* 192: 111-120.

Uauy, C., Distelfeld, A., Fahima, T., Blechl, A. and Dubcovsky, J. 2006. A NAC gene regulating senescence improves grain protein, zinc, and iron content in wheat. *Science.* 314: 1298-1301.

Varier, A., Vari, A.K. and Dadlani, M. 2010. The sub cellular basis of seed priming. *Cur. Sci.* 99: 450-456.

Vasil, I.K. 2007. Molecular genetic improvement of cereals: transgenic wheat (*Triticum aestivum* L.). *Plant Cell Rep.* 26: 1133-1154.

Wardlaw, I.F., Blumenthal, C., Larroque, O. and Wrigley, C.W. 2002. Contrasting effects of chronic heat stress and heat shock on kernel weight and flour quality in wheat. *Funct. Plant Biol.* 29: 25-34.

Wollenweber, B., Porter, J.R. and Schellberg, J. 2003. Lack of interaction between extreme high temperature events at vegetative and reproductive growth stages in wheat. *J. Agron. Crop Sci.* 189: 142-150.

Warrington, I.J., Dunstone, R.L. and Green, L.M. 1977. Temperature effects at three development stages on the yield of the wheat ear. *Australian Journal of Agricultural Research.* 28(1): 11-27.

Welch, R.M. and Graham, R.D. 2004. Breeding for micronutrients in staple food crops from a human nutrition perspective. *J. Exp. Bot.* 55: 353-364.

White, P.J. and Broadley, M.R. 2005. Biofortifying crops with essential mineral elements. *Trends Plant Sci.* 10: 586-593.

WHO. 2002. The World Health Report. Reducing Risks, Promoting Healthy Life. World Health Organization, Geneva.

Ye, X., Babili, A.S., Kloeti, A., Zhang, J., Lucca, P., Beyer, P. and Potrykus, I. 2000. Engineering the provitamin A (â-carotene) biosynthetic pathway into (carotenoid-free) rice endosperm. *Science.* 287: 303-305.

Yin, L.J., Lun, S., Jun-feng, G., Feng-yun, Q. and Shu-shen, Y. 2009. Effect of drought stress on photosynthesis and some other physiological characteristics in flag leaf during grain filling of wheat. *Agricultural Research in the Arid Areas* 2003-02.

Zagdanska, B. and Winievski, K. 1996. Endoproteinase activities in wheat leaves upon water deficit. *Acta Biochim. Pol.* 43(3): 515-519.

Ziello, C., Sparks, T.H., Estrella, N., Belmonte, J., Bergmann, K.C. and Bucher, E. *et al.,* 2012. Changes to airborne pollen counts across Europe. *PLoS One.* 7: e34076.

8

Strategies for Improving Soil Health Under Current Climate Change Scenario

Pradeep K. Rai, G.K. Rai, Bhav Kumar Sinha and Reena

Soils are the basis of food production. Ninety five percent of our food is directly or indirectly produced on our soil. A healthy soil maintain a diverse community of organisms that helps in controlling pests (insects, weeds, and fungus) but also form beneficial symbiotic association with plant roots, recycle essential plant nutrients and improve soil structure. A healthy soil can also be a strategic ally in mitigating and adapting to climate change, as soil sequesters CO_2 and prevent to escape into the atmosphere. Beside this, it contributes to mitigate climate change by maintaining or increasing its carbon content. This can be said that the proportion of organic carbon is available more in the soil than combining both atmosphere and ground vegetation. Most of the soil organic carbon (SOC) stored in the first metre of the soil in the form of organic matter. However, organic matter degrades due to deforestation; deplete soil biodiversity, loss of nutrients as consumed by crop plants, soil compaction due to excessive use of agricultural machineries, soil erosion, water logging conditions, and urbanization, which release greenhouse gases like CO_2, CH_4, and N_2O into the atmosphere causing global warming and climate change. One third of all CO_2 emissions come from changes in land use (deforestation, shifting cultivation, and intensification of agriculture) whereas two-thirds of CH_4 and majority of N_2O emitted through agricultural practices (Kotschi and Müller-Sämann, 2004).

Expected Climate by 2060

- Several degrees warmer, on average (3 to 5°C)
- Generally higher expected rainfall along the East Coast
- Greater rainfall extremes

    - More floods
    - More droughts
- Possibly lower rainfall in some areas
- Increase in sea levels by up to 60 cm
- Likely impact on C sources and sinks

Climate plays an important role in determining the soil type in an area. Furthermore, the climatic conditions have a strong bearing upon the type of activity occurring within the soils. Of the climatic parameters temperature and precipitation are perhaps the most important. Climate change is occurring, both in terms of air temperature and precipitation.

Climatic change may alter the hydrologic cycle and, consequently, it will impact the available water resources, thus resulting frequent floods in some case and drought conditions in other. Either way the agricultural productivity will be impacted. By changing the proportion of soil organic and mineral constituents and altering soil structure, soil degradation will usually decrease a soil's water-holding capacity, with a consequent increase in runoff and the magnitude of nutrient and water stress.

Among the potentially most important characteristics of expected climate change in relation to agriculture, are changes in climate extremes, warming of high latitudes, shift of monsoon rainfall areas toward the poles, and reduction in soil humidity. Possible combination of increased temperature with drought or flood is perhaps the biggest risk for agriculture in many areas during the expected climate change. With the changing climate, many challenges will come in the way viz., soil process will alter.

These variations in soil processes may occur in many ways.

Physical degradation

Soil erosion

Climate change can either result in increased or decreased rainfall. The change in mean precipitation may take place by a change in storm frequency alone, intensity alone, or a combination of both. Higher precipitation may result in an increasing rate of erosion. Lower precipitation generally reduces the rate of erosion, but it can be counterbalanced by less intensive soil conservation, influence on poor vegetation due to the non-adequate water supply for plants. This can be the consequence of increasing temperature, as well. Lower precipitation (or higher temperature) may intensify wind erosion.

Climate change is also likely to bring about changes in the management practices. In their quest to adapt to changes in the climatic conditions farmers may switch to alternate crops or may change the dates of planting. This may result in altering the timing of and duration of soil cover and such will influence the soil erosion rates. Further the increasing CO_2 levels in the atmosphere will impact the biomass production affecting canopy cover thus influencing the soil erosion.

Alteration in soil structure

Soil structure is affected by variation in temperature and rainfall. In particular, during hotter, dryer summers there is an increased tendency for subsoil to become strong, making it more difficult for roots to penetrate. Some soils are likely to form impenetrable caps, increasing the risk of run-off and subsequent pollution events and flooding. Others may form cracks through which any rainfall will pass, reducing the trapping effect of the surface layers, further increasing risk of drought in the following year and also reducing the filtering effect of soil and increasing pollution risk. Increased intensity of rainfall can destroy soil structure and aggregation. The damage will be directly proportional to the intensity of the rainfall.

Chemical degradation

Increasing precipitation may result in increased infiltration resulting in increased leaching of bases as in the laterization process resulting in acidification of soils. Lower precipitation and higher temperature will intensify salinization / sodification processes. Higher temperatures will result in increased rate of evapo-transpiration, leading to capillary rise in water bringing the salts to the surface.

Soil desertification

Higher temperatures and evapo-transpiration combined with less summer rainfall make conditions for drought more likely.

Changes in soil organic matter dynamics

Temperature also impacts microbial activity levels, and hence residue decomposition rates. Increasing temperatures will result in increased decomposition rates thus increasing the release rate of nutrients as well as increased production of CO_2.

Table 1. Summary of expected effects of individual climate change variables on soil processes

Increasing temperature	Loss of soil organic matter Reduction in labile pool of SOM Reduction in moisture content Increase in mineralization rate Loss of soil structure Increase in soil respiration rate
Increasing CO_2 concentration	Increase in soil organic matter Increase in water use efficiency More availability of carbon to soil microorganisms Accelerated nutrient cycling.
Increasing rainfall	Increase in soil moisture or soil wetness Enhanced surface runoff and erosionIncrease in soil organic matter Nutrient leaching Increased reduction of Fe and nitrates Increased volatilization loss of nitrogen Increase in productivity in arid regions
Reduction in rainfall	Reduction in soil organic matter Soil salinization Reduction in nutrient availability

Table 2: Soil quality indicators and soil processes with high relevance to assess climate change impacts

Indicator	Soil processes affected
Soil organic matter fractions	Residue decomposition, organic matter storage and quality
Mineralizable C and N	Metabolic activity of soil organisms, mineralization-immobilization turnover
Total C and N	C and N mass and balance
Soil respiration, soil microbial biomass	Microbial activity
Microbial quotients	Substrate use efficiency
Microbial diversity	Nutrient cycling and availability
Porosity	Air capacity, plant available water capacity
Available water	Field capacity, permanent wilting pointing, water flow

Table 3: Soil quality indicators and soil processes with medium relevance to assess climate change impacts

Indicator	Soil processes affected
Soil structure	Aggregate stability, soil organic matter turnover
pH	Biological and chemical activity thresholds
EC	Plant and microbial activity thresholds
Available N, P and K	Plant available nutrient and potential for loss

Potential positive and negative impacts on plants and soil under warmer high and low rainfall scenario are discussed hereunder.

Potential positive impacts under a warmer high rainfall scenario

- Potentially higher yields from increased stem growth, diameter and tillering.
- Increased volume of trash & root material with increased C/N ratios.
- Increased C sequestration in soils and potential for N release through mineralisation.
- Increased K availability through improved diffusion especially in heavy vertic soils.

Potential negative impacts under a warmer high rainfall scenario

- Increased CO_2 release with higher temperatures + tillage.
- Increased runoff and stream flow will result in soil erosion and flooding events.
- Increased water logging potential & fertiliser N loss by denitrification.
- Loss and of nutrients (N, Ca, Mg, K, Si) through leaching favouring greater acidification.
- High C/N in trash will result in temporary fertiliser N loss by immobilisation.
- Increased pest and disease risk.
- In coastal alluvial floodplains increased risk of acid sulphate soils.

Potential negative impacts under a warmer low rainfall scenario

- Higher rates of evaporation with increased incidence of moisture stress and lower dry matter production.
- Decrease in soil moisture, runoff, stream flow, water table and aquifer discharge.
- Reduced SOM and microbial turnover.
- N volatilisation from urea is temperature dependent leading to increased risk of N loss.
- In low lying areas prone to water logging increased incidence of saline sodic conditions.

Table 4. Permissible concentration of elements in irrigation water

Element	Concentration(mg/l)	Remarks
Manganese	0.20	Toxic to a number of crops but usually only in acid soils.
Zinc	2.0	Toxic to many plants at widely varying concentrations; reduced toxicity at pH > 6-0 in fine textured or organic soils
Cadmium	0.01	Toxic to beans, beets and turnip at concentration as low as 0.1 mg/l in nutrient solutions. Conservative limits recommended due to its potential for accumulation in plants and soils to concentrations that may be harmful to humans
Copper	0.20	Toxic to number of plants at 0.1 to 1.0 mg/l in nutrient solutions
Fluoride	1.0	Inactivated by neutral and alkaline soils
Iron	5.0	Not toxic to plants in aerated soils, but can contribute to soil acidification and loss of availability of essential phosphorus and molybdenum. Overhead sprinkling may result in unsightly deposits on plants, equipments and buildings.
Aluminum	5.0	Can cause non-productivity in acid soils (pH less than 5.5) but more alkaline soils at pH > 7 will precipitate the ions and eliminate any toxicity.
Arsenic	0.10	Toxicity to plants varies widely, ranging from 12 mg/l for Sudan grass to less than .05 mg/l for rice
Molybdenum	0.01	Not toxic to plants at normal concentrations in soil and water. Can be toxic to livestock if forage is grown in soils with high concentrations of available molybdenum.
Lead	5.0	Can inhibit plant cell growth at very high concentrations.
Selenium	0.02	Toxic to plants at high concentrations as low as 0.025 mg/l and toxic to livestock if forage is grown in soils with relatively high levels of added selenium.
Nickel	0.20	Toxic to a number of plants at 0.5 mg/l to 1mg/l: reduced toxicity at neutral or alkaline pH.

Acclimatization, adaptation and mitigation strategies for improving soil health

Acclimatization is essentially adaptation that occurs spontaneously through self-directed efforts. Adaptation to climate change involves deliberate adjustments in natural or human systems and behaviours to reduce the risks to people's lives and livelihoods. Mitigation of climate change involves actions to reduce greenhouse gas emissions and sequester or store carbon in the short term, and development choices that will lead to low emissions in the long term.

The major aim of these strategies is to attempt a gradual reversal of effects caused by climate change and sustain development under the inescapable effect of climate change. Mitigation and adaptation are related to temporal and spatial scales on which they are effective. The benefits of mitigation activities carried out today will be evidenced in several decades because of the long residence time of greenhouse gases in the atmosphere, whereas the effects of adaptation measures should be apparent immediately or in the near future (Kumar and Parikh, 2001). Besides, mitigation has global in addition to local benefits, whereas adaptation typically takes place on a local or regional scale (Srinivasarao *et al.*, 2014).

Technically, adaptation measures are related with change in production systems like adjusting planting or fishing dates, rotations, multiple cropping/species diversification, crop-livestock pisciculture systems, agroforestry, soil, water and biodiversity conservation and development by building soil biomass, restoring degraded lands, rehabilitating rangelands, harvesting and recycling water, planting trees, developing adapted cultivars and breeds, protecting aquatic ecosystems to maintain long-term productivity Adaptation measures also include disaster risk management plans and risk transfer mechanisms, such as crop insurance and diversified livelihood systems.

Mitigation strategies

Mitigation options include carbon sequestration in agriculture and forestry. Mitigation of climate change is a global responsibility. Agriculture, forestry, fisheries/aquaculture provide in principle, a significant potential for greenhouse gases mitigation (Venkateswarlu and Shanker, 2009; Srinivasa Rao *et al.*, 2014).

Agro ecology

It is a system approach based on a variety of diverse technologies, practices and innovations including local and traditional knowledge and modern science.

Organic farming

Organic agriculture is a holistic production management system, which promotes sustainable agriculture and enhances agro-ecosystem health. It ensures qualitative development of soil, water and environment on sustainable basis. The experimental evidences could prove that a holistic soil and crop management had similar productivity as compared to conventional agriculture and often higher yields in the regions of the world where the production environment is much fragile and tough (rainfed and hilly areas) (Scialabba and Müller-Lindenlauf, 2010). On this note, organic agriculture may prove to be an alternating strategy to avert climate change with basic principles as mentioned by Kotschi and Müller-Sämann (2004) in organic agriculture with regard to mitigation of climate change, and it include to:

- Encourage and enhance biological cycles within the farming system
- Maintain and increase long-term fertility in soils
- Use as far as possible, renewable resources in locally organized production systems
- Minimize all forms of pollution

Beside this, organic agriculture give priority to the optimal (recycled and reuse) use of inputs with an aim to achieve maximum output.

Agroforestry

It includes both traditional and modern land-use system where trees are managed together with crops and/or animal production systems in agricultural setting. Agroforestry systems buffer farmers against climate variability, and reduce atmospheric loads of greenhouse gases. Agroforestry can both sequester carbon and produce a range of economic, environmental, and socio-economic benefits. For example, trees in agroforestry systems improve soil fertility through control of erosion, maintenance of soil organic matter and physical properties, increased nitrogen aeration, extraction of nutrients from deep soil horizons, and promotion of more closed nutrient cycling. In India, reducing emissions from deforestation and forest degradation is a major event that needs to evaluate to rectify it.

Tank silt application

Community tanks and taals may be used to collect rain water along with nutrient-rich top soil eroded from catchment areas. This tanks' silt can supply organic carbon and several nutrients besides improving soil physical, chemical and biological properties, if applied in the field.

Mulching-cum-manuring

In permanent cropping system, soil organic carbon improves by use of organic manures added through plant residues, mixed cropping, legume based crop rotations, or agroforestry (Drinkwater *et al.*, 1998). On the opposite side, sole use of synthetic nitrogen fertilizer application increases oxidation of organic matter, thereby reducing organic carbon from the soil. It improves soil surface conditions to increase infiltration, and water holding capacity, and reduce evaporation losses from the field. Beside this, it also provides additional nitrogen into the soil thereby improving soil health. It reduces temperature fluctuation in the soil and lowered down the canopy temperature at reproductive stage in many crops giving higher yield and test weight levels. It can be done through the use of crop residues, green manure crops, green leaf manure crops, brown manuring etc.

Mitigating N_2O emission from soils

N_2O emission contributes 38% of agricultural GHGs emissions (Smith *et al.*, 2007) of which 1% as direct N_2O emission from applied nitrogenous fertilizers. These emissions can be reduced by adding catch or cover crops that have the capacity to extract stored nitrogen from the soil that was not used by the previous crops. However, study suggests higher N_2O emission after manure application compared to mineral fertilizer application due to higher oxygen consumption for decomposition of organic matter (Flessa and Beese, 2000). These can be reduced by improving soil aeration either by lowering bulk density, rescue incorporation of legumes, and increase tillage practices (as no-tillage cause low aeration with more release of N_2O gas.

Soil carbon sequestration

Soil carbon sequestration is the process of transferring carbon dioxide from the atmosphere into the soil through crop residues and other organic solids, and in a form that is not immediately reemitted. This transfer or "sequestering" of carbon helps off-set emissions from fossil fuel combustion and other carbon-emitting activities while enhancing soil quality and long-term agronomic productivity. Soil carbon sequestration can be accomplished by management systems that add high amounts of biomass to the soil, cause minimal soil disturbance, conserve soil and water, improve soil structure, and enhance soil fauna activity. Continuous no-till crop production is a prime example. The term 'carbon sequestration' is commonly used to describe any increase in soil organic carbon (SOC) content caused by a change in land management, with the implication that increased soil carbon (C) storage mitigates climate change. However, this is only true if the management practice causes an additional net transfer of C from the atmosphere to land. Limitations of C sequestration for climate change mitigation include the following constraints: (i) the quantity of C stored in soil is finite, (ii) the process is reversible and (iii) even if SOC is increased there may be changes in the fluxes of other greenhouse gases, especially nitrous oxide (N_2O) and methane. Removing land from annual cropping and converting to forest, grassland or perennial crops will remove C from atmospheric CO_2 and genuinely contribute to climate change mitigation. However, indirect effects such as conversion of land elsewhere under native vegetation to agriculture could negate the benefit through increased CO_2 emission. Re-vegetating degraded land, of limited value for food production, avoids this problem. Adding organic materials such as crop residues or animal manure to soil, whilst increasing SOC, generally does not constitute an additional transfer of C from the atmosphere to land, depending on the alternative fate of the residue. Increases in SOC from reduced tillage now appear to be much smaller than previously claimed, at least in temperate regions, and in some situations increased N_2O emission may negate any increase in stored C. The climate change benefit of increased SOC from enhanced crop growth (for example from the use of fertilizers) must be balanced against greenhouse gas emissions associated with manufacture and use of fertilizer. An over-emphasis on the benefits of soil C sequestration may detract from other measures that are at least as effective in combating climate change, including slowing deforestation and increasing efficiency of N use in order to decrease N_2O emissions.

During the last decade much has been written about the possibility of slowing climate change through the sequestration of carbon (C) in soil. In view of the large quantity of C held in organic matter in the world's soils, it is entirely appropriate to consider how its management might either mitigate or worsen climate change. It is estimated that the global stock of soil organic C (SOC) is in the range 684–724 Pg to a depth of 30 cm and 1462–1548 Pg to a depth of 1m (Batjes, 1996). Thus the quantity of SOC in the 0–30 cm layer is about twice the amount of C in atmospheric carbon dioxide (CO_2) and three times that in global above-ground vegetation. It was estimated in the Fourth Assessment Report (AR4) of the Intergovernmental Panel on Climate Change (IPCC) that the annual release of CO_2 from deforestation (coming from both vegetation and soil) is currently some 25% of that from fossil fuel burning (IPCC, 2007a). The large number of publications on soil C sequestration indicates the degree of interest, at least in the scientific community, of combating climate change by increasing the quantity of C stored in soil and vegetation, to some extent seeking to reverse the current trend. To quote just two examples: Freibauer *et al.* (2004) give quantitative estimates of the potential for soil C sequestration through changes in the management of agricultural soils in Europe; Smith *et al.* (2007) give global estimates for greenhouse gas (GHG) mitigation within agriculture, taking account of wider considerations in addition to soil C sequestration. Carbon sequestration removes carbon, in the form of CO_2, either directly from the atmosphere or at the conclusion of combustion and industrial processes. One type of sequestration is the long-term storage of carbon in trees and plants (the terrestrial biosphere), commonly referred to as terrestrial sequestration. CO_2 removed from the atmosphere is either stored in growing plants in the form of biomass or absorbed by oceans. Sequestering carbon helps to reduce or slow the build up of CO_2 concentrations in the atmosphere. Increases in atmospheric CO_2 concentration may be generating increases in average global temperature and other climate change impacts. Although some of the effects of increased CO_2 levels on the global climate are uncertain, most scientists agree that doubling atmospheric CO_2 concentrations may cause serious environmental consequences. Rising global temperatures could raise sea levels, change precipitation patterns and affect both weather and climate conditions.

Using trash blankets is one of the best methods of increasing Carbon Sequestration.

Benefits of trash blankets

- Greatly reduced soil and water run-off
- Increased infiltration and moisture conservation
- Weed suppression
- Decreased agricultural chemical run-off
- Improved soil health (Increased organic matter, increased micro organism activity, nutrient recycling)
- Reduced soil compaction

Carbon sequestration for mitigating climate change

Lal (2004) implies soil carbon sequestration is the transfer of atmospheric CO_2 into long-lived soil pools and storing it securely so that it may not be reemitted back instantly. It means increasing soil organic and inorganic carbon stocks through judicious land use with best recommended management practices. The global soil carbon pool of 2500 gigatons (Gt) include 1550 Gt of soil organic carbon and 950 Gt of soil inorganic carbon (Lal, 2004). Depletion of soil organic carbon (SOC) degrade soil quality, reduces biomass productivity, and adversely impact water quality, and this depletion increase the cause of global warming (Lal, 2004). Cultural practices like no-tillage improves carbon sequestration however it also increases N_2O emissions. Beside this, carbon stored through no-tillage is released by single ploughing due to its labile quality (Stockfisch *et al.,* 1999) that showed that removal of GHGs from the atmosphere through carbon sequestration is limited.

Improving and sustaining SOC particularly in rainfed agro-ecosystem is a major agronomic challenge. Their concentration improves or is maintained at a higher level in the soil with INM practices with the application of organics in conjunction with the fertilizers (Srinivasarao *et al.*, 2013). The SOC stock in the soil profile varies with soil order with more SOC accumulate in vertisol > Inceptisol > Alfisol > Aridisol. Their rate of depletion ranges from 0.15 Mg C/ ha/yr under rice system to 0.92 Mg C/ ha/yr in groundnut-finger millet system (Srinivasarao *et al.*, 2014). To arrest this depletion, carbon input of 1.10 – 3.47 Mg C/ ha/yr is required as a maintenance dose. Further, potential of tropical soils to sequester more carbon can be harnessed by identifying appropriate production systems and management practices for sustainable development and improved livelihood in the tropics (Srinivasarao *et al.*, 2014).

Biomass recycling

Indian agriculture produce around 500 550 million tonnes (Mt) of crop residues annually, which can be used as animal feed, soil mulch, manure, thatching for rural homes and fuel for domestic and industrial purposes. However, large portion of around 90-140 Mt residues burnt on-farm annually to clear the field for the next succeeding crop (NAAS, 2012). Burning causes release of smoke, deleterious particles, emission of green house gases, and loss of plant nutrients like N, P, K, S, and carbon from the soil, which are beneficial for soil health.

A large amount of energy is used in cultivation and processing of crops like sugarcane, food grains, vegetables and fruits, which can be recovered by utilizing residues for energy production (Srinivasarao *et al.*, 2014). This can be a major strategy of climate change mitigation by avoiding burning of fossil fuels and recycling crop residues (Venkateswarlu, 2010). Lack of availability equipment that helps to incorporate soil is one of the major reasons for biomass wastage in India. Other issues include high labour and transport costs that cause lack of interest in utilizing the biomass. Many technologies like briquetting, anaerobic digestion, vermicomposting and biochar etc exist, but they have not been commercially exploited among farmers and growers (Srinivasarao *et al.*, 2014). CRIDA (2014) suggest biochar application to soil improves soil properties and crop yield.

Application of biochar for soil carbon sequestration

The current availability of biomass in India (2010-2011) is estimated at about 500 million ton/ year. Studies sponsored by the Ministry of New and Renewable Energy, Govt. of India have estimated surplus biomass availability of about 120-150 million ton/annum (MNRE, 2009). Of this, about 93 million ton of crop residues are burnt each year. Generation of crop residues is highest in Uttar Pradesh (60 million ton) followed by Punjab (50 million ton) (IARI, 2012). Efficient utilization of this biomass by converting it as a valuable source of soil amendment is one approach to manage soil quality, fertility, mitigate GHGs emission and increase carbon sequestration (Srinivasarao *et al.*, 2013b).

Biochar is a fine-grained, carbon-rich, porous product remaining after plant biomass has been subjected to thermo-chemical conversion process (pyrolysis) at low temperatures (~350–600°C) in an environment with little or no oxygen. Biochar is not a pure carbon, but rather mix of carbon (C), hydrogen (H), oxygen (O), nitrogen (N), sulphur (S) and ash in different proportions. Biochar is a fine-grained charcoal high in organic carbon and largely resistant to decomposition. It is produced from pyrolysis of plant and waste feed stocks. As a soil amendment, biochar creates a recalcitrant soil carbon pool that is carbon-negative, serving as a net withdrawal of atmospheric carbon dioxide

Fig. 1: Biochar

stored in highly recalcitrant soil carbon stocks. Char-amended soils have shown 50 - 80 percent reductions in nitrous oxide emissions and reduced runoff of phosphorus into surface waters and leaching of nitrogen into groundwater. As a soil amendment, biochar significantly increases the efficiency of and reduces the need for traditional chemical fertilizers, while greatly enhancing crop yields. The carbon in biochar resists degradation and can hold carbon in soils for hundreds to thousands of years.

Biochar has a condensed aromatic structure that makes it a stable solid rich in carbon content which is known to be highly resistant to microbial decomposition, thus it can be used to lock carbon in the soil. Biochar application has received a growing interest as a sustainable technology to improve highly weathered or degraded tropical soils (Lehmann and Rondon, 2006; Ogawa *et al*., 2006; and Woolf, 2008). Biochar has capacity to reduce N_2O emission from soil which might be due to inhibition of either stage of nitrification and/or inhibition of denitrification, or encouragement of the decrease of N_2O, and these impacts could occur simultaneously in a soil (Berglund *et al*., 2004; DeLuca *et al*., 2006). Several workers have reported that applications of biochar to soils have shown positive responses for yield of several crops (Chan *et al*., 2008; Chan and Xu, 2009; Major *et al*., 2009 and Spokas *et al*., 2009). Similarly, biochar has also been found to have significant positive interaction with plant growth promoting rhizobacteria for improving total dry matter yield of rice (Singh *et al*., 2015).

Resource conservation based technologies

The key resource conservation-based technologies are in situ moisture conservation, rainwater harvesting and recycling, efficient use of irrigation water, conservation agriculture, energy efficiency in crop production and irrigation and use of poor quality water. Other strategies include characterization of bio-physical and socio-economic resources utilizing GIS and remote sensing; integrated watershed development; developing strategies for improving rainwater use efficiency through rainwater harvesting, storage, and reuse; contingency crop planning to minimize loss of production during drought / flood years (Kapoor, 2006).

Integrated Nutrient Management (INM) and Site-Specific Nutrient Management (SSNM) techniques have the potential to mitigate effects of climate change by reducing carbon dioxide emissions and improving crop yield. One of the key emerging technologies to reduce GHG emissions from paddy fields is the use of zymogenic bacteria, acetic acid and hydrogen-producers, methanogens, methane oxidizers, and nitrifiers and denitrifiers in rice paddies which help in maintain the soil redox potential in a range where both nitrous oxide and methane emissions are low (Venkateswarlu and Shanker, 2009). The application of urease inhibitor, hydroquinone (HQ), and a nitrification inhibitor, dicyandiamide (DCD) together with urea also is an effective technology for reducing nitrous oxide and methane from paddy fields. Use of neem-coated urea is another simple and cost effective technology.

Conservation agriculture

In the climate change scenario with nine billion mouths to feed by 2050, conservation agriculture is a key to future food security. It observes three major principles / pillars as

- Direct seeding of growing crops without mechanical field preparation and with minimal soil disturbance since the harvest of previous crop.
- A permanent coil cover to protect soil against deleterious effect of exposure to rain and sun; addition of micro and macro organisms in the soil with a constant supply of 'food'; and alter the microclimate in the soil for optimal growth and development of soil organisms, including plant roots.
- Crop rotation provides different root depths exploring different soil layers for nutrients, and also provides diverse 'diet' to the soil microorganisms.

With the increasing organic matter under conservation agriculture, soil can retain carbon from carbon dioxide and act as a carbon repository for longer time period fighting climate change problem. It provides small-scale farmers with diversification opportunities, reduced labour requirements for tillage, land preparation and weeding. More time availability offers real opportunities for diversification options such as for example poultry farming or on-farm sales of produce, or other off-farm small enterprise developments.

Conservation agriculture lowers farm power. Reduced requirements for farm power and energy for field production by up to 60% compared to conventional farming. Additionally equipment investment, particularly the number and size of tractors, is significantly reduced. Zero tillage reduces soil compaction causing more carbon to retain in the soil, thereby improving soil health.

Conservation agriculture for carbon sequestration

Conservation agriculture, in broader sense includes all those practices of agriculture, which help in conserving the land and environment while achieving desirably sustainable yield levels. Tillage is one of the important pillars of conservation agriculture which disrupts inter dependent natural cycles of water, carbon and nitrogen. Conservation agriculture enhances the carbon sequestration potential of the soil, which helps in maintaining and/ or enhancing the carbon content of the soil. There are several reports on the influence of conservation agricultural management practices of tillage, residue recycling, application of organic manures, green manuring and integrated use of organic and inorganic sources of nutrients, soil water conservation treatments, integrated pest management, organic farming, etc., on soil quality. Improved soil quality parameters create additional muscle power to soil to combat the ill effects of climate change (Srinivasarao *et al.*, 2013a).

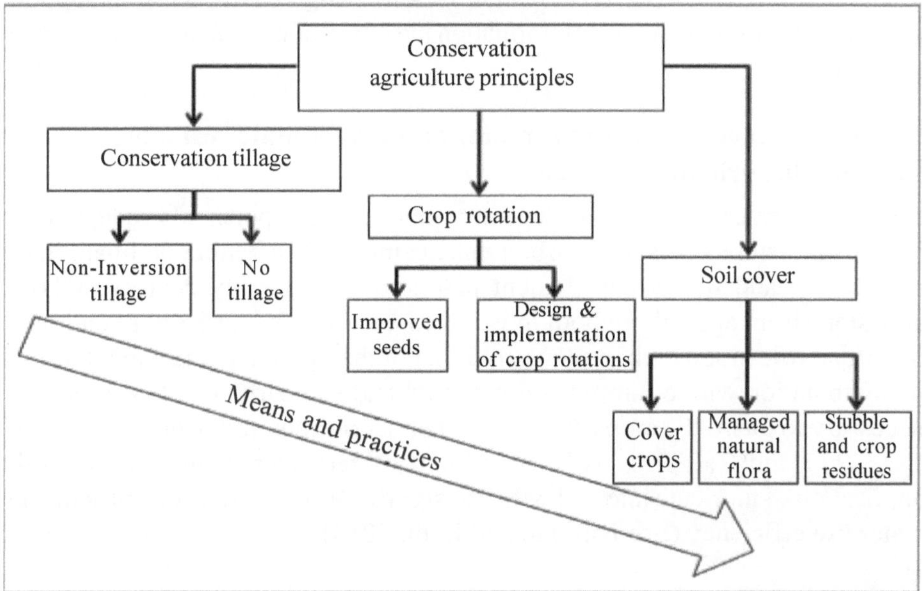

Fig. 2: Three principal components of conservation agriculture (Srinivasarao *et al.*, 2013a)

Climate resilient sustainable agriculture

Climate Resilient Sustainable Agriculture is an initiative that revolves around the concept and practices of sustainable agriculture. It represents an effort to incorporate in our work the new challenges posed by climate change and its impacts on poor people's lives. It is based on the identification of the major risks and challenges local communities face, and/or are likely to face in the near future, and on the design and implementation of site-specific adaptation strategies aimed at reducing vulnerabilities and increasing the resilience of the smallholder production systems. (www.actionaid.org).

Adaptation strategies for resilient agro-climatic system

Chary *et al*., (2013) gave the main adaptation strategies which include: (a) development of new genotypes, (b) intensifying search for genes for stress tolerance across plant and animal kingdom, (c) development of heat and drought tolerant genotypes, (d) attempting conversion of C_3 plants to C_4 plants, (e) development of new land use systems, (f) to explore opportunities for restoration of soil health, (g) use of multipurpose adapted livestock species and breed, (h) development of spatially differentiated operational contingency plans for weather related risks (i) supply management through market and non-market interventions in the event of adverse supply changes, (j) enhancement in research on applications of short, medium and long range weather forecasts for reducing production risks, and (k) development of knowledge based decision support systems for translating weather information into operational weather management sources.

Soil management strategies for adaptation of climate resilient sustainable agriculture system

Better land, water and tillage management practices have potential in adaptations towards moisture stress, soil carbon management and a better combination of soil, tillage and water management practices considerably decrease GHGs emissions from agriculture both in irrigated and dry land agriculture and play important role in soil carbon sequestration. For that reason, management of soil in combination with optimum soil moisture and soil carbon management are necessary to guard the crops during weather aberrations and overall decline of CO_2, N_2O and CH_4 from soil. Placement of fertilizer materials and split application of nutrients into soil will also significantly improve the nutrient and water use efficiency (Srinivasarao and Rani, 2013).

Soil fertility management

Proper fertility management of soil is an important part of soil conservation technique. The conservation of soil implies utilization without misuse so as to maintain higher crop productivity as well as improving simultaneously the environmental quality. Soil conservation, in practice refers to the protection of surface deposits as well as subsurface deposits. Some of the methods used to maintain soil fertility consist of the in-situ moisture conservation practices, balanced use of fertilizer nutrients, efficient use of fertilizers, site specific nutrient management, taking care of nutrient needs of cropping systems, optimum fertilizer rates, following the principles of crop rotation, mixed cropping, use of biofertilizers, sowing of cover crops, use of organic manures, green manuring and efficient utilization of either rainwater or irrigation water given from harvested rain water etc (Nayyar and Sudhir, 2009).

Site- specific nutrient management

Site- specific nutrient management (SSNM) can be prescriptive, corrective or a combination of both (Dobermann and Cassman, 2004). SSNM has potential to mitigate adverse effects of climate change (Srinivasarao *et al.*, 2008). SSNM approach follows the principles of participatory soil sampling and development of soil health cards, which is very helpful for recommendation of nutrients. Based on the expected crop demand, targeted yield, soil test data, fertilizer efficiency and crop grown in each field, SSNM sheet could be developed in each farmer's field. With this, farmers will invest only on deficient nutrients and omit nutrient application which was in sufficient range in soils. Thus, it reduces the input cost and improves the use efficiency of nutrients significantly. Nitrogen content of the standing crop is measured by employing certain diagnostic tools such as chlorophyll meter and leaf colour chart, which helps in deciding the most suitable time for application of nitrogen during period of crop growth. Various benefits of SSNM practice include lowering input cost, higher nutrient use efficiency, higher water use efficiency and reducing GHGs particularly N_2O (Mohanty *et al.*, 2009).

Integrated nutrient management

One of the very useful findings of the long term fertilizer experiments under different agro-climatic conditions points out that addition of organic manures along with chemical sources of nutrients (NPK), results in higher yields over period of time as compared to a decrease in yield over time when only chemical fertilizers were applied, besides improving organic carbon content of the soil both in normal rainfall years and deficient rainfall years. INM also provides overall resilience to soil system against mid-season droughts (Srinivasarao and Rani, 2013).

Conclusions

Global soil and water resources are limited, exposed to exploitation and negligence and strongly attached with the processes that also govern climate change. Soil quality is the key that determines the resilience of crop production under changing climate. A number of interventions are designed to build carbon content of soil, prevent soil loss due to erosion and improve water holding capacity of soils, all of which will collectively build resilience in soil. Mandatory soil testing is required to be done in all the farmers' fields, to make sure balanced use of chemical fertilizers, improved methods of fertilizer application, matching with crop requirement to reduce nitrous oxide emission. Conservation agriculture, biochar application to the soil, watershed management, enhancing water use efficiency, precision agriculture and techniques of carbon sequestration in soils and ecosystems, site specific nutrient management, and integrated nutrient management are proven technologies of soil management for climate resilient agriculture.

References

Aggarwal, P. K. 2003. Impact of climate change on Indian agriculture. *J. Plant Biology* 30(2), 189–198.

Berglund, L.M., DeLuca, T.H. and Zackrisson, O. 2004. Activated carbon amendments to soil alters nitrification rates in Scots pine forests. Soil *Biology & Biochemistry*, 36: 2067-2073.

Biochars as soil amendments. *Australian Journal of Soil Research*, 46: 437-444.

Chan, K.Y. and Xu, Z. 2009. Biochar: nutrient properties and their enhancement. In: Biochar for environmental management (J. Lehmann and S. Joseph eds.), *Science and Technology, Earthscan*, London. pp 67-84.

Chan, Y., Van Zwieten, L., Meszaros, I., Downie, A. and Joseph, S. 2008. Using poultry litter

Chary, G. R., Srinivasarao, Ch., Srinivas, K., Marurhi Sankar., G.R., Kumar, R.N. and Venkateswarlu, B. 2013. Adaptation and Mitigation Strategies for Climate Resilient Agriculture. Central Reasearch Institute for Dryland Agriculture, ICAR, Hyderabad, India.

CRIDA. 2014. Annual Report 2013-14, Central Research Institute for Dryland Agriculture, Hyderabad, India. pp 37-38.

DeLuca, T.H., MacKenzie, M.D., Gundale, M.J. and Holben, W.E. 2006 Wildfire-produced charcoal directly influences nitrogen cycling in forest ecosystems. *Soil Science Society America Journal*, 70: 448-453.

Dobermann, A. and Cassman, K.G. 2004. Environmental dimensions of fertilizer nitrogen. What can be done to increase nitrogen use efficiency and ensure global food security? In: Agricultural and Nitrogen Cycle: Assessing the Impacts of Fertilizer Use of Food production (Mosier, A.R., Syers, J.K. and Freney, J.R., Eds), Scope 65, Paris, France, pp.261-278.

Drinkwater, L.E., Wagoner, P. and M. Sarrantonio, M. 1998. "Legume-Based Cropping Systems Have Reduced Carbon and Nitrogen Losses," *Nature* 396: 262–65.

FAO. 2014. Target and indicators for the post 2015 development agenda and the sustainable development goals. Available at www. Fao.org/file admin/user_upload/post2015/ Target_and_indicator_RBA_joint_proposal.pdf

Flessa, H. and Beese, F. 2000. Laboratory estimates of trace gas emissions following surface application and injection of cattle slurry. *Journal of Environmental Quality* 29:262– 268.

Gebbers, R.; Adamchuk, V.I. Precision agriculture and food security, 2010. Science, 327: 828–831

IARI 2012. Crop residues management with conservation agriculture: Potential, constraints and policy needs. Indian Agricultural Research Institute, New Delhi, 32 p.

IPCC. 2007a. Climate Change: The Physical Science Basis. Contribution of Working Group I to the Fourth Assessment Report of the Intergovernmental Panel on Climate Change. Eds: S.Solomon, D. Qin, M. Manning, Z. Chen, M. Marquis, K.B. Averyt, M. Tignor and H.L. Miller. Cambridge University Press, Cambridge, UK. 996 p.

IPCC. 2001. Climate Change 2001: The Scientific Basis. Contribution of Working Group I to the Third Assess-ment Report of the Intergovernmental Panel on Climate Change [Houghton, J.T.,Y. Ding, D.J. Griggs, M. Noguer, P.J. van der Linden, X. Dai, K. Maskell, and C.A. Johnson (eds.)]. Cambridge University Press, Cambridge, United Kingdom and New York, NY, USA, 881 pp.

IPCC. 2007b. Contribution of Working Group 1 to the Fourth Assesment Report of the Inter Governmenttal Panel on Climate Change. pp 996. Solomom S, Qin D, Manning M, Chen Z, Marquis M, Averyt KB, Tignor M and Miller HC (Eds).Cambridge University Press, Cambridge, UK and New York, USA.

Kapoor, A. 2006. Mitigating natural disasters through preparedness measures. Proc. International Conference on Adaptation to Climate Variability and Change, 5-7 January 2006, New Delhi, organized by Institute for Social and Environmental Transition and Winrock International India. 184 p.

Kotschi, J., and Müller-Sämann, K. 2004. The Role of Organic Agriculture in Mitigating Climate Change: A Scoping Study. IFOAM, Bonn, Germany. Available at http://www.ifoam.org/press/positions/Climate_study_green_house-gasses.html. Accessed February 12, 2009.

Kumar, K.S.K. and Parikh, J. 2001. Indian agriculture and climate sensitivity. Glob. *Environ. Chang*. 11(2): 147-154.

Lal, R., 2004. Soil carbon sequestration impacts on global climate change and food security. Science 304, 1623–1627.

Lehmann, J. and Rondon, M. 2006. 'Bio-char soil management on highly weathered soils in the humid tropics', in N. Uphoff (ed) Biological Approaches to Sustainable Soil Systems, CRC Press, Boca Raton, pp 517–530.

Major, J., Steiner, C., Downie, A. and Lehmann, J. 2009. Biochar effects on nutrient leaching. In: Biochar for environmental management (J. Lehmann and S. Joseph eds.), Science and Technology, Earthscan, London. pp 271-287.

MNRE, 2009. Ministry of New and Renewable Energy Resources, Govt. of India, New Delhi. www.mnre.gov.in/biomassrsources.

Mohanity, S.K., Singh, T.A. and Aulakh, M.S. 2009. Nitrogen In: Fundamental of Soil Science. Second edition: Indian Society of Soil Science.

NAAS, 2012. Management of crop residues in the context of conservation agriculture. Policy Paper No. 58, National Academy of Agricultural Sciences, New Delhi. pp. 12.

Nayyar, V.K. and Sudhir, K. (2009) Soil Fertility Management. In: Fundamental of Soil Science. Second edition: Indian Society of Soil Science.

Ogawa, M., Okimori, Y. and Takahashi, F. 2006. Carbon sequestration by carbonization of biomass and forestation: Three case studies. *Mitigation and Adaptation Strategies for Global Change*, 11: 429-444.

Pandey, S, *Behura,* D.D, *Villano,* R. and *Naik,* D. 2000. *Economic costs* of drought and farmers' coping mechanisms: A study of rainfed rice systems in eastern India. Discussion Paper Series 39, IRRI, Phillipines. 35 p.

Pathak H, Aggarwal, P. K. and Singh, S. D. 2012. Climate Change Impact, Adaptation and Mitigation in Agriculture: Methodology for Assessment and Applications. Indian Agricultural Research Institute, New Delhi. pp xix + 302.

Lal, R. 2014. Climate Strategic Soil Management, Challenges, 5, 43-74 doi:10.3390/challe5010043

Samra, J.S., Sharma, U.C. and Dadhwal, K.S. 2009 Soil Erosion and Soil Conservation. In: Fundamental of Soil Science. Second edition: Indian Society of Soil Science.

Scialabba, N.E. and Müller-Lindenlauf, M. 2010.Organic agriculture and climate change. *Renewable Agriculture and Food Systems*: 25(2); 158–169.

Singh, A., Singh, A.P., Singh, S.K. and Singh, C.M. 2015. Effect of biochar along with plant growth promoting rhizobacteria (PGPR) on growth and total dry matter yield of rice. *Journal of Pure and Applied Microbiology*. 9(2):1627-1632

Smith, P., Martino, D., Cai, Z., Gwary, D., Janzen, H.,Kumar, P., McCarl, B., Ogle, S., O'Mara, F., Rice, C., Scholes, B., and Sirotenko, O..2007. Agriculture. In B. Metz, O.R. Davidson, P.R. Bosch, R. Dave, and L.A. Meyer (eds). Climate Change 2007. Mitigation. Contribution of Working Group III to the Fourth Assessment Report of the Intergovernmental Panel on Climate Change. Cambridge University Press, Cambridge, UK.

Spokas, K.A., Koskinen, W.C., Baker, J.M. and Reicosk, D.C. 2009. Impacts of woodchip biochar additions on greenhouse gas production and sorption/degradation of two herbicides in a Minnesota soil. *Chemosphere*, 77: 574-581.

Srinivasarao, Ch, Venkateswarlu, B, Rattan Lal, Singh, A.K. and Kundu, S. 2013. Sustainable management of soils of dryland ecosystems of India for enhancing agronomic productivity and sequestering carbon. In: 'Advances in Agronomy' (Ed. Donald L. Sparks). Vol. 121, pp. 253-329. (Academic Press, Burlington).

Srinivasarao, Ch., and Rani, Y.S. 2013. Soil Management Strategies for Adaptation and Mitigation. In: In Adaptation and Mitigation Strategies for Climate Resilient Agriculture. Central Reasearch Institute for Dryland Agriculture, Hyderabad.

Srinivasarao, Ch., Chary G R. and Pravin B. T. 2013a. Conservation Agriculture Practices for Soil Carbon Sequestration and Sustainbility of Rainfed Production Systems of India. In: Adaptation and Mitigation Strategies for Climate Resilient Agriculture. Central Reasearch Institute for Dryland Agriculture, ICAR, Hyderabad, India.

Srinivasarao, Ch., Gopinath, K.A., Venkatesh, G., Dubey, A.K., Wakudkar, H, Purakayastha, T.J., Pathak, H., Pramod Jha, Lakaria, B.L., Rajkhowa, D.J., Mandal,S., Jeyaraman, S., Venkateswarlu, B., and Sikka, A.K. 2013b. Use of Biochar for Soil Health Enhancement and Greenhouse Gas Mitigation in India: Potential and Constraints, Central Research Institute for Dryland Agriculture, Hyderabad, Andhra Pradesh. 51p.

Srinivasarao, Ch., Vittal K.P.R., Gajbhiye P.N., Sumanta Kundu and Sharma, K.L. 2008. Distribution of micronutrients in soils in rainfed production systems of India. *Indian Journal of Dryland Agricultural Research and Development*, 23:29-35

Srinivasrao,Ch, Gopinath, K.A., Venkatesh, G., and Jain, M.P. 2014. Climate change adaptation and mitigation strategies in rainfed agriculture. In Souvenir: National Seminar on 'Technologies for sustainable production through climate resilient agriculture', JNKVV, Jabalpur, August8-9, 2014.L-13.

Status of Indian Agriculture. 2013. Directorate of Economics and Statistics. Department of Agriculture and Cooperation. Ministry of Agriculture, Government of India, New Delhi.

Stockfisch, N., Forstreuter, T., and Ehlers, W. 1999. Ploughing effects on soil organic matter after twenty years of conservation tillage in Lower Saxony, Germany. *Soil and Tillage Research* 52:91–101.

Venkateswarlu, B. 2013. Climate Change Scenario in India and its Impact on Agroecosystems. In: Adaptation and Mitigation Strategies for Climate Resilient Agriculture. Central Reasearch Institute for Dryland Agriculture, Hyderabad.

Venkateswarlu, B. 2010. Climate change: Adaptation and mitigation strategies in rainfed agriculture. The 21st Dr. S.P. Raychaudhuri Memorial Lecture, delivered on 25th September, 2010 at BAU, Ranchi.

Venkateswarlu, B. and Rao, V.U.M. 2014. Agriculture in Madhya Pradesh: Impacts of climate change and adaptation strategies. Available at www.skmccc.net.

Venkateswarlu, B. and Shanker, A.K. 2009. Climate change and agriculture: Adaptation and mitigation strategies. *Indian J. Agron.* 54(2): 226-230.

Woolf, D. 2008. Biochar as a soil amendment: A review of the environmental implications. (Accessed online at http://orgprints.org/13268/1/Biochar_ as_ a _soil_ amendment_- _a_ review. pdf).

www.actionaid.org, Climate Resilient Sustainable Agriculture, Experiences from ActionAid and its partners.

9

Abiotic Stress Management in Pulse Crops

Madhuri Gupta, Pankaj Kumar, Jitender Singh, Shivani Khanna and Mini Sharma

Pulses belongs to the family *Fabaceae* (earlier known as *leguminosae*) comprises more than 600 genera and about 18,000 species of cultivated plants. It is the second largest family after *Poaceae* (earlier known as *Gramineae*), in terms of food and vegetable protein source and of fodder. The sub-family *Papilionoideae* consist of 480 genera and about 12,000 species, of which only a few species are cultivated for human nutrition. Endowed with excellent food and fodder qualities, these crops also restore soil fertility by scavenging atmospheric nitrogen, adding organic matter, enhancing phosphorous availability and improving physical, chemical and biological properties of soil (Graham and Vance, 2003; Dita *et al.*, 2006).

The word 'pulse' is derived from latin word 'puls' meaning pottage i.e. seeds used to make porridge or thick soup. Pulses or grain legumes in general are indispensible source of supplementary proteins to daily vegetarian diets (Table 1) these are regarded as "poor man's meat". Pulse proteins are chiefly globulins and contain low concentrations of sulphur containing amino acids such as methionine and cysteine, but higher concentration of lysine than cereals, pulses provide a perfect mix of essential amino acids with high biological value and also contain higher calcium and iron than cereals.

Table 1: Nutritive value of Pulse (Singh *et al.*, 2015)

Constituents	Magnitudes
Protein	>20 %
Carbohydrate	55-60 %
Fat	>1.0 %
Fibre	3.2 %
Phòsphorus	300-500 mg/100 g
Iron	7-10mg/100 g
Vitamin C	10-15 mg/100 g
Calcium	69-75mg/100g
Calorific value	343
Vitamin A	430-489 IU

A variety of pulse crops are grown in the world among which India is the largest producer, consumer and the largest importer of pulses in the world as shown in (figure 1). India accounts for about 33 percent of world area and about 22 percent of world production. About 90 percent of the total global area under pigeonpea, 65 percent under chickpea and 37 percent under lentil is contributed by India, with a corresponding share of production of 93 percent, 68 percent and 32 percent, respectively (Reddy, 2004). Ironically, the country's pulse production has been hovering around 14–15 MT, coming from a near stagnated area of 22– 23 Mha, since 1990–91 (Singh *et al.*, 2013). Major areas under pulses are in the States of Madhya Pradesh (20.3%), Maharashtra (13.8%), Rajasthan (16.4), Uttar Pradesh (9.5%), Karnataka (9.3%), Andhra Pradesh (7.9%), Chhattisgarh (3.8%), Bihar (2.6%) and Tamil Nadu (2.9%). Pulse productivity which was 441 kg/ha in 1950 increased up to 689 kg/ha during 2011, registering 0.56% annual growth rate (Singh *et al.*, 2015). Among different pulses, the leading contributors are chickpea and pigeonpea. In India, pulses are grown in around 24-26 million hectares area producing 17-19 million tonnes of pulses annually accounts for over one third of the total world area and over 20 per cent of total world production. India primarily produces bengal gram (chickpeas), red gram (tur), lentil (masur), green gram (mung) and black gram (urad).

For majority of vegetarian population in India, pulses are the major source of protein. Pulses and pulse crop residues are also major sources of high quality livestock feed. The pulses are grown across the country in which Chickpea to total pulses area is 35 %, Pigeonpea 16%, Urdbean 12%, Moongbean 13%, Lentil 7%, Fieldpeas 3%, Horsegram 2% and Lathyrus 2%. Chickpea is grown by 22 states and 02 UTs of Daman & Nagar, Haveli and Delhi, Pigeonpea by 24 states and 03 UTs of Andaman & Nicobar Island, Daman & Nagar, Haveli and Delhi, Urdbean by 20 states and 01 UT i.e. Daman & Nagar, Moongbean by 19 states, Lentil by 15 states, Fieldpeas by 18 states and Lathyrus by 5 states (Anon., 2012). Pulses got strengthened in 1967 with the initiation of All-India Co-ordinated Research Improvement Programme. Through this programme, the varieties suitable for Northern Hills Zone, North West Plain Zone, North East Plain Zone, Central Zone and South Zone are evolved. Pulses are grown since ages in different parts of the country and these are well suited to different environments and fit in various cropping systems owing to their wide adaptability, low input requirements, fast growth, nitrogen fixing and weed smothering ability. Harvesting threshing and storage of pulses is also kept in mind carefully as Moongbean, urdbean, fieldpeas, lentil, chickpea, cowpea, mothbean, khesari and horsegram are harvested manually using sickle. Pigeonpea crop is harvested by sickle as well as gadasa. At maturity of pigeonpea crop, the field is irrigated and after 3-4 days crop is uprooted by the farmers. This practice helps farmers to sow the seeds of succeeding crop in time. Most of the pulse crops are dried in the fields for 2-3 days and thereafter, threshed by normal thresher and this is also done by beating the dried plant by heavy sticks. Pigeonpea grains are separated by beating the plants on some hard objects like stone and heavy wood. To avoid storage losses in pulses, these should have 9-10% moisture at the time of placing in storage. Farmers, who have small quantity of pulses, store them in storage bins (Tin made) mixing them with neem leaves. If the quantity is big, then it may be kept in scientific storage of Central Warehousing Corporation (CWC) and Food Corporation of India (FCI) to minimize the storage losses by pests.

Chickpea contribution at first place with 26.85 and 38.81 per cent share with respect to area and production in India, followed by pigeonpea and mungbean respectively. Lentil occupies only 4.94 % area and corresponding contribution in national pulse production is 6.96% (Reddy, 2009) as shown in Table 2.

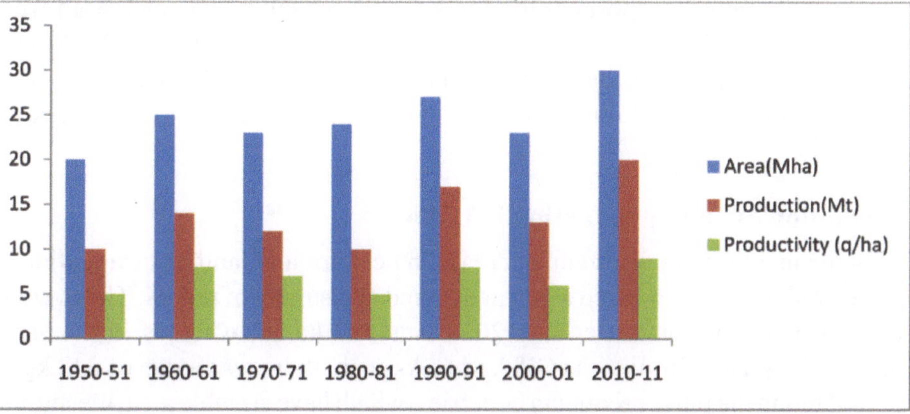

Fig. 1: Decadal growth in All-India Area, Production and Productivity of pulses (Singh *et al.*, 2015).

Name of crops	Area (Mha)	Area (% of Total)	Production (MT)	Production (% of Total)
Chickpea	5.81	26.85	5.69	38.81
Pigeonpea	3.62	16.73	3.07	20.94
Mungbean	2.98	13.77	1.61	10.98
Urdbean	3.18	14.70	1.46	9.96
Lentil	1.07	4.94	1.02	6.96
Other Pulses	6.30	29.11	1.75	11.94
Total	21.64	26.85	14.66	100.00

Availability of pulses

According to research report on marketing of pulses in India (2011-12), the share of pulses in the total food grain production of the country fell from 17 percent in 1950-51 to about 6 percent in 2008-09. The same has been reflected on the net availability of cereals and pulses. The net availability of pulses during the same period came down from 61 to 37 grams/ person/ day against the ICMR norms of 40 grams/ person/ day. During the same period the consumption of cereals increased from 334 to 407 grams/ person/ day.

The situation has shown some signs of recovery during the last decade. The availability of pulses during last decade i.e. between 1999- 2000 (TE) and 2009-10 (TE) has increased from 14.22 MT to 17.43 MT. During the same period, net availability of pulses increased from 35.47 grams per person per day to 38.10 grams per person per day. Owning to continuous increase in population, an increase of about 23 percent in total availability of pulses could result into an increase of just 7 percent in per capita net availability of pulses. Imports of pulses have played an important role in the increase in net availability of pulses. The contribution of imports in total availability of pulses has increased from 4.27 percent during 1999- 2000 (TE) to 16.75 percent during 2009-10 (TE). More than 72 percent of the increase in availability of pulses is on account of increase in imports and remaining 27 percent is contributed by an increase in production.

Promotion of pulse production in India

With the increase in pulses and decrease in per capita of land and stagnation in pulse production has created a gap in demand and supply of pulses. The average yield of world triennium ending 2010 was 890 kg/ha, whereas in the same period, the yield of India was 648 kg/ha. As such, there was a gap of 242 kg or 27%. Among the pulse producing countries, which have sizeable area, the highest productivity at triennium ending 2010 was of France (4219 kg / ha) followed by

Canada (1936), USA (1882), Russia (1643) and China (1596). Higher yields in these countries might be due to prevailing environmental condition and crop management practices. In India, government has made various efforts to bridge this gap like launching of Technology Mission in 1986 in order to reduce import and achieve self sufficiency in production of pulse crops covered under Mission. Pulses were brought within the purview of the Mission with the introduction of National Pulses Development Project in 1990-91(Anon., 2011).

The Government of India has introduced the National Food Security Mission in 2007-08, to increase production of pulses through area expansion and productivity enhancement in sustainable manner in the identified states. Accelerated Pulses Production Programme (A3P) is another step forward for vigorous implementation of the pulse development under the NFSM-Pulses. A3P has been conceptualized to take up the active propagation of key technologies such as Integrated Nutrient Management (INM) and Integrated Pest Management (IPM). The centrally sponsored National Pulses Development Project was implementated in 30 states/UTs in the country upto the year 2003-2004 covering 356 districts. Various schemes under the technology mission including National Pulses Development Project were merged into one Centrally Sponsored Integrated Scheme of Oilseeds, Pulses, Oil palm and Maize (ISOPOM) during 2004 for providing flexibility to the states in implementation of these schemes.

Over last 100 years, population of India has seen a fivefold increase and projections have been made by the government that with the present growth rate of 1.45%, population will increase to 1613 million by 2050, surpassing China (1417 million). The consumption requirements alone for the 1613 million people, on normative requirement of 14.60 kg pulses/capita/yr as recommended by National Institute of Nutrition, Hyderabad, works out to be 23.55 Mt, which in terms of production requirements would be about 26.50 Mt, assuming seed, feed and wastage as 12.5% of the gross output. As against total pulse requirements of 18.33 Mt for 2009–10, the domestic production is only 14.60 Mt. The annual import of 2–3 Mt provides only partial relief and checks escalation in the market price. By 2050, the domestic requirement would be 26.50 Mt (Ali and Gupta, 2012) as shown in table 3. These tasks have to be accomplished under more severe production constraints, abrupt climatic changes, abiotic stresses, emergence of new species/ strains of insect-pests and diseases, and increasing deficiency of secondary and micronutrients in the soil.

Table 3: Requirement, production and import of pulses in India (Ali and Gupta, 2012)

Year	Population (Million)	Requirement (Mt)	Production (Mt)	Import (Mt)
2000-01	1027	16.02	11.08	0.35
2004-05	1096	17.10	13.13	1.31
2009-10	1175	18.33	14.60	2.83
2020-21	1225	19.10		
2050-51	1613	26.50		

The yield levels of pulses have remained low and stagnant in comparison to other countries. Number of districts harvesting more than 0.8 or 1 t/ha yield of *kharif* pulses is very less (Anon., 2013). Situation of *rabi* pulses is better in this regard. The gap between demand and supply has been widening and has necessitated import of pulses of 2.8 million tons in 2007-08. Perusal of data presented in table revealed that Uttar Pradesh contributes significantly to the pulses production and its share to the national pulses security is 21.8% with 3.196 MT. Other two leading states are Madhya Pradesh and Rajasthan with 19.5% and 13.6% production share of India. Only eight states contributing 90 % of total pulses production. State wise contribution to the total area and production is shown in pie diagram (figure 2) respectively (Anon., 2012).

However, in recent years fluctuations in pulse yield and frequent crop failure in several regional environments caused by biotic and abiotic stresses are threatening the future expansion of the pulse industry in India. Even with the best efforts, pulses production and productivity has been stagnant. Due to the low productivity low input nature, pulses are grown as residual/alternate crops on marginal lands after taking care of food/income needs from high productivity-high input crops like paddy and wheat by most farmers.

The low priority accorded to pulse crops may be related to their relatively low status in the cropping system. As a crop of secondary importance, in many of these systems, pulse crops do not attract much of the farmer's crop management attention. In addition to this, these crops are adversely affected by a number of biotic and abiotic stresses, which are responsible for a large extent of the instability and low yields.

Pulses encounter a number of abiotic and biotic stresses during various stages and phases of their life cycle. The nature of abiotic stresses may vary depending upon species, prevailing weather conditions and the type of soil. Each type of stress hampers the growth of the plant by disturbing the normal physiology and morphology (Sultana *et al.,* 2014). For example, winter grain legumes (cool season pulses) such as chickpea, lentil, peas and faba beans, which relatively

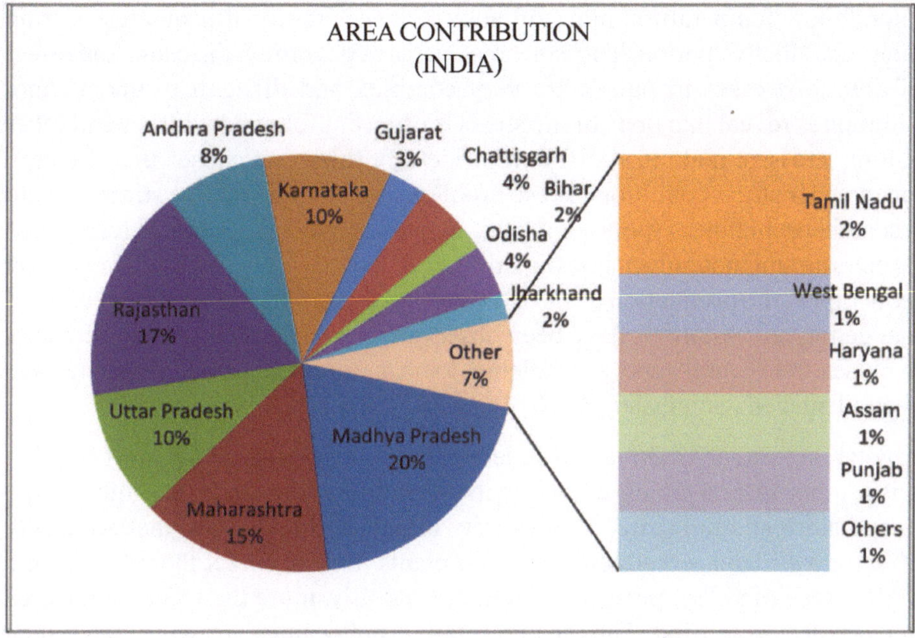

A

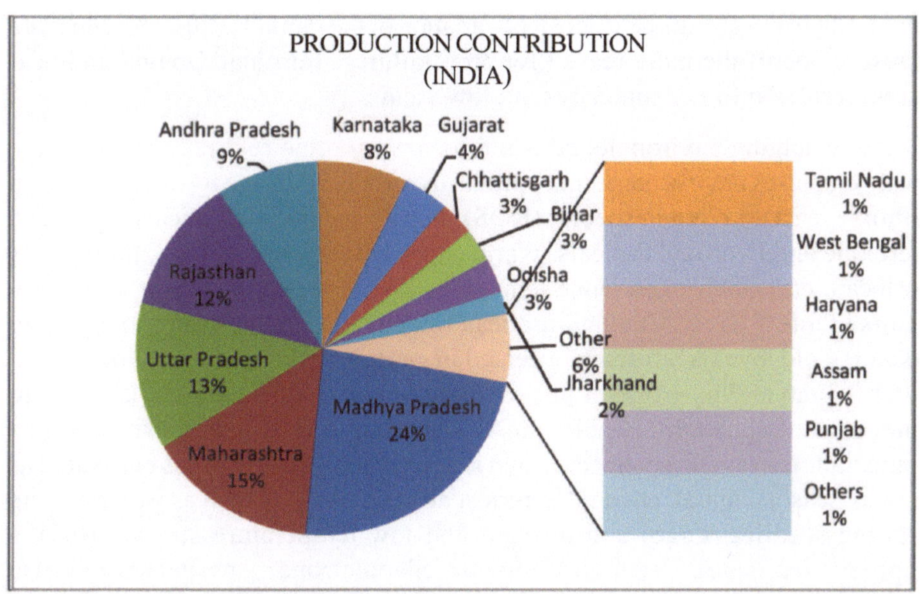

B

Fig. 2: (A) Area contributing to total pulse production in India. (B) Total production contribution of pulses in India. (Anon, 2012)

tolerate low temperature, more often experiences terminal heat stress during their reproductive period. The exact mechanisms governing the cause and effect of abiotic stresses in pulses are very complex and difficult to understand. Literatures reveal that one abiotic stress is often confounded with several other abiotic stresses, making it difficult to identify the exact causes that the crop faced under stress condition. These problems require a proactive strategy from researchers, planners, policy-makers, extension workers, market forces and farmers aiming not only at boosting the per unit productivity of land, but also at reduction in the production costs. However, recent empirical evidence suggests that genotypic variations have been observed for almost all the abiotic stresses in pulses and several genotypes tolerant to heat, drought and waterlogging have been identified (Sultana *et al.,* 2014).

Abiotic stresses can be explained as deviation from optimum production condition arising due to non-living components of environment (such as high or low temperature and moisture, high salinity or acidity in soil, etc) that adversely affect growth and reproduction of crop plants. Due to these, there has been a high degree of risk in pulses production. Presently, more than 87% of the area under pulses is rainfed. The mean rainfall of major pulse growing states such as Madhya Pradesh (MP), Uttar Pradesh (UP), Gujarat and Maharashtra is about 1,000 mm and the coefficient of variation of the rainfall is 20-25%. Moisture stress is one of the main reason for crop failures. Terminal drought and heat stress results in forced maturity with low yields.

Due to changing environmental conditions, very often referred to as 'climate change', pulses have become more prone to oxidative damage by overproduction of toxic reactive oxygen species (ROS) such as superoxide radicals, hydrogen peroxide and hydroxyl radicals. Warm (rainy) season pulses (e.g., mungbean, urdbean, etc) often experience temporary waterlogging that may vary from hours to a few days. Although urdbean tolerates excess moisture to a greater extent, yield level is adversely affected in case excess water is not drained out after 2-3 days. This group of pulses also encounter moisture deficit owing to uneven rainfall pattern. Besides these, this group is relatively sensitive to low temperature stress. Pigeonpea, which is perennial by nature but is cultivated as a rainy season annual, encounter almost all such stresses such as waterlogging (during seedling stage) and drought and low temperature stresses (during reproductive stage). Depending upon the edaphic conditions, pulses may also face stresses imposed by salinity/alkalinity (high pH), Al toxicity and sodicity (Chaudhary, 2007).

Constraints related to pulse production

Pulses are mainly grown under rainfed conditions except in few districts of Karnataka, Uttar Pradesh, Madhya Pradesh, Rajasthan and Bihar. As a consequence area under pulses and their productivity are dependent on amount and distribution of rainfall. Rainfall intensity and distribution leads to vulnerability of kharif pulses to water stagnation (oxygen stress) and that of rabi pulses to water stress. Occurrence of mid-season cold waves and terminal heat during winter season has also been causing losses to crop productivity of rabi pulses in many regions.

Drought stress

It is the most universal and significant environmental stress affecting plant growth and productivity worldwide. The optimal temperatures for the cool season pulses range between 10°C and 30°C. Therefore, understanding crop response to this stress is the basis for regulating crops appropriately and achieving agricultural water savings. There are significant differences in the tolerance of plants to drought stress depending upon intensity and duration of stress, plant species and the stage of development (Singh *et al.,* 2012). Drought stress alone may reduce seed yields by 50% in the trop-ics as it causes a series of physiological, biochemical and morphological responses of crops which finally results in low yield of crops (Baroowa *et al.,* 2012). A plant can resist drought condition through reduced water loss from aerial portions, increased water uptake from deep layers of the soil or by giving more yields at low water potentials. The maintenance of water uptake under drought condition is related to several properties concerning roots of plants such as root size and efficiency, root density and the size of xylem vessel. Many studies have been conducted on drought tolerant genes in legumes.

The genetic and molecular basis of drought resistance in legumes has been deeply dissected via QTL or gene discovery approaches through association mapping and linkage technologies mostly in soybean (Charlson *et al.,* 2009; Du *et al.,* 2009) and in *M. trancatula* (Badri *et al.,* 2011).

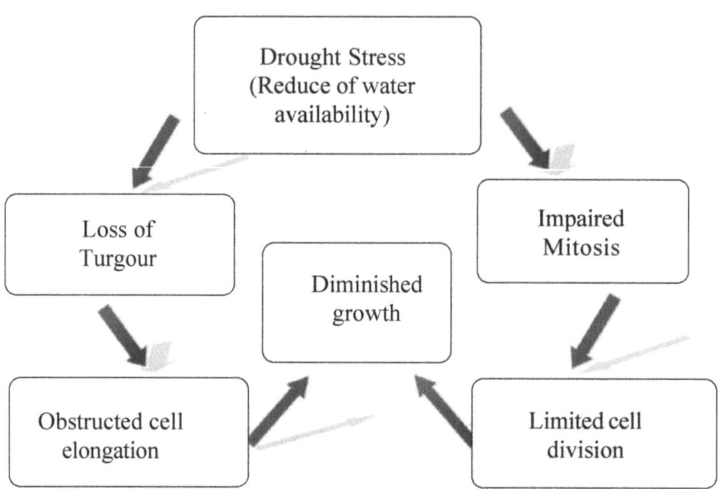

Fig. 3: Description of possible mechanisms of growth reduction under drought stress. Under drought stress conditions, cell elongation in higher plants is inhibited by reduced turgor pressure. Reduced water uptake results in a decrease in tissue water contents. As a result, turgor is lost. Likewise, drought stress also trims down the photoassimilation and metabolites required for cell division. As a consequence, impaired mitosis, cell elongation and expansion result in reduced growth (Farooq *et al.,* 2008).

During prolonged water stress plants must be able to survive with low water content and maintain a minimum amount of water, through water uptake and retention. To cope with prolonged drought stress plants respond with energy demanding processes that alter the growth pattern, chemical content of the plants and the up or down regulation of genes. When the water availability is reduced, plants change the biochemistry to be able to retain as much water as possible and take up whatever water they can. During water stress plants produce and accumulate compatible solutes such as sugars, polyols and amino acid to lower the osmotic potential in the cells to facilitate water absorption and retention. Some of the compatible solutes also contribute to maintaining the conformation of macromolecules by preventing misfolding or denaturation (Xiong and Zhu, 2002). Plants also produce higher levels of the plant stress hormone ABA during water stress and this affects their growth pattern and stress tolerance. A group of proteins called late embryogenesis abundant like (LEA) proteins are also produced during water stress. These LEA like proteins are highly hydrophilic, glycine-rich and highly soluble and have been found to be regulated by ABA (Xiong and Zhu, 2002). The LEA-like proteins are thought to act as chaperones, protecting enzymatic activities (Reyes *et al.*, 2005) and preventing misfolding and denaturation of important proteins (Xiong and Zhu, 2002).

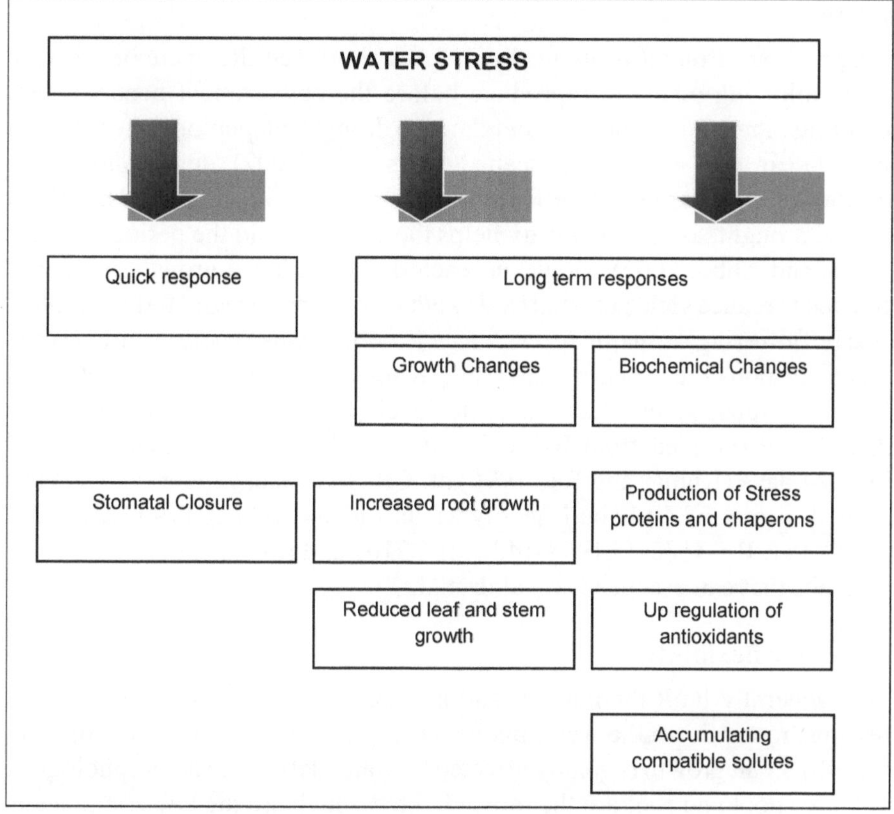

Fig. 4: Plant responses to water stress (Xiong and Zhu, 2002)

Drought resistance mechanisms

Plants respond and adapt under drought stress by different mechanisms as by the induction of various morphological, biochemical and physiological responses plants survive under harsh conditions. Mechanisms of drought tolerance at different levels are explained in the following sections.

Morphological mechanisms

Plant drought tolerance involves changes at various physiological and molecular levels. Induction of a single or a combination of inherent changes determines the ability of the plant to sustain itself under limited moisture supply. Various morphological mechanisms taking place under drought conditions are explained in brief below:

Escape

"Escape" from drought is attained through a shortened life cycle or growing season, allowing plants to reproduce before the environment becomes dry. Flowering time is an important trait related to drought adaptation, where a short life cycle can lead to drought escape (Araus *et al.,* 2002), developing short-duration varieties has been an effective strategy for minimizing yield loss from terminal drought, as early maturity helps the crop to avoid the period of stress (Kumar and Abbo, 2001). However, each day of reduced growth cycle was estimated to reduce yield potential by 74 kg/ha (White and Singh,1991) In context of early flowering, a major recessive gene 'efl-1' was reported in chickpea which is responsible for early flowering (Kumar and van Rheenen 2000), and this finding subsequently facilitated the development of super early genotype ICCV 96029 (derived from ICCV 2 9 ICCV 93929 cross) which flowered within 24 days (Kumar and Rao 1996) at ICRISAT. Sabaghpour *et al.* (2006) screened a total of 40 kabuli genotypes and identified ILC1799, ILC3832, FLIP98-141, ILC3182, FLIP98-142C, ILC3101 and ILC588 as superior early genotypes that can escape terminal drought.

Phenotypic flexibility

Plants generally limit the number and area of leaves in response to drought stress just to cut down the water budget at the cost of yield loss (Schuppler *et al.,* 1998). Plant growth is greatly affected by water deficit and at morphological level, the shoot and root are the most affected which are the key components of plant adaptation to drought. Since roots are the only source to acquire water from soil, the root growth, its density, proliferation and size are key responses of plants to drought stress (Kavar *et al.,* 2007). It has long been established that plants bearing small leaves are typical of xeric environments. Such plants withstand drought very well, albeit their growth rate and biomass are relatively low (Ball *et al.,* 1994).

Avoidance

Drought avoidance explains the mechanisms regulating reduce water loss from plants, due to stomatal control of transpiration, and also maintains water uptake through an extensive and prolific root system (Turner *et al.,* 2001; Kavar *et al.,* 2007) This mechanism is also reflected in stomatal control, by reducing evapotranspiration during its vegetative phase and conserving moisture for grain filling period. The root characters such as length, density, depth and biomass are one of the main drought avoidance traits that contribute to final yield under drought environments (Subbarao *et al.,* 1995; Turner *et al.,* 2001). A deep and thick root system is most helpful trait for extracting water from considerable depths (Kavar *et al.,* 2007). This mechanism was recognized in soybean (Sinclair *et al.,* 2000; Sinclair, 2000).

Physiological mechanisms

Osmotic adjustment, osmoprotection, antioxidation and a scavenging defense system have been the most important bases responsible for drought tolerance. The physiological basis of genetic variation in drought response is yet not clear; because more intricate mechanisms have been suggested. Some of these mechanisms include cell and tissue water conservation, membrane stability, response to growth regulators etc.

Cell and tissue water conservation

Under drought stress, sensitive pea genotypes were more affected by a decline in relative water content than tolerant ones (Upreti *et al.*, 2000). In faba bean, determination of leaf water potential was useful for describing the drought effect, but was not suitable for discriminating tolerant from sensitive genotypes. This suggested that water potential was not the defining feature of the tolerance (Riccardi *et al.,* 2001). Nevertheless, in other studies determination of leaf water status in the morning and water content in leaves in the afternoon was potentially useful for screening drought tolerance in chickpea (Pannu *et al.,* 1993). Osmotic adjustment allows the cell to decrease osmotic potential and, as a consequence, increases the gradient for water influx and maintenance of turgor. Tissue water improvement status may be achieved through osmotic adjustment and/or changes in cell wall elasticity. This is essential for maintaining physiological activity for extended periods of drought (Kramer and Boyer, 1995).

Antioxidant defense

The antioxidant defense system in the plant cell constitutes both enzymatic and non-enzymatic components. Enzymatic components include superoxide dismutase, catalase, peroxidase, ascorbate peroxidase and glutathione reductase. Non-enzymatic components contain cysteine, reduced glutathione and ascorbic acid (Gong *et al.,* 2005). In environmental stress tolerance, such as drought, high activities of antioxidant enzymes and high contents of non-enzymatic constituents are important. The reactive oxygen species in plants are removed by a variety of antioxidant enzymes and/or lipid-soluble and water soluble scavenging molecules (Hasegawa *et al.,* 2000); the antioxidant enzymes being the most efficient mechanisms against oxidative stress (Farooq *et al.,* 2008). Apart from catalase, various peroxidases and peroxiredoxins, four enzymes are involved in the ascorbate-glutathione cycle, a pathway that allows the scavenging of superoxide radicals and H_2O_2. These include ascorbate peroxidase, dehydroascorbate reductase, monodehydroascorbate reductase and glutathione reductase (Fazeli *et al.,* 2007).

Cell membrane stability

Biological membranes are the first target of many abiotic stresses. It is generally accepted that the maintenance of integrity and stability of membranes under water stress is a major component of drought tolerance in plants (Bajji *et al.,* 2002). The causes of membrane disruption are unknown; notwithstanding, a decrease in cellular volume causes crowding and increases the viscosity of cytoplasmic components. This increases the chances of molecular interactions that can cause protein denaturation and membrane fusion. Cell membrane stability, reciprocal to cell membrane injury, is a physiological index widely used for the evaluation of drought tolerance (Premachandra *et al.,* 1991). For model membrane and protein systems, a broad range of compounds have been identified that can prevent such adverse molecular interactions. Some of these are proline, glutamate, glycinebetaine, carnitine, mannitol, sorbitol, fructans, polyols, trehalose, sucrose and oligosaccharides (Folkert *et al.,* 2001). Another possibility of ion leakage from the cell may be due to thermal induced inhibition of membrane-bound enzymes responsible for maintaining chemical gradients in the cell (Reynolds *et al.,*2001).

Plant growth regulators

Plant growth regulators, when applied externally, and phytohormones, when produced internally, are substances that influence physiological processes of plants at very low concentrations (Morgan, 1990). Under drought, endogenous contents of auxins, gibberellins and cytokinin usually decrease, while those of abscisic acid and ethylene increase (Nilsen and Orcutte, 1996). Nevertheless, phytohormones play vital roles in drought tolerance of plants. Auxins induce new root formation by breaking root apical dominance induced by cytokinins. As a prolific root system is vital for drought tolerance, auxins have an indirect but key role in this regard. Drought stress limits the production of endogenous auxins, usually when contents of abscisic acid and ethylene increase (Nilsen and Orcutte, 1996).

Molecular mechanisms

Plant cellular water deficit may occur under conditions of reduced soil water content. Under these conditions, changes in gene expression (up- and down-regulation), genes induced responses to drought at the transcriptional level, and gene products are thought to function in tolerance to drought (Kavar *et al.,* 2007). Gene expression may be triggered directly by the stress conditions or result from secondary stresses and/or injury responses. Nevertheless, it is well established that drought tolerance is a complex phenomenon involving the concerted action of many genes (Agarwal *et al.,* 2006; Cattivelli *et al.,* 2008).

Aquaporins

Aquaporins have the ability to facilitate and regulate passive exchange of water across membranes. They belong to a highly conserved family of major intrinsic membrane proteins (Tyerman *et al.,* 2002). In plants, aquaporins are present abundantly in the plasma membrane and in the vacuole membrane. The structural analysis of aquaporins has revealed the general mechanism of protein-mediated membrane water transport. Studies on aquaporins and plant water relations have been carried out for many years. Mercury is a potential inhibitor of aquaporins. This was evident from a number of reports on mercury-induced decline in root hydraulic conductivity, which substantiated that aquaporins play a major role in overall root water uptake (Javot and Maurel, 2002), and play a role in cellular osmoregulation of highly compartmented root cells (Maurel *et al.,* 2002; Javot *et al.,* 2003). Recently, efforts have been concentrated on investigating the function and regulation of plasma membrane intrinsic protein aquaporins. The aquaporins play a specific role in controlling transcellular water transport. For instance, they are abundantly expressed in roots where they mediate soil water uptake (Javot and Maurel, 2002) and transgenic plants downregulating one or more prolactin-inducible protein genes had lower root water uptake capacity (Javot *et al.,* 2003).

Signalling and drought stress tolerance

The complexity of signalling events associated with the stress and the activation of defense and acclimation pathways is believed to involve reactive oxygen species, calcium, calcium regulated proteins, mitogen-activated protein kinase cascades and cross-talk between different transcription factors (Kovtun *et al.,* 2000; Chen *et al.,* 2002). Chemical signals, e.g., reactive oxygen species, calcium and plant hormones are involved in inducing stress tolerance by acting via transduction cascades and activate genomic re-programming (Joyce *et al.,* 2003).Various chemical signals transduced under drought stress activate an array of genes, leading to the synthesis of proteins and metabolites, conferring drought tolerance in a number of plant species.

Stress proteins

Synthesis of stress proteins is a ubiquitous response to cope with prevailing stressful conditions including water deficit. Most of the stress proteins are soluble in water and therefore contribute towards the stress tolerance phenomena by hydration of cellular structures (Wahid *et al.,* 2007). Membrane-stabilizing proteins and late embryogenic abundant proteins are important protein group responsible for conferring drought tolerance. These increase the water-binding capacity by creating a protective environment for other proteins or structures, referred to as dehydrins. They also play a major role in the separation of ions that are concentrated during cellular dehydration (Gorantla *et al.,* 2006). Dehydrins, also known as a group of late embryogenesis abundant proteins, accumulate in response to both dehydration and low temperature (Close, 1997). Synthesis of a variety of transcription factors and stress proteins is exclusively implicated in drought tolerance (Taiz and Zeiger, 2006).

Waterlogging stress

It refers to soil saturation with water. Some crop plants including rice tolerate this stress by virtue of their special character (for example, presence of aerenchymatous cells). However, other crop plants are prone to waterlogging stress especially at seedling stage (Choudhary and Vijayakumar, 2012). Under waterlogged condition oxygen diffusion rates (ODR) in flooded soil is about 100 times lower than air and respiration of plant roots, soil micro-flora and fauna leads to rapid exhaustion of soil oxygen, eventually causing anaerobiosis. However, proximate causes of plant injury can be oxygen deficit or mineral nutrient imbalances, a decrease in cytokinins or other hormones released from the roots, a decrease in available soil nitrogen and/or nitrogen uptake, an increase in toxic compounds in soil such as methane, ethylene, ferrous ions or manganese, an increase in toxic compounds (in the plant) such as ethanol or ethylene, and an increase in disease causing organisms (Choudhary *et al.,* 2011).

In general, warm season grain legumes are relatively tolerant to heat stress and sensitive to low temperature. Winter season pulses, on the other hand, tolerate low temperature, but are sensitive to heat stress (> 35°C). Amongst rainy season pulses, traditional pigeonpea encounters low temperature stress during winter months in North India. The stress adversely affects growth, survival and reproductive capacity of plants if the minimum temperature falls below 5°C. Genotypic variations for cold tolerance are well documented in pigeonpea (Yong *et al.,* 2002; Sandhu *et al.,* 2007) especially for survival traits. Low temperature primarily affects development and growth and opening of flower buds (Choudhary, 2007). Winter season pulses (chickpea, field pea and lentil) experience terminal heat stress especially during pod formation and grain-filling stages in North India. Screening techniques to identify heat tolerant chickpea genotypes have been developed (IIPR Annual Report, 2011-12).

Salinity stress

Salinity is an ever increasing problem constraining production of both cool season and tropical and sub-tropical grain legumes in many parts of the world. It affects about 80 million hectares of arable lands worldwide (Munns and Tester, 2008). Crop response to salinity also changes with crop age. For example, lentil, faba bean and field pea are more sensitive at germination than at subsequent growth stages and the converse is true for chickpea. To quantify the effect of salinity on plant growth it is necessary to establish critical values, relating salt concentrations and their effect on growth and yield reduction. In chickpea, germination is less sensitive to salinity than early vegetative stage, and reproductive phase is considered to be even more sensitive than vegetative phase. Under saline condition, symptoms of leaf necrosis, presumably related to the destruction of chlorophyll in leaf cells resulting from ion toxicity when Na+ and/or Cl- exceed threshold level in tissues have been observed; 'visual scores' of necrosis could be used as an index of salinity tolerance in chickpea (Flowers *et al.*, 2010). Salinity also causes physiological drought so that chickpea is unable to remove as much water from saline soil as from non-saline soil. Although chickpea shows osmotic adjustment (OA), its role in salt-sensitive compared to salt resistant genotypes is not conclusive, and requires further study. Salinity has been shown to decrease number of pods/plant, seeds/pod and size of seeds; nevertheless, size is relatively less affected. The global impact of salt stress as well as the capacity to recover from the salt treatment was investigated in *M. truncatula* at transcriptional level (Merchan *et al.*, 2007).Recent efforts to elucidate the salinity stress tolerant mechanisms were focussed on a transcriptomic profiling analysis of seedling roots and leaves performed using high-throughput Illumina sequencing technology of salt stress tolerant line HJ-1 in soybean (Araujo *et al.*, 2015). Also aluminium (Al) toxicity is a well known problem limiting crop production in 30% of arable lands (Campbell *et al.*, 1988). Both cool and warm season pulses are sensitive to Al toxicity.

Abiotic stresses play a major role in determining crop and forage productivity (Boyer, 1992; Rao, 2013), and also affects the differential distribution of the plant species across different types of environments (Chaves *et al.*, 2003). Climate change exacerbates abiotic stress on a global scale, with increased irregularity and unpredictability, and as a result, adaptation strategies need to be developed to target crops to specific environments (Beebe *et al.*, 2011). A remarkable feature of plant adaptation to abiotic stresses is the activation of multiple responses involving complex gene interactions with many pathways at the physiological, biochemical, cellular and molecular levels (Grover *et al.*, 2001; Le *et al.*, 2006; Atkinson *et al.*, 2012).

Cold stress

Cold stress represents a major limiting factor in production especially in North India, Canada and some parts of Australia. Low temperature stress is becoming more prevalent in temperate region creating a serious threat to vegetative growth by several means like creating chlorosis, necrosis of leaf tip and curling of whole leaf (Jha *et al.,* 2013). Plants are grouped into two broad categories based on their level of sensitivity to sub-optimal temperatures among which some are sensitive to temperatures ranging from 0 to 15ÚC (chilling stress), while others are capable of withstanding freezing temperatures. Further, low temperature injuries are also categorised under two types: (i) Chilling injury (>0ÚC) and (ii) freezing injury (temperature below freezing point i.e. 0ÚC. Regarding the response to cold stress most of the studies were performed on *Medicago truncatula.* At the molecular level, its tolerance response has been studied through the expression of CBF/DREB1 gene family (Zhang *et al.,* 2010; Li *et al.,* 2011).

Although plant physiology provided a general overview of plant responses, identifying stress tolerance-related traits or the generation of better performing cultivars through breeding, further elucidation of the genetic basis of these important traits, integrating molecular biology and genomics approaches, is needed to further dissect, and eventually profit from, the mechanisms underlying plant adaptation to abiotic stresses (Mir *et al.*, 2012).

Abiotic stress responses in model legumes

The study of the unique biological mechanisms used by legumes in response to stress conditions has been facilitated by the establishment of several model species. *Medicago truncatula* and *Lotus japonicus* have been the primary models developed to investigate plant-microbe interaction and nitrogen fixation (Araujo *et al*., 2015). Being the most important legume crop, the use of *G. max* as a model species presents several advantages over the *M. truncatula* or *L. japonicus*, providing valuable outputs in questions related to yield or grain production (e.g., grain filling or pod abortion), susceptible to be transferred among others cultivated legume. The release of almost complete genome sequences for these species (Young and Bharti, 2012) combined with the existence of genetic transformation protocols (Stewart, 1996; Aoki *et al.*, 2002; Araujo *et al.*, 2004) make them important tools to dissect the molecular mechanisms underlying legume adaptation to abiotic stresses.

Abiotic stress response of legume crops

Legume improvement should focus on a myriad of challenges for grain legumes and forages, namely concerning improvement towards abiotic stress resistance or tolerance. Common bean, chickpea (*Cicer arietinum*), pea and faba bean (*Vicia faba*) are some examples of the most cultivated staple food legumes for direct human consumption in the world. Soybean is the world's leading economic oilseed crop and vegetable protein for food and feed (Manavalan *et al.*, 2009). However, due to its characteristics and genomic or genetic resources available, it is now rightly considered as well a model species (Cannon, 2013) and thus it was addressed on the previous model legume section. Others, such as cowpea (*Vigna unguiculata*), pigeon pea (*Cajanus cajan*), lentil (*Lens culinaris*), and grass pea (*Lathyrus sativus*) also have also an important role as staple crops mainly in some of the most marginal and harsh regions of the world.

Soil related constraints

Pulses crops are generally very sensitive to acidic, saline and alkaline soil conditions. North-western states have extensive areas with high soil pH whereas eastern and north eastern states have chronically acidic soils. The problem has been compounded by rising deficiency of micronutrients such as zinc, iron, boron and molybdenum and that of secondary nutrients like sulphur particularly in traditional pulse growing areas. This emerges to an extent from the fertilizer subsidy policies. Recent incentives to speciality fertilisers ameliorate this stress. Deep black cotton soils in the states of Madhya Pradesh, Maharashtra, Gujarat, Andhra Pradesh, and Tamil Nadu get inundated during kharif season thereby causing serious damage to pigeonpea, urdbeans and mungbeans. On the other hand, shallow and coarse textured soils in north and western states have low water rententivity and require irrigation for supporting a good rabi pulse crop.

Input quality and availability related constraints

Nutrient requirement of pulses is much lower than cereals mainly because of biological nitrogen fixation and relatively low productivity levels although pulse crops respond favourably to higher doses of fertilizer nutrients than generally applied or even recommended. But, since pulses are invariably subjected to abiotic stresses leading to sub-optimal nutrient uptake, farmers tend to use low doses of fertilizer nutrients. Further, nutrient use is unbalanced and seldom based on soil-test values. Timely availability of quality chemical fertilizers continues to be a problem in many pulses growing area. Inadequate availability of gypsum or pyrites as a cheap source of sulphur remains a serious impediment in many states/regions. Availability of pesticides (including herbicides) in most of the states has been comfortable but their quality in terms of effectiveness and eco-friendliness has been an issue in spite of a well designed regulatory mechanism put in place.

Pests and diseases

Although pulse crops are prone to many insect pests and seed borne diseases, pod-borer in chickpea and pigeonpea has been a major cause of concern as its incidence, if not controlled, devastates the crop. Podfly and Maruca also cause serious damage to pigeonpea. Fusarium wilt is wide spread in chickpea, pigeonpea and lentil growing regions. Urd and mungbean crops are often damaged by yellow mosaic virus and powdery mildew. In addition, heavy damage to pulses grain is caused by pests during storage.

Blue-bull menace

Pulses are vulnerable to attack by Blue-bulls in the Indo-Gangetic Plains. Because of the widespread menace particularly in Uttar Pradesh, Bihar, Madhya Pradesh, Rajasthan and Chhattisgarh the potential area suitable for taking pulses crops is left uncultivated by the farmers. There is no viable strategy available in the country to effectively the menace.

Technological constraints

Pulses are grown under varied agro-climatic conditions (soil types, rainfall and thermal regime) in the country. This calls for region specific production technology including crop varieties with traits relevant to prevailing biotic and abiotic stresses. Even biological fertilizers and pesticides used should be based on strains isolated from regions with similar agro-climatic conditions for them to be effective. Our research and development programme in pulses has yet to appreciate and address this issue adequately. Production technology for a pulse crop has to be soil type/region specific. So is true for tillage and seeding device/gadgets. Non-availability of a dependable ridge planter for kharif pulses in black soil region (for which ridge planting is most relevant and recommended) has left farmers with no option but to grow kharif pulses on flat beds following conventional practices. The country has lagged in state of the art biotechnology research in pulses, now common in some of the countries exporting pulses to India.

Infrastructural constraints

Rainfall received during maturity of kharif pulses, causes losses in yields and grain quality when farmers usually do not have pakka and covered threshing floor. Farmers also lack awareness and means for safe storage of grain/seed of pulses. Many areas are approachable only during fair weather. Warehousing facilities are either inadequate or inaccessible.

Table 4: Important Biotic and Abiotic Stresses Identified in Major Pulse Crops of India (Reddy (2006).

Crop	Seasons	Stress	
		Biotic	Abiotic
Chickpea pod-borer	Timely sown	FW, root rot, chickpea stunt, BGM,	Low temperature
	Early sown	FW, root rot, AB, or chickpea stunt, pod borer	Terminal drought, salt stress
	Late sown	FW, pod borer	Terminal drought, cold
Pigeonpea	Kharif-early	FW, PB, pod-borer complex	Waterlogging
	Medium late	FW, SM, pod-borer complex	Cold, terminal drought, waterlogging
	Pre-rabi	FW,ALB, Pod fly	Cold, terminal drought
Moong	Kharif	MYMV, CLS, WB, sucking insect pests	Pre-harvest sprouting, terminal drought
	Zaid	MYMV, root and stem rot, stem agromyza, sucking insect pests	Pre-harvest sprouting temperature drought stress
	Rabi	PM, Rust, CLS	Terminal drought
Urad	Kharif	MYMV, anthracnose, WB, LCV	Terminal drought
	Zaid	MYMV, root and stem rot, stem agromyza	Pre-harvest sprouting, temperature stress, drought
	Rabi/rice fallow	Spot	Terminal drought
Lentil		FW, root rot, rust	Moisture, temperature

FW= Fusarium wilt, PB= Phytophthora blight, SM= Sterility mosaic, ALB= Alternaria leaf blight, MYMV= Moongbean yellow mosaic virus, BGM= Botrytis gray mould, AB= Ascochyta blight.

Credit and marketing related constraints

Farmers engaged in cultivation of pulses are mostly small and marginal. A majority are in areas with poor banking infrastructure. They have poor resource base and lack risk-bearing capacity. They therefore either lack access to credit or turn defaulters. There is lack of marketing network in remote areas. Procurement of produce by a dedicated agency is virtually non-existent or in-effective.

Policy related issues

System of regulating quality of inputs though in place in all the states, needs to be made more effective. Delivery of improved technology, inputs, credits need to be stream lined through appropriate policy interventions. Benefit of crop insurance need to be extended to pulses farmers.

Genetic potentiality advancement

Genetic break through for yield improvement

ICAR/SAUs/ International organizations have been screening the accessions / germplasmlines since inception of research for yield and other improvements. A number of good improved varieties of pulse crops have been evolved using traditional and non traditional techniques of plant breeding like selection, back cross and cytoplasmic male sterility (CMS). Old varieties of pulses are working well on farmers' field compared to newly evolved / notified and therefore, farmers preferold variety seeds. ICRISAT has evolved ICPH 2671 hybrid, which has been notified by the State Varietal Release Committee of Madhya Pradesh for cultivation in Madhya Pradesh. GTH- 1 cytoplasmic hybrid of pigeonpea has also been evolved by Gujarat Agricultural University for cultivation in Gujarat (Anon., 2012).

Status of transgenic

Efforts are being made in Indian Public Research Institutions since early eighties to develop transgenic crops. The Government of India has been very supportive of the efforts to develop transgenic crops and invested liberally through the Department of Biotechnology, Department of Science, Department of Technology and Indian Council of Agriculture Research. As a result many transgenic crops have been developed and are being tested by various public and private institutions. The crops covered are Brinjal, Castor, Groundnut, Potato, Rice, Tomato, Chickpea, Sorghum, Watermelon, Papaya, Sugarcane, Mustard, Cabbage, Cauliflower, Maize and Okra including pulses. At present, the Institutions working in respect of pulses are ICRISAT, Hyderabad dealing with Abiotic stress tolerance in chickpea, NRCPB: IIPR, Kanpur working on

Resistant to pod borer in chickpea and pigeonpea. The work for development of transgenic pulses is in progress at the aforesaid centres.

Suggestions for improvement in scenario of pulses

Pulse production is affected by a number of biotic and a biotic factors including others like inadequate marketing facilities and less recovery of pulses due to use of obsolete processing machines. About 20-25% area is sown with certified / quality seeds and the remaining is sown with seeds of farmers. It is understood that about 10 - 20% pulses production may be increased by increasing distribution of certified / quality seeds covering 50 % of the total area under pulses. To sow this area, 50 lakh quintals of certified seed is required. It will be a positive approach in the direction of increasing pulse production by increasing supply of quality seeds maintaining multiplication chain involving nucleus, breeder and foundation seeds. Pigeonpea and chickpea are major pulses, which contribute about 60% of total pulse production. Among the insect pests and diseases, *Helicoverpa armigera* and wilt are the major pests, damaging about 20-30% of the productivity. To minimize the losses by the above and other pests to pulses, aggressive implementation of IPM and INM technologies is required. In addition, development of pulses varieties resistant to pests, especially podborer and wilt of Arhar and gram are urgently required. Many rainfed rice fallow lands in Chhattisgarh, Madhya Pradesh, Jharkhand, Bihar, Odisha and Andhra Pradesh remain uncultivated during rabi season due to lack of cultivation knowledge of field crops in non availability of irrigation water.

Hence, the farmers of such areas are required to be guided to grow pulses in Rabi season on residual moisture, lentil in upland, chickpea / Batry in medium and lowlands as pulse crops provide better production in the aforesaid conditions. Mechanization of pulse production, processing and handling is very important in order to increase production and saving of losses. It also helps in timeliness of operations, better utilization of resources, reduction of drudgery, increasing production and productivity leading to economic benefits. At present, more than 80% area of pulses is rainfed and therefore, arranging irrigation at critical stage by micro irrigation devices (Sprinkler set and Raingun etc.) may increase production by about 10-15%. About fourteen Pulse crops are cultivated across the country by major and minor states and marketing facility is available for buying Arhar, Moongbean, Urdbean, gram and Lentil on Minimum Support Price (MSP) under Price Support Scheme in Andhra Pradesh, Assam, Bihar, Chhattisgarh, Gujarat, Haryana, Jharkhand, Karnataka, Madhya Pradesh, Maharashtra, Odisha, Punjab, Rajasthan, Tamil Nadu, Uttar Pradesh and West Bengal. Creation / development of such facility in minor pulse producing states and inclusion of some more important crops under price support scheme (PSS) may help in increasing production. Inter cropping of pulses with suitable crops like sugarcane, tomato, soybean, cotton and with other main crops may increase production by expansion in area. Better recovery of Dal by processing grain of pulses with modern machines, besides minimizing losses in storage by utilizing scientific storage may also help improving pulses scenario. Pulse crops require well drained soils. During south west monsoon, it has been observed that most of the crops affected are in lowlying areas. This may be minimized by growing pulses in well drained soils or making well drainage system.

This apart, crops are also affected by high and low temperatures, especially when crop of pigeonpea, chickpea and lentil are in flowering to pod development stages. Farmers are to be guided to cultivate varieties tolerant / resistant to these problems. It has also been observed that most of the farmers do not follow proper crop rotations, besides, growing pulses in less fertile lands. There appears a need for creating awareness among the farmers to grow pulses following crop rotations for increasing production by restoration of soil fertility and biological nitrogen for long life of soil. The states, which produce pulses as Inter /mixed crops, do not estimate area, production and yield. As such, estimation of area, production and yield of inter crops separately may also help in better planning to achieve goal of improving pulses scenario. Production of pulses in the off seasons that is summer / rabi is affected by stray cattle and Blue Bull, which damage pulse crops such as Arhar, Moongbean and Urdbean more than any other crop. Proper management of the aforesaid may definitely help to increase the magnitude of pulses production. Maintenance of genetic purity of old popular high yielding varieties of pulses may also support in increasing production. Development of short duration varieties of Arhar are required with synchronized maturity, especially for central zone states cultivating Arhar- Gram / wheat cropping sequence. Varieties of pulses suitable for harvesting by harvester need to be developed. It is observed that in some of the states extreme cold & heat are affecting production of Gram, Arhar and lentil and therefore, it has become necessary to develop tolerant and resistant varieties.

Overcoming socio-economic constraints

Farmers' awareness on improved varieties and seed availability of varieties are the key factors in awareness and access to new technologies and in the spread of improved varieties. The television will be the most popular media for increasing awareness; FPVS trials and farmers fairs/field days will also be helpful. The identified technology needs to be subsidized for wider adoption.

Semi-formal seed systems explains that yet there is a good number of High Yielding Varieties (HYVs) released for all major pulses in India, and there is enough Breeder seed and Foundation seed produced, there is a shortage of

Certified/Truthful seed at farmers' level. Both public and private agencies are working on this but yet they have not been able to meet the requirement of quality seed and also the seed replacement ratio is very low. There has been some success in establishing semi-formal seed systems to produce Truthfully Labelled seed, in which linkages were established between the formal and informal seed sectors through supply of basic quality seed by the NARS, and quality of seed production is monitored by universities/non-governmental organizations/farmers' associations (Reddy, 2013).

Cash is a key element for enabling smallholder farmers to shift from low input-low output to high-input-high-output agriculture. But access to credit by these farmers is low because of their low asset base, low risk bearing ability and high risk environments. This can be effectively tackled by the insurance-linked credit to pulse crops without any collateral security. The scale of finance should be sufficient enough to cover all the costs of the recommended practices (Reddy 2009).

Farm mechanization is one of the reasons for success of expansion of area under chickpea in Andhra Pradesh and farm mechanization can further be enhanced by developing varieties suitable for harvesting by combine harvesters. Hence, farm mechanization in peak season activities such as harvesting and threshing needs to be encouraged through the distribution of subsidized farm machinery to cope with labor shortage and higher wage rates.

With the expansion of irrigation facilities through groundwater and also through canal irrigation systems, there is a scope for expansion of irrigated area under pulse crops, especially summer, rabi and spring season crops, as yield response is higher. Harvesting and management of rainwater through watersheds rather than exploitation of costly groundwater needs to be emphasized.

Markets for legumes are thin and fragmented due to scattered production and consumption across states. Farmers sell their marketed surplus immediately after harvest, while some large traders/wholesalers trade between major markets and hoard pulses to take advantage of speculative gains in the off-season. Due to this, farmers do not benefit from the higher market prices of pulses. Investments in market infrastructure, warehouses, market information systems both in public and private sectors through Public-Private Partnership (PPP) models and economic viability gap funding models need to be encouraged in SAT India (Reddy, 2013).

Management of abiotic stresses in pulse crops

Pulse crops reported huge losses due to biotic (pests and diseases) and abiotic (drought, high temperature, etc) stresses. Some of the studies estimated the losses in the range from 15% to 20% of normal production (IIPR 2011). This means, India can increase pulses availability by 15% to 20% with investments in appropriate crop protection R&D. With abiotic stress, the emphasis shifts to the here-and-now. The focus is on those margins where such stresses currently limit crop production and threaten farmer livelihoods. We want genes for today, rather than genes for tomorrow. Tolerance or resistance is commonly a complex of characters combining mechanisms of both direct tolerance and indirect avoidance. These different mechanisms can also differ in their associated yield costs. Thus a diversity of sources of stress tolerances is abiotic stress needed as the raw material for combining or pyramiding to achieve improved tolerance, and for crossing with high-yielding but sensitive populations. Recent studies from overseas suggest that wild species of pulses posses desired traits for a number of abiotic stresses. Future effort is required to identify desirable genes from this germplasm for transfer to adapted cultivars by conventional and/or biotechnological approaches to develop abiotic stress resistant cultivars. In short, to increase area and production of pulse crops, we need crop specific and region specific approaches. Already ICAR and ICRISAT, with the support of state and central governments, are involved in the development of short duration, photo-thermo insensitive varieties for different agro-ecology, development of hybrids in pigeonpea, development of efficient plant architecture in major pulse crops, development of bio-intensive Integrated Pest Management modules, design of improved machines to cope with labor shortage, production of Breeder seed of the latest released varieties and in organizing frontline demonstrations in farmers' fields. The efforts under NFSM-Pulses and R&D under NARS needs to be further strengthened (Anon., 2012).

Abiotic stresses can also be reduced by choosing the most appropriate pulse species and adjusting agronomy (sowing time, plant density, soil management) to ensure sensitive crop stages occur at the most favourable time in the season. For example choosing the optimum sowing time, species and cultivars with appropriate phenology can reduce the effect of frost and drought in pulse crops in dry land environments with terminal stress. One of approach of dealing with stresses caused by extremes in the abiotic environment is to develop cultivars resistant to specific stresses. Breeding and selection for resistance to these stresses is often considered difficult because of the unpredictability of climatic conditions. 'Escape' mechanism has been invariably and widely utilized to mitigate the effects of abiotic stresses in almost all pulses. However, the same strategy cannot be applied for mineral stresses (Allard, 1999). By extrapolation, grain legumes are likely to accumulate a specific combination of genes (alleles) if exposed separately to above-mentioned stresses. Therefore, a genotype showing tolerance to drought may react differently if exposed to water-logging and vice-versa. The best breeding strategy appears to be selection of superior genotypes of grain legumes under actual field condition. We are ultimately concerned with the yield (a measure of relative reproductive capacity) of genotypes. Therefore, selection should be based on yield and sometimes on its component traits (Flowers *et al.*, 2010). It is often argued that there may be temporal and spatial variation under field condition; therefore, screening should be done under controlled condition. However, under field conditions a number of variables interact to produce final outcome, thus field testing of genotypes cannot be ignored. Therefore, we suggest assessing genotypes for such abiotic stresses under actual field condition, and the results may be reconfirmed under controlled condition and vice versa. However, selection must finally be practiced for high yielding genotypes.

The developmental stage of the plants and the duration of the stress have a strong influence on the effect of the abiotic stress, together with the ability of the plants to tolerate the stress themselves. Drought and heat escape through earliness in flowering and maturity is probably the characteristic most widely used by breeders for pulses and other crops to escape drought especially in low rainfall terminal drought environments. Currently lack of simple and accurate screening procedures to screen parental genotypes and breeding population for various abiotic stresses is the major bottleneck in the development of stress tolerant pulse crops. One of the solution to compete with the increasing problems can be to focus on kharif pulse crops to reduce cost of production as kharif pulses (pigeonpea, mungbean and urdbean) are cheaper in their cost of production than for rabi pulse crops (chickpea and lentil) due to cultivation under more uncertain conditions (Reddy, 2013).

References

Agarwal, P. K., Agarwal, P., Reddy, M. K., Sopory, S. K. 2006. Role of DREB transcription factors in abiotic and biotic stress tolerance in plants. *Pl. Cell Rep.* 25: 1263–1274.

Ali, M. and Gupta, S. 2012.Carrying capacity of Indian Agriculture: pulse crops. *Current Sci.*: 102(6).

Allard, R. W. 1999. Principles of Plant Breeding (2nd Edition). John Wiley and Sons, New York.

Anonymous 2011. Research report on marketing of pulses in India. CCS, National Institute of Agriculture Marketing, Jaipur, pp. 11-13.

Anonymous 2012. Status paper on pulses: Government of India Ministry of Agriculture (Department Of Agriculture & Cooperation), pp. 9-15.

Anonymous 2013. Report of Expert Group on Pulses. Department of Agriculture & Cooperation Government of India, Ministry of Agriculture.pp.139.

Aoki, T., Kamizawa, A. and Ayabe, S. 2002. Efficient Agrobacterium-mediated transformation of Lotus japonicus with reliable antibiotic selection. *Pl. Cell Rep.* 21: 238–243.

Araujo, S. D. S., Duque, A. S. R. L. A., Santos, D. M. M. F., and Fevereiro, M. P. S. 2004. An efficient transformation method to regenerate a high number of transgenic plants using a new embryogenic line of *Medicago truncatula* cv. Jemalong. *Pl. Cell Tiss. Organ Cult.* 78: 123–131.

Araus, J. L., Slafer, G. A., Reynolds, M. P. and Royo C. 2002. Plant breeding and drought in C_3 cereals: what should we breed for? *Ann. Bot.* 89: 925–940.

Atkinson, N. J. and Urwin, P. E. 2012. The interaction of plant biotic and abiotic stresses: from genes to the field. *J. Exp. Bot.* 63: 3523–3544.

Badri, M., Chardon, F., Huguet, T. and Aouani, M. E. 2011. Quantitative trait loci associated with drought tolerance in the model legume *Medicago truncatula. Euphytica.* 181: 415–428.

Bajji, M., Kinet, J. and Lutts, S. 2002. The use of the electrolyte leakage method for assessing cell membrane stability as a water stress tolerance test in durum wheat, *Pl. Growth Regul.* 36: 61–70.

Ball, R.A., Oosterhuis, D.M., Mauromoustakos, A. 1994. Growth dynamics of the cotton plant during water-deficit stress. *Agron. J.* 86: 788–795.

Baroowa, B., Gogoi, N., Paul, S. and Sarma, B. 2012. Morphological responses of pulse (Vigna spp.) crops to soil water deficit. *J. Agri. Sci.* 57(1): 31-40.

Beebe, S., Ramirez, J., Jarvis, A., Rao, I. M., Mosquera, G., Bueno, G. and Blair, M. 2011. Genetic improvement of common beans and the challenges of climate change. In: *Crop Adaptation to Climate Change*. pp. 356–369. 1st ed. Yadav, S. S., Redden, R. J., Hatfield, J. L., Lotze-Campen, H., and Hall, A. E., Eds., Wiley, New York.

Campbell, T. A., Foy, C. D., Mc Murty, E. and Elgin, J. E. 1988. Selection of alfalfa for tolerance to toxic levels of Al. Canadian. *J. Pl. Sci.* 68:743-753.

Cannon, S. B. 2013. The model legume genomes. *Methods Mol. Biol.* 1069: 1–14.

Cattivelli, L., Rizza, F., Badeck, F. W., Mazzucotelli, E., Mastrangelo, A. M., Francia, E., Mare, C., Tondelli, A., Stanca, A. M. 2008. Drought tolerance improvement in crop plants: An integrative view from breeding to genomics. *Field Crop. Res.* 105: 1–14.

Charlson, D. V., Bhatnagar, S., King, C. A., Ray, J. D., Sneller, C. H., Carter, T. E. and Purcell, L. C. 2009. Polygenic inheritance of canopy wilting in soybean [*Glycine max* L. Merr.]. *Theor. Appl. Genet.* 119: 587–594.

Chaves, M. M., Maroco, J. P., and Pereira, J. S. 2003. Understanding plant responses to drought-from genes to the whole plant. *Funct. Pl. Biol.* 30: 239–264.

Chen, W., Provart, N.J., Glazebrook, J., Katagiri, F., Chang, H.S., Eulgem, T., Mauch, F., Luan, S., Zou, G., Whitham, S.A., Budworth, P.R., Tao, Y., Xie, Z., Chen, X., Lam, S., Kreps, J.A, Harper, J.F., Si-Ammour, A., Mauch-Mani, B., Heinlein, M., Kobayashi, K., Hohn, T., Dangl, J.L., Wang, X. and Zhu, T. 2002. Expression profile matrix of Arabidopsis transcription factor genes suggests their putative functions in response to environmental stresses. *Pl. Cell.* 14:559–574.

Choudhary, A.K. 2007. Selection criteria for low temperature tolerance in long-duration pigeonpea (p: 266). Abstract published in the National Symposium on "Legumes for Ecological Sustainability: Emerging Challenges and Opportunity, November 3-5, 2007". Indian Institute of Pulses Research, Kanpur.

Choudhary, A.K. and Singh D. 2011. Screening of pigeonpea genotypes for nutrient uptake efficiency under aluminium toxicity. *Physiol. and Mol. Bio. of Pl,* 17: 145-152.

Choudhary, A.K. and Vijayakumar, A.G. 2012. Glossary of Plant Breeding, A Perspective. LAP LAMBERT Academic Publishing, Germany (ISBN 978-3-659-21039-6).

Close T. J. 1997. Dehydrins: a commonality in the response of plants to dehydration and low temperature, *Physiol. Pl.* 100: 291–296.

Dita, M. A., Rispail, N., Prats, E., Rubiales, D. and Singh, K. B. 2006. Biotechnology approaches to overcome biotic and abiotic stress constraints in legumes. *Euphytica.* 147: 1–24.

Du, W., Wang, M., Fu, S. and Yu, D. 2009. Mapping QTLs for seed yield and drought susceptibility index in soybean *Glycine max* L. across different environments. *J. Genet. Genomics.* 36: 721–31.

Farooq, M., Aziz, T., Basra, S. M. A., Cheema, M. A., Rehamn, H. 2008. Chilling tolerance in hybrid maize induced by seed priming with salicylic acid. *J. Agron. Crop Sci.* 194: 161–168.

Fazeli, F., Ghorbanli, M., Niknam, V. 2007. Effect of drought on biomass, protein content, lipid peroxidation and antioxidant enzymes in two sesame cultivars. *Biol. Pl.* 51: 98–103.

Flowers, T. J., Gaur, P. M., Gowda, C. L. L., Krishnamurthy, L., Srinivasan, S. and Siddique, K. H. M. 2010. Salt sensitivity in chickpea. *Pl. Cell Environ.* 33: 490–509.

Folkert, A. H., Elena, A. G., Buitink, J. 2001. Mechanisms of plant desiccation tolerance. *Trends Pl. Sci.* 6: 431–438.

Gong, H., Zhu, X., Chen, K., Wang, S., Zhang, C. 2005. Silicon alleviates oxidative damage of wheat plants in pots under drought. *Pl. Sci.* 169: 313–321.

Gorantla, M., Babu, P.R., Lachagari, V. B. R., Reddy, A. M. M., Wusirika, R., Bennetzen, J. L., Reddy, A. R. 2006. Identification of stress responsive genes in an indica rice (*Oryza sativa* L.) using ESTs generated from drought-stressed seedlings. *J. Exp. Bot.* 58: 253– 265.

Graham, P. H. and Vance, C. P. 2003. Legumes: importance and constraints to greater use. *Pl. Physiol.* 131: 872–877.

Grover, A., Kapoor, A., Laksmi, O. S., Agarwal, S., Sahi, C., Katiyar-Agarwal, S., Agarwal, M., and Dubey, H. 2001. Understanding molecular alphabets of the plant abiotic stress responses. *Curr. Sci.* 80: 206–216.

Hasegawa, P. M., Bressan, R. A., Zhu, J. K., Bohnert, H.J. 2000. Plant cellular and molecular responses to high salinity. *Annu. Rev. Pl. Phys.* 51: 463–499.

IIPR Annual Report 2011-12, Indian Institute of Pulses Research, Kanpur, India.

Javot, H., Lauvergeat, V., Santoni, V., Martin-Laurent, F., Guclu, J., Vinh, J., Heyes, J., Franck, K. I., Schaffner, A. R., Bouchez, D. and Maurel C. 2003. Role of a single aquaporin isoform in root water uptake. *Pl. Cell.*15: 509–522.

Javot H. and Maurel C. 2002.The role of aquaporins in root water uptake. *Ann. Bot.* 90: 301–313.

Jha, U. C., Chaturvedi, S. K., Bohra, A., Basu, P., Khan, M. S. and Barh, D. 2013. Abiotic stresses, constraints and improvement strategies in chickpea. *Pl. Breeding.* 133:163–178.

Joyce, S. M., Cassells, A. C., Mohan, J. S. 2003. Stress and aberrant phenotypes *in vitro* culture. *Pl. Cell Tiss. Orga.* 74: 103–121.

Kavar, T., Maras, M., Kidric, M., Sustar-Vozlic, J., Meglic, V. 2007. Identification of genes involved in the response of leaves of *Phaseolus vulgaris* to drought stress. *Mol. Breed.* 21:159–172.

Keating, B. A. and Fisher, M. J. 1985. Comparative tolerance of tropical grain legumes to salinity. *Australian J. Agri. Res.* 36:373-383.

Kovtun, Y., Chiu, W.L., Tena, G. and Sheen, J. 2000. Functional analysis of oxidative stress-activated mitogen-activated protein kinase cascade in plants, *Proc. Natl Acad. Sci.* 97: 2940–2945.

Kramer, P. J. and Boyer, J. S. 1995. Water relations of Plants and Soils Academic Press, San Diego.

Kumar, J. and Abbo, S. 2001. Genetics of flowering time in chickpea and its bearing on productivity in semiarid environments. *Adv. Agron.* 72: 107–138.

Kumar, J. and Van Rheenen, H. A. 2000. A major gene for time of flowering in chickpea. *J. Hered.*, 91: 67–68.

Li, D., Zhang, Y., Hu, X. Shen, X., Ma, L., Su, Z.,Wang, T. and Dong, J. 2011. Transcriptional profiling of *Medicago truncatula* under salt stress identified a novel CBF transcription factor MtCBF4 that plays an important role in abiotic stress responses. *BMC Pl. Biol.* 11: 109.

Li, W. Y. F., Wong, F. L., Tsai, S. N., Phang, T. H., Shao, G. and Lam, H. M. 2006. Tonoplast-located GmCLC1 and GmNHX1 from soybean enhance NaCl tolerance in transgenic bright yellow (BY)-2 cells. *Pl. Cell Environ.* 29: 1122–1137.

Farooq, M., Wahid, A., Kobayashi, N., Fujita, D. and Basra, S. M. A. 2009. Plant drought stress: effects, mechanisms and management. *Agronomy for Sustainable Development, Springer* Verlag (Germany). 29 (1): pp.185-212.

Manavalan, L. P., Guttikonda, S. K., Tran, L. S. and Nguyen, H. T. 2009. Physiological and molecular approaches to improve drought resistance in soybean. *Pl. Cell Physiol.* 50: 1260–1276.

Merchan, F., de Lorenzo, L., Rizzo, S. G., Niebel, A., Manyani, H., Frugier, F., Sousa, C. and Crespi, M. 2007. Identification of regulatory pathways involved in the reacquisition of root growth after salt stress in *M. truncatula. Pl. J.* 51: 11–17.

Mir, R. R., Zaman-Allah, M., Sreenivasulu, N., Trethowan, R., and Varshney, R. K. 2012. Integrated genomics, physiology and breeding approaches for improving drought tolerance in crops. *Theor. Appl. Genet.* 125: 625–645.

Morgan, P.W. 1990. Effects of abiotic stresses on plant hormone systems, in: Stress Responses in plants: adaptation and acclimation mechanisms. Wiley-Liss, Inc., pp. 113–146.

Munns, R., and Tester, M. 2008. Mechanisms of salinity tolerance. *Annu. Rev. Pl. Biol.* 59: 651–681.

Nilsen, E.T. and Orcutte, D.M. 1996. Phytohormones and plant responses to stress, in: Nilsen E.T., Orcutte D.M. (Eds.), *Physiology of Plant under Stress*: Abiotic Factors, John Wiley and Sons, New York, pp. 183–198.

Pannu, R. K., Singh, D. P., Singh, P., Chaudhary, B. D. and Singh, V. P. 1993. Evaluation of various plant water indices for screening the genotypes of chickpea under limited water environment. *Haryana J. Agron.* 9: 16–22.

Premachandra, G. S., Saneoka, H., Kanaya, M. and Ogata, S. 1991 .Cell membrane stability and leaf surface wax content as affected by increasing water deficits in maize. *J. Exp. Bot.* 42: 167–171.

Rao, D. L. N., Giller, K. E., Yeo, A. R. and Flowers, T. J. 2002. The effects of salinity and sodicity upon nodulation and nitrogen fixation in chickpea. *Annu. of Bot.* 89: 563-570.

Rao, I. M., Beebe, S. E., Polania, J., Ricaurte, J., Cajiao, C., Garcia, R. and Rivera, M. 2013. Can tepary bean be a model for improvement of drought resistance in common bean? *Afr. Crop Sci. J.* 21: 265–281.

Reddy, A. A. 2004."Consumption Pattern, Trade and Production Potential of Pulses", *Economic & Political Weekly*, 30 October, pp 4854-60.

Reddy, A. A. 2006. "Impact Assessment of Pulses Production Technology", Research Report No 3, Indian Institute of Pulses Research, Kanpur.

Reddy, A. A. 2009. Pulses Production Technology: Status and Way Forward. *Economic & Political Weekly.* 44 (52): 73-80.

Reddy, A. A., Bantilan, M. C. S., Mohan, G. 2013. Pulses production scenario: Policy and technological options.

Reynolds, M. P., Oritz-Monasterio, J. I., Mc Nab, A., 2001. Application of physiology in wheat breeding, CIMMYT, Mexico.

Riccardi, L., Polignano, G.B. and de Giovanni, C. 2001. Genotypic response of faba bean to water stress, *Euphytica.* 118: 39–46.

Sandhu, J.S., Gupta, S.K., Singh, S. and Dua, R.P. 2007. Genetic variability for cold tolerance in pigeonpea. *E-Jour. of SAT Agri. Res.* 5:1-3.

Schuppler, U., He, P. H., John, P. C. L., Munns, R. 1998. Effects of water stress on cell division and cell-division-cycle-2-like cell-cycle kinase activity in wheat leaves. *Pl. Physiol.* 117: 667–678.

Sinclair, T. R., Purcell, L. C., Vadez, V., Serraj, R., King, C. A. and Nelson, R. 2000. Identification of soybean genotypes with N_2 fixation tolerance to water deficits. *Crop Sci.* 40: 1803–1809.

Singh, A. K., Manibhushan , Bhatt, B. P., Singh, K.M. and Upadhyaya, A. 2013. An Analysis of Oilseeds and Pulses Scenario in Eastern India during 2050-51. *J. Agril. Sci.* 5 (1): 241- 9.

Singh, A. K., Singh, S. S., Prakash, V., Kumar, S. and Dwivedi, S. K. 2015. Pulses Production in India: Present Sta-tus, Bottleneck and Way Forward. *J. of Agrisearch.* 2(2): 75-83.

Singh, S., Gupta, A. K., Kaur, N. 2012. Differential responses of antioxidative defence system to long-term field drought in wheat (*Triticum aestivum* L.) genotypes differing in drought tolerance. *J. Agro. and Crop Sci.* 198(3):185-195.

Stewart, C. 1996. Genetic transformation, recovery, and characterization of fertile soybean transgenic for a synthetic *Bacillus thuringiensis* cryIAc gene. *Pl. Physiol.* 112: 121–129.

Subbarao, G. V., Johansen, C., Slinkard, A. E., Rao R. C. N., Saxena N. P. and Chauhan Y. S. 1995. Strategies and scope for improving drought resistance in grain legumes. *Crit. Rev. Plant Sci.* 14: 469–523.

Sultana, R., Choudhary, A. K., Pal, A. K., Saxena, K. B., Prasad, B. D and Singh, R. 2014. Abiotic stresses in major pulses: Current status and strategies. *Springer*, Pp: 173-190.

Araujo, S. S., Beebe, S., Crespi, M., Delbreil, B., Esther, M. G., Gruber, V., Lejeune-Henaut, I., Link, W., Maria, J. M., Prats, E., Rao, I., Vadez, V. and Patto, M. C. V. 2015. Abiotic Stress Responses in Legumes: Strategies Used to Cope with Environmental Challenges. *Critical Reviews in Plant Sciences*. 34:1-3, 237-280.

Taiz, L. and Zeiger, E. 2006. Plant Physiology, 4th Ed., Sinauer Associates Inc. Publishers, Massachusetts.

Turner, N. C., Wright, G. C., Siddique, K. H. M. 2001. Adaptation of grain legumes (pulses) to water-limited environments, *Adv. Agron.* 71: 123–231.

Upreti, K.K., Murti, G.S.R., Bhatt, R.M. 2000. Response of pea cultivars to water stress: changes in morpho-physiological characters, endogenous hormones and yield. *Veg. Sci.* 27: 57–61.

Wahid, A. 2007. Physiological implications of metabolites biosynthesis in net assimilation and heat stress tolerance of sugarcane (*Saccharum officinarum*) sprouts. *J. Plant Res.* 120: 219–228.

White, J.W. and Singh, S. P. 1991. Sources and inheritance of earliness in tropically adapted indeterminate common bean. *Euphytica.* 55: 15–19.

Xiong, L. M., Schumaker, K. S. and Zhu, J. K. 2002. Cell signaling during cold, drought, and salt stress. *Pl. Cell.* 14: S165–S183.

Yong, Gu., Zhenghong, Li., Chaohong, Zhou., Saxena, K.B. and Kumar, R.V. 2002. Field studies on genetic variation for frost injury in pigeonpea. *International Chickpea and Pigeonpea Newsletter.* 9: 39-42.

Young, N. D. and Bharti, A. K. 2012. Genome-enabled insights into legume biology. *Annu. Rev. Pl. Biol.* 63: 283–305.

Zhang, G., Chen, M., Chen, X., Xu, Z., Li, L., Guo, J. and Ma, Y. 2010. Isolation and characterization of a novel EAR-motif-containing gene GmERF4 from soybean (*Glycine max* L.). *Mol. Biol. Rep.* 37: 809–818.

10

Mitigation Strategies of Abiotic Stress in Fruit Crops

Parshant Bakshi, Amit Jasrotia and V.K. Wali

Stress, is usually defined as an external factor that exerts disadvantageous influence on the plant. In most cases, stress is measured in relation to plant survival, crop yield, growth (biomass accumulation), or the primary assimilation processes (CO_2 and mineral uptake), which are related to overall growth. Some environmental factors (such as air temperature) can become stressful in just a few minutes; other may take days to become stressful or even months (some mineral nutrients) to become stressful. Different types of abiotic stresses that affect fruit production are; water stress, heat stress, chilling and freezing stress, salinity stress, flooding stress. Although, it is convenient to examine each of these factors separately, many are interrelated. For example, water deficit is often associated with salinity in the root zone/ or with heat stress in the leaves. Plants often display cross resistance, or resistance to one stress induced by acclimatization to another. This behaviour implies that mechanism of resistance to various stresses share many common features.

Plants encounter adverse environmental stresses during their life cycle, which have negative impacts on growth and greatly affect crop productivity. As perennial crops, fruit trees are exposed to an array of stresses for a long time once they are planted. If the trees are severely injured by environmental stresses, it would be hard for them to recuperate, leading to retarded growth and reduced fruit production. Furthermore, the negative effects of the stresses are not limited to the fruit production in the current year but can also extend to the next year(s). Therefore, it is of critical importance to develop techniques for reducing stress injury and/or improving stress tolerance in fruit trees for sustainable cropping, which can be fulfilled by the cultivation method or genetic engineering (Holmberg and Bulow, 1998).

Water stress

Soil moisture in fruit orchards is deeply involved in tree growth, productivity, and moreover fruit size. In fruit orchards, water absorption content from the soil is less than that of evaporation through the stomata of the leaves under water stress condition. Subsequently, cell enlargement is depressed.

Morphological and physiological responses of plants to water deficit

Decreased leaf area

As the water content of the plant decreases, the cells shrink and cell walls relax. As the water loss progresses and the cells contract further, the solutes in the cells become more concentrated. The plasma membranes become thicker and more compressed. Because turgor loss is the earliest significant biophysical effect of water stress, turgor dependent activities are most sensitive to water deficit.

Cell expansion is turgor-driven process which is defined by equation:

$$GR = m\,(\Psi p - y)$$

GR is growth rate, Ψp is turgor, y is the yield threshold (pressure below which cell wall resists plastic or non-reversible deformation), and m is wall extensibility (response of wall to pressure). In this equation, decrease in turgor causes a decrease in growth rate. Ψp needs decrease only to the value of y, not to zero, to eliminate expansion. In normal conditions, y is usually 0.1 to 0.2 MPa less than Ψp, so growth rate takes place over a very narrow range of turgor. Inhibition of cell expansion results in slowing of leaf expansion. The smaller leaves transpire less water, and conserves limited water in soil for use over a longer period.

In intact leaves, stress not only decreases turgor, but also decreases m and increases y. Under unstressed conditions, wall extensibility (m) is greatest when cell wall solution is slightly acidic. Stress decreases m in part because proton transport across plasma membrane is inhibited resulting in raising of cell wall pH. Cell wall changes are very important because unlike turgor, they are slow to recover after re-hydration. Water deficient plants tend to become re-hydrated at night, as a result substantial leaf growth occurs at night. However, because of changes in m and y the rate of growth is still less than the growth of unstressed plants at the same turgor. Water stress not only limits the size of individual leaf but also number of leaves on indeterminate plants, because it decreases both number and growth rate of branches.

Stimulates leaf abscission

The total leaf area of the plant does not remain constant after all the leaves have matured. If the plants become water stressed after a substantial leaf area has developed leaves will senesce and eventually fall off. This leaf area adjustment is an important long term change that improves plant's fitness for water limited environment. Many drought-deciduous desert plants drop all their leaves during drought and sprout new ones after a rain. Abscission during water stress results largely from enhanced synthesis of and responsiveness to the endogenous plant hormone ethylene.

Enhances root extension into deeper, moist soil

Mild water deficits also affect the development of root system. Root-shoot relations appear to be governed by functional balance between water uptake by roots and photosynthesis by the shoots. A shoot will grow until it becomes so large that water uptake by roots becomes limiting to further growth; conversely, roots will grow until their demand for photosynthates from the shoots equals the supply. This functional balance is shifted when water supply decreases.

When water uptake is curtailed, leaf expansion is affected very early, but photosynthetic activity is much less affected. Inhibition of the leaf expansion reduces the consumption of carbon and energy, and the greater proportion of plant's assimilates can be distributed to the root system where they can support further growth. At the same time, root apices in the dry soil loose turgor. All these factors lead to root growth preferentially into soil zone that remains moist. As the water deficit progresses the upper layers of soil dry first, the roots proliferate deeper into moist layers. Enhancement of root growth into moist soil zones during stress depends on the allocation of assimilates to the growing root tip. As, alternative sinks for assimilates, fruits predominate over roots, and assimilates are directed to the fruits and away from the roots. This competition between roots and fruits is one explanation for the fact that plants are generally more sensitive to water stress during reproduction.

Stomata close during water deficit in response to abscissic acid

Under first two responses we focused on changes in plant development during slow, long-term dehydration. When the onset of stress is more rapid or the plant has reached its full leaf area before initiation of stress, other responses protect the plant against immediate desiccation. Under these conditions, the stomata close to reduce evaporation from the existing leaf area. Because guard cells are exposed to the atmosphere, they lose water directly by evaporation and so lose turgor, causing stomata to close by a mechanism called hydro-passive closure. A second mechanism, called hydro-active closure, closes the stomata when the whole leaf or the roots are dehydrated and depends upon metabolic processes in the guard cells. A reduction in the solute content of the guard cells results in water loss and decreased turgor, causing the stomata to close.

Water deficit limits photosynthesis within the chloroplast

Water stress usually affects both stomatal conductance and photosynthetic activity in the leaf. Upon stomatal closure during early stages of water stress, water use efficiency may increase (more CO_2 is taken up per unit of water transpired) because stomatal closure inhibits transpiration more than it decreases intercellular CO_2 concentrations. Additionally dehydration of mesophyll cells also inhibits photosynthesis. As stress becomes severe, however, water use efficiency usually decreases, and the inhibition by the mesophyll cells becomes stronger.

Water deficit increases wax deposition on the leaf surface

A common developmental response to water stress is production of thicker cuticle that reduces water loss from the epidermis (cuticular transpiration). A thicker cuticle also decreases CO_2 permeability, but leaf photosynthesis remains same because epidermal cells underneath the cuticle are non-photosynthetic.

Water deficit may induce CAM

It is a plant adaptation in which stomata open at night and close during the day. The leaf-to-air vapour pressure deficit that derives transpiration is much reduced at night, when both leaf and air are cool. As a result the water use efficiency of the CAM plants are among the highest measured in all higher plants.

Response of fruit plants to water stress and its management

In fruit orchards, water absorption content from the soil is less than that of evaporation through the stomata of the leaves under water stress conditions. Physiological and morphological changes in plant structure, growth rate, tissue osmotic potential, and stomatal conductance occur due to internal drought tolerance mechanisms in response to external environmental factors (Ruiz Sanchez *et al.*, 1997; Save *et al.*, 1993).

The effect of water stress on plant hormones (GAs, IAA and ABA) level in the leaves and flower bud formation of the Satsuma mandarin (*Citrus unshiu* Marc.) trees was investigated to determine the relationship between flower-bud induction and the level of endogenous plant hormones as a result of water stress. Severe water stress (-1.5 to -2.0 MPa) in autumn, which causes heavy leaf fall, reduced the percentage of flowering nodes by one-third of the moderately water stressed ones (-0.5 to -1.0 MPa). The $GA_{1/3}$ levels in the leaves was higher in the less flower producing severely water stressed trees during flower inducing period, and involvement of endogenous GA_3 in the leaves in flower bud formation of Satsuma mandarin is suggested. The content of IAA in the leaves of the trees under moderate water stress was higher in late February (Koshita and Takahara, 2004). Some soil bacteria among plant growth-promoting rhizobacteria (PGPRs) can stimulate plant growth even under stressful conditions by reducing ethylene levels in plants, hence the term "stress controllers" for these bacteria (Kang *et al.*, 2010).

It is well known that stomatal function is closely related to leaf water status and especially to leaf turgor. The latter could be mediated during water stress conditions by two distinct strategies: (i) by lowering the osmotic potential due to active solute accumulation and (ii) by increasing the elasticity of the cell walls. Chartzoulakis *et al.* (2002) reported that water stress affected avocado cultivars to a different degree. 'Hass' seems more affected by water stress, since at the same RWC (relative water content) turgor potential was lower than that of 'Fuerte'.

Different mechanisms are involved in the plant's response to water limitation. One of the most common mechanisms is the stomatal closure, which reduces water loss and regulates plant water potential. Leaf water potential has been used as a sensitive indicator of plant water stress. It is believed that hormone signals from the roots are involved in the initial response to drought. Zhu *et al.* (2004) investigated leaf water potential and endogenous cytokinins in xylem sap of young apple trees treated with or without paclobutrazol under drought stress conditions. The results showed that leaf water potential of the plants decreased significantly when drought was applied, while for the paclobutrazol treatment leaf water potential was significantly less. The paclobutrazol treated plants generally showed a higher concentration of zeatin riboside compared with the drought treatment alone. This might be due to a reduction of cytokinin oxidation prevented by paclobutrazol.

Lee *et al.* (2006) reported that the formation of stone cells in pear seems to be elevated by water stress condition. Water stress in pear trees results in decreased root activity and decreased leaf water potential. The decrease in water potential appeared to suppress the absorption of calcium due to reduced transpiration, increasing the activity of peroxidase. Increased peroxidase activity increased cell wall lignification and the formation of stone cells.

Droughted and non-droughted trees showed similar gas exchange values at the end of the droughting period when measured at 25°C, but gas exchange of droughted trees remained repressed when measured at 30°C due to increased non-stomatal limitation. This study showed that post-harvest stomatal control was more attuned to ambient atmospheric evaporative demand in the orchard than to slowly developing soil moisture stress. So, both atmospheric factors and changing sink demand influence stomatal control of apple leaves post-harvest and that this should be taken into account when determining irrigation strategy (Pretorius and Wand, 2003). The wider adaptability of papaya to tropical and subtropical environments has bestowed much significance to its cultivation in Indian subcontinent. However, the abiotic stresses particularly, higher ambient temperature (>35°C), photosyn*thetic photon flux density (>1650 ìmolm*$^{-2}$s^{-1}) and water deficit (<82 % leaf relative water content) experienced by the plants during summer months (March-May) influence many eco-physiological processes and chlorophyll fluorescence. Net CO_2 assimilation (P_n) and stomatal conductance (g_s) are the key physiological indices, that declined when the plants are exposed to abiotic stresses. Papaya, being a heavy yielder shows significant reduction in fruit and latex yield when there is a marginal reduction in P_n. In the present study, eco-physiological parameters *viz*., P_n, g_s, transpiration (E) and water use efficiency (P_n /E) were observed during summer months in selected papaya cultivars. Chlorophyll stability and fluorescence under dark- adapted conditions were also assessed. Excessive light intensity and high temperature or water deficit in the plant system sensitized the physiological processes to slow down, and the data on chlorophyll fluorescence revealed the influence of abiotic stress on photosynthetic processes. The papaya cv. Co-7 had higher net CO_2 assimilation (P_n = *15.7 ìmol m*$^{-2}$ s^{-1}) coupled with better physiological efficiency as evidenced by higher Fv/Fm (0.802). The stress tolerance was seen with higher Fm/Fo (4.05). The data on g_s, E and WUE were also found more associated to stress tolerance. The results of the present study might aid not only in identifying papaya cultivars those could tolerate summer months, but also in developing crop improvement and management systems (Jeyakumar *et al.* 2007).

The protein osmotin is involved in water stress. Xiong *et al.* (2002) reported on cell signalling following cold, drought and salt stress. The problems of drought and salinization could be solved by using drought- and salt-resistant cultivars, which are also appropriate for high-density plantings.

Mitchell *et al*. (1989) showed a decrease in the number of flowers per tree in European pear in response to deficit irrigation early in the season. Severe post-harvest water stress delayed flower bud development in stone fruits (Naor *et al*., 2005). Southwick and Devenport (1986) reported that both continuous and cyclical water-stress treatment induces flowering.

In-situ water harvesting is one of management practice used in orchards to protect trees from water stress. The term *in- situ* water harvesting refers to measures of storing water in soil profile under the tree canopy during rain spells to augment the seasonal requirement of the plant (Pareek *et al.*, 1996). In fruit trees under rain fed condition critical stage of water requirement of a plant species is of paramount importance and therefore the *in-situ* water harvesting and moisture conservation measures should aim to provide the maximum moisture to the plants at the critical stage of its requirement (Evenari *et al.*, 1968). Nath *et al.* (2005) reported that full moon terracing followed by mulching with paddy straw resulted in early panicle initiation, higher fruit set and final fruit retention in litchi. Minimum fruit drop and fruit cracking was also recorded in the same treatment.

Ber (*Ziziphus mauritiana*) is considered as the king of arid zone fruits, because of its adaptations to tolerate the biotic and abiotic stresses prevailing under rainfed conditions. Crop diversification through arid fruits like ber ensured some production during the drought year when traditional crops fail. Among the cultivars evaluated, Kaithali provided consistently significant higher yields compared to other cultivars. As regards the *in situ* water harvesting, increasing the catchments area at 2.5 times of the normal area with 5% slope exhibited higher yield compared to 1.5 and 2.0 times of the normal area, normal area with and without slope. The yield contributing factors were also positively improved (Anbu, *et al.*, 2009).

The use of drought-tolerant rootstocks would minimise the immediate effects of dry conditions and enable the variety to recover quickly. Dog Ridge (*Vitis champinii*), 110 R (*Vitis berlandieri* × *Vitis rupestris*), 1103 P (*Vitis berlandieri* × *Vitis rupestris*), 99 R (*Vitis berlandieri* × *Vitis rupestris*), St. George (*Vitis rupestris*) are some of the drought tolerant rootstocks for grapes. Dog Ridge was the only popular rootstock used by grape growers prior to late 1990s. However, Satisha, *et al.*, (2010) reported that during the initial years, Thompson Seedless grafted on Dog Ridge produced the highest yield, with good quality fruit. Over the years we could observe uneven bud sprouting, gaps on the cordon due to dead wood formation, and reduced yield in vines grafted on Dog Ridge rootstocks. In contrast, Thompson Seedless grafted on 110R performed well in terms of moderate vigour, increased fruitfulness and consistently higher yield. Dog Ridge and St. George produced a lower yield, owing to increased vigour measured in terms of pruning weight, total shoot length and cane diameter. Rootstocks 110R, 1103 P and 99 R have good potential in the tropical and subtropical climate of India due to their drought tolerant characteristics, such as increased water-use efficiency and high proline content, which was established in our earlier studies (Satisha *et al.*, 2008).

Molecular breeding methods have recently been widely applied to improve the stress- and disease resistance of different fruit crops. To this end the suitability of a ferritin gene derived from *Medicago sativa* (*MsFerr*) was tested in grapes.

It is proposed that by sequestering the intracellular iron involved in generation of the very reactive hydroxyl radicals through Fenton-reaction, the increased overall ferritin concentration results in increased protection of plant cells from oxidative damage induced by a wide range of stresses. The existing regeneration and transformation protocols were improved. The number of regenerated plants were raised due to the application of 1 μM benzyladenine (BA). The plant regeneration was supported by the maintenance of plant material on "half MS" medium without selection agent (after two-year selection), and by cut of abnormal embryos under the hypocotyl. The independent regenerated transformants were tested by PCR, qPCR, Western blot, and were used for various abiotic stress tolerance experiments (Zok, *et al.*, 2009). A strawberry genomic clone containing an osmotin-like protein (OLP) gene, designated *FaOLP2*, was isolated and sequenced. *FaOLP2* is predicted to encode a precursor protein of 229 amino acid residues, and its sequence shares high degrees of homology with a number of other OLPs. Genomic DNA hybridization analysis indicated that *FaOLP2* represents a multi-gene family. The expression of *FaOP2* in different strawberry organs was analyzed using real-time PCR. The results showed that *FaOLP2* expressed at different levels in leaves, crowns, roots, green fruits and ripe red fruits. In addition, the expression of *FaOLP2* under different abiotic stresses was analyzed at different time points. All of the three tested abiotic stimuli, abscissic acid, salicylic acid and mechanical wounding, triggered a significant induction of *FaOLP2* within 2-6 h post-treatment. Moreover, *FaOLP2* was more prominently induced by salicylic acid than by abscissic acid or mechanical wounding. The positive responses of *FaOLP2* to the three abiotic stimuli suggested that strawberry *FaOLP2* may help to protect against osmotic-related environmental stresses and that it may also be involved in plant defense system against pathogens (Zhang and Shih, 2007).

Chilling and freezing stress

Sensitive species are injured by chilling at temperatures that are too low for normal growth but not low enough for ice to form. Typically tropical or sub tropical species are susceptible to chilling injury. Freezing injury on the other hand, occurs at temperatures below the freezing point of the water.

Plant responses to chilling and freezing injury

Membrane properties change in response to chilling injury

Leaves from plants injured by chilling show inhibition of photosynthesis and carbohydrate translocation, slower respiration, inhibition of protein synthesis, and increased degradation of existing proteins. All these responses probably depend on a common primary mechanism involving loss of membrane function during chilling.

Why are the membranes of chill sensitive plants affected by chilling?

Plant membranes consist of lipid bi-layer interspersed with proteins and sterols. The physical properties of the lipids greatly influence the activities of the integral membrane proteins including H^+ ATPases, carriers and channel forming proteins that regulate the transport of ions, and solutes, and of enzymes on which metabolism depends.

In chill sensitive plants, the lipids in the bi-layer have a high percentage of saturated fatty acid chains, and this type of membrane tends to solidify into a semi-crystalline state at a temperature well above 0°C. As the membranes become less fluid, their protein components can no longer function normally. The result is inhibition of H^+- ATPase activity, of solute transport into and out of cells, of energy transduction and of enzyme-dependent metabolism. In addition, chill-sensitive leaves exposed to high photon fluxes and chilling temperatures are photo inhibited causing acute damage to photosynthetic machinery. Membrane lipids from chill-resistant plants have a greater proportion of unsaturated fatty acids than those from chill-sensitive plants. During acclimation to cool temperatures the activity of desaturase enzyme increases and the proportion of unsaturated lipids rise. This modification allows membranes to remain fluid by lowering the temperature at which membrane lipids begin a gradual phase change from fluid to semi-crystalline.

Freezing kills cells by forming intracellular ice crystals or by dehydrating the protoplast

When tissue is cooled under natural conditions, ice usually forms first within the intercellular spaces, and in the xylem vessels, along which the ice can quickly propagate. This ice formation is not lethal to hardy plants, and the tissue recovers fully if warmed. However, when plants are exposed to freezing temperatures for an extended period, the growth of extracellular ice crystals results in the movement of liquid water from the protoplast to the extracellular ice, causing excessive dehydration. Several hundred molecules are needed for an ice crystal to begin forming. The process whereby these hundreds of ice molecules start to form a stable ice crystal is called ice nucleation. Some large polysaccharides and proteins facilitate ice crystal formation, and are called ice nucleators. In plant cells, ice crystals begin to grow from endogenous ice nucleators, and the resulting relatively large intracellular ice crystals cause extensive damage to the cell and are usually lethal.

Management of chilling/freezing stress

Efforts to protect crops from freezing injury began at least 2000 years ago, when Roman farmers protected grapes by burning scattered piles of dead vines and prunings (Blanc *et al.* 1963). Over the past 100 years, several different methods of freeze protection have emerged from considerable research efforts devoted to reduction of freezing injury in horticultural crops. Before understanding the methods of freeze protection, we need to know some of the terminology associated with freeze protection.

Freeze vs. Frost

The primary usage of "frost" given in the Glossary of Meteorology (Huschke, 1959) is a synonym for hoarfrost: "a deposit of interlocking ice crystals (hoar crystals) formed by direct sublimation on objects such as tree branches. Plant stems and leaf edges". The secondary usage is given as "The condition which exists when the temperature of the earth's surface and earthbound objects falls below freezing, 0°C or 32°F." Hoarfrost may or may not be formed during a frost, and when it is not formed the event is termed as "Black frost". The primary usage of "freeze" denotes the phase change from liquid to solid, whereas the secondary usage is given as "The condition that exists when, over a widespread area, the surface temperature of the air remains below freezing for a sufficient time to constitute the characteristic feature of the weather".

Radiative vs. advective

The term "radiative" (radiational, radiation, radiation type) and "advective" (advection) are commonly used to modify freeze (or frost) and convey information on the intensity, duration and general meteorological conditions during a period of subfreezing temperatures. A radiative freeze is characterized by clear skies, calm or light wind, and relatively high subfreezing temperature and dewpoint. During advective freezes, subfreezing temperatures result largely from an influx of an arctic or polar air mass into a region. Such air mases typically have a relatively low air temperature and dewpoint and high wind speed, and unlike radiative freezes may have cloud cover. Radiative freezes often follow advective freezes because cold, dry air masses are conducive to radiant heat loss and wind velocity usually decreases after passage of the leading edge of the air mass (cold front). An important distinction is that vegetation and earthbound objects cool the air during radiative freezes, whereas an influx of air cools the vegetation and earthbound objects under advective freeze conditions. However, there are no rigid threshold values of air temperature, wind speed and dew point above or below which a freeze is classified as radiative or advective.

Cold Hardiness of plants is affected by following factors

Environmental

Light is important in cold acclimation since hardiness is related to carbohydrate content (Flore and Howell, 1987). Citrus seedlings did not harden when exposed to acclimating temperatures in darkness (Young 1969). Lack of adequate light due to shading by nearby plants, improper pruning and training, etc., can therefore result in increased incidence of winter injury. Decreasing photoperiod induces dormancy and increases cold hardiness in temperate zone woody plants (Weiser, 1970). Therefore, long photoperiods during the acclimation process may be expected to reduce cold hardiness.

Water stress has been shown to enhance hardiness of many plants. Davies *et al* (1981) reported that 'Orlando' tangelo trees not receiving irrigation in the fall had less leaf and fruit damage than fall irrigated trees following a radiative freeze. Conversely, Koo (1981) found that trees irrigated in the fall had less leaf and fruit damage following a freeze than those unirrigated during the fall. The conflicting results with citrus suggest that the level of water stress is important in determining the plant cold hardiness response.

Pruning and crop load

Pruning and crop load affect the cold hardiness of certain woody perennials. Pruning fruit trees late in the fall can reduce cold hardiness and increase incidence of winter injury. Fall pruning is often cited as a contributing factor in peach tree short life complex (Reilly *et al.* 1986) which is closely linked to freezing injury. Timing and severity of pruning also affect cold hardiness of grapevines. A common practice to delay pruning of vines until the danger of freezing temperatures has passed so that bud break will not be induced (Howell and Dennis, 1981).

Nutrition

The effect of nutritional status on cold hardiness of plants has been reviewed thoroughly by Pellett and Carter (1981). One report (Savage, 1970) not cited by Pellet and Carter indicates that late fall nitrogen application nitrogen application to peaches delayed bloom up to 6 days on twig samples taken from the trees the following spring and placed in a heated greenhouse. However the trees from which the twigs were taken did not show any difference in bloom date due to fall nitrogen application under orchard conditions.

Rootstock

The rootstock can influence the hardiness of the scion cultivar of several fruit tree species (Westwood 1970). Selection of a cold hardy rootstock is important where midwinter temperatures approach the killing temperature of the scion cultivar. Apple trees on M.5, M.7 and M.9 rootstocks had greater trunk injury than those on M.1, M.4, M.16 or seedling rootstocks after exposure to -23^{0}C (Westwood and Bjornstad, 1981)

The influence of rootstock on scion hardiness of pear, cherry, peach, and other deciduous fruit trees has been thoroughly reviewed by Westwood (1970). It has long been recognized that hardiness of citrus trees is affected by rootstock (Young and Olson, 1963).

Chemicals

Numerous studies on the use of growth regulators and various other chemicals to increase cold hardiness and/or delay spring development of crop plants have been reviewed by Howell and Dennis (1981). It is now established that chemicals have the potential to provide freeze protection. A summary of data on chemicals used to increase cold hardiness or delay budbreak of horticultural crops is given in Table 1 and 2.

Table 1: Reports of increased hardiness induced by chemical substances in horticultural crops.

Crop	Chemical substance	Effect on hardiness	Reference
Citrus paradisi	ABA	0-	Young 1971
(grapefruit)	Chloremquat	0	Young 1971
	Daminozide	0	Young 1971
	Kinetin, Benzyladenine	+	Stewart & Leonard 1960
	Maleic hydrazide	+	Hendershott 1965
C. Sinensis	DMSO	0	Burns 1970
	Maleic hydrazide	+	Hendershott 1965
		0	Burns 1970
Strawberry	Daminozide	+	Freeman and Carne 1970
	Maleic hydrazide	0	White and Kennard 1955
Apple	ABA	+	Holubowicz and Boe 1969
	AMO-1618	+	Raese 1977
	Chlormequat	+	Raese 1983
	Daminozide	+	Raese 1977
		0	Modilbowska 1968
	Ethephon + NAA + Daminozide	+	Raese 1977
	Ethephon	0	Dennis 1976
Apricot	Ancymidol	0	Dennis 1976
	Paclobutrazol	-	Proebsting & Mills 1985

Contd.

Sweet cherry	Daminozide	-	Proebsting & Mills 1969
	Ethephon	+	Proebsting & Mills 1973
		+	Dennis 1976
	GA	-	Dennis 1976
Sour cherry	Ancymidol	0	Dennis 1976
	GA	-	Dennis 1976
Peach	Daminozide	0	Edgerton 1966
	Ethrel	+	Funt & Ferree 1985
	GA	+	Edgerton 1966
		+	Corgan and Widmoyer 1971
	Paclobutrazol	-	Proebsting & Mills 1985
Pear	Chlormequat	+	Modlibowska 1968
	GA	-	Modlibowska 1968
	Glycerol, Ethylene	+	Ketchie & Murren 1976
Grape	Maleic hydrazide	0	White & Kennard 1955

+ = increase, 0 = no effect, - = decrease

Table 2. Reports of delay of budbreak induced by chemical substances in fruit crops.

Crop	Chemical substance	Delay of budbreak	Reference
Orange	Maleic hydrazide	+	Hendershott 1965
Strawberry	Maleic hydrazide	0	White & Kennard 1955
Apple	Daminozide	0	Dennis 1976
		+	Sullivan & Widmoyer 1970
	Ethephon	0	Dennis 1976
	Maleic hydrazide	0	White & Kennard 1955
	NAA	+	Hitchcock & Zimmerman 1943
Almond	Ethephon	+	Browne *et al.* 1978
	GA	+	Hicks & Crane 1968
Apricot	Ancymidol	0	Dennis 1976
	Ethephon	+	Dennis 1976
	Paclobutrazol	-	Proebsting & Mills 1985
Sweet cherry	Ethephon	+	Proebsting & Mills 1973
		+	Dennis 1976
	GA	0	Dennis 1976
		+, 0	Proebsting & Mills 1974
	NAA	+	Hitchcock & Zimmerman 1943
	Paclobutrazol	-	Proebsting & Mills 1985
Plum	AVG	+	Dennis *et al.* 1977
	Ethephon & GA3	+	Webster 1985
	NAA	+	Hitchcock & Zimmerman 1943
Peach	AVG	+	Dennis *et al.* 1977
	CGA-15281	+	Coston 1985
	Daminozide	+,0	Coston 1985
	Ethephon	+	Coston 1985
	Ethrel	+	Funt & Ferree 1985
		+	Corgan & Widmoyer 1971

Contd.

	NAA	+	Hitchcock & Zimmerman 1943
	Paclobutrazol	-	Proebsting & Mills 1985
Pear	Chlomequat	-	Modilbowska 1968
	NAA	+	Hitchcock & Zimmerman 1943

+ : delay, 0 : no effect, - : advance

Karlidag *et al.*, (2009) conducted a study to determine the effect of foliar Salicylic acid (SA) applications on fruit-quality characteristics and yield of strawberry under anti-frost heated greenhouse conditions. Spraying of 1 mM SA (1 mM) was done once (SA1), twice (SA2), three times (SA3), or four times (SA4) during the vegetation period with 7 d intervals. The early yield and total yield of strawberry were significantly affected by SA applications, among which SA3 and SA4 resulted in the highest early and total yields. They suggested that SA3 and SA4 treatments can ameliorate the deleterious effects of low temperatures on strawberry plants and that SA application may offer an economical and simple method for low-temperature protection.

Sprinkling to delay spring budbreak

Phenological development of flower buds on fruit trees is related to the number of degree hours above about 4.5^{0}C accumulated after the chilling requirement is completed. Cold hardiness of flower buds decreases by 10^{0}C or more as bud development proceeds from the dormant to full bloom condition (Ballard and Proebsting 1978). Hence, evaporative cooling of buds by sprinkling during the period of degree hour accumulation in late winter can slow the rate of bud development and delay bloom sufficiently to reduce the risk of freezing injury.

Heating

Heating has been employed for centuries as a method of freeze protection, and is still widely practiced throughout the world today. Heating is regarded as one of the most reliable methods of freeze protection.

Heating provides freeze protection by raising air temperature within the planting and through radiant heat transfer from the heater stack or flames directly to plant surfaces (Perry *et al.*, 1977). As warm air rises from the heater, it cools rapidly and stops rising where its density equals that of the surroundings. An important and widely accepted generalization stems from this concept: several small fires provide better protection than fewer large fires. Large fires cause air to rise above the warmest level of the inversion, "puncturing" the inversion and allowing much of the heated air to rise above the level of the crop.

Irrigation

Irrigation is one of the most popular methods of freeze protection used today. In general, surface or furrow flooding provides 1 to 3°C protection during radiative freezes only. Sprinkling for freeze protection can be beneficial or deleterious because freezing of water on the plant releases heat, while evaporation extracts heat.

Wind machines

On radiative freeze nights when low-level temperature inversions occur, warm air above the planting can be mixed with cooler air at ground level to provide a limited amount of temperature moderation. Wind machines are most often used for this purpose. Advantages of wind machines over other freeze protection methods include reduced labour requirement, low operating cost, low fuel input per °C of protection, and proven effectiveness under radiative freeze conditions for over 60 years. Disadvantages include high initial costs and lack of effectiveness under advective conditions or unusually cold radiative freeze conditions.

Fogging

Fog is essentially a thin stratus cloud at ground level, containing water droplets with radii ranging from 2 to 40 µm, and liquid water contents of 0.05 to 0.1 gm^{-3}. Such clouds are effectively "black" or totally opaque to thermal radiation in the 8 to 12 µm range of wavelength (known as "atmospheric window"). Therefore radiative cooling of the earth and plants is generally reduced on cloudy or foggy nights. In addition, condensation of water on plants surrounded by fog replaces any heat lost by radiation through thin or low density fog layers. Fog increased survival of citrus trees during a severe freeze (-9°C) in Arizona, although protection decreased with distance from the fog emitting lines (Finch, 1977). In other field trials with citrus, fog typically increased air temperature 1 to 1.5°C (Brewer *et al*., 1973).

Over the past two decades, significant advances have been made in understanding the genetic regulation of cold hardiness, as well as resistance to other abiotic stresses. At first, research focused on isolating and characterizing cold-regulated (*cor*) genes and then advanced to the discovery of cold-induced transcription factors and the characterization of cold-induced changes in whole genomes using microarray technologies. Metabolomics and proteomics have also provided a wealth of new information on the metabolism and biochemistry of cold acclimation. In fruit trees, and woody plants in general, cold hardiness is a complex trait and a thorough understanding of the physiology of freezing tolerance is needed if biotechnology is going to be used effectively to improve environmental stress resistance. The factors that limit cold hardiness in mid-winter are very different from those that are responsible for frost susceptibility in the spring. In contrast to what is observed in herbaceous crops, tissues in woody plants that are in very close proximity to each other can differ dramatically in cold hardiness and the mechanisms by which they cold acclimate. For example, xylem tissues are generally less cold hardy than bark tissues, and the flower buds of some fruit crops exhibit deep super-cooling, a freeze-avoidance mechanism that relies on the biophysical properties of the bud tissues. Therefore, different approaches will be needed to influence cold resistance depending on the type of injury (mid-winter or spring) and tissue-type (buds vs. stems) that is being targeted for improvement. Despite this complexity, significant opportunities exist for improving resistance to environmental stress in fruit crops. The targeting of dehydrin proteins to flower buds and the use of transcription factors to regulate suites of genes are two approaches. Additionally, the over-expression of antioxidant enzyme (APX or SOD) genes is also a viable approach to improving resistance to several environmental stresses. Recent genomic and proteomic research on stress response in fruit trees is also being used to develop a more comprehensive understanding of environmental stress resistance (Wisniewski *et al.*, 2007).

Climatic changes create problems in olive culture due to frost damage. Most olive cultivars can resist temperatures no lower than –12°C after acclimation due to freezing-tolerance mechanisms (Thomashow, 1999). However, in certain olive-growing areas lower temperatures are common. Cold-hardiness is a very important trait for olive improvement (Bartolozzi and Fontanazza, 1999). Improved cold tolerance could involve one or more of the following:

- The superoxide dismutase gene (*SOD*), which is responsible for repair of ozone-damaged cells (Van Camp *et al.*, 1993).
- The *CBF1* gene, which increases tolerance to cold (Jaglo-Ottosen *et al.*, 1998).
- The involvement of the cryoprotective proteins COR/LEA/dehydrin (Late Embryogenesis Abundant proteins). Among the genes that are the most highly induced during cold acclimation are: (i) the 'classical' *COR* (coldregulated) genes, alternatively designated *KIN* (cold-induced); (ii) *RD*(responsive to dehydration); (iii) *LTI* (low temperature-induced); and (iv) *ERD* (early responsive to dehydration) (Thomashow, 1998). The proteins encoded by these genes are extremely hydrophilic and are either novel or members of the dehydrins (Close, 1997).

Salinity stress

Soil salinity is a serious problem in arid and sub-arid climates, such as in the Mediterranean region, where plants are subjected to high temperature regimes and extreme water deficits during the dry season. Under these climatic conditions salts tend to accumulate in the soil because of the high evaporative demand and insufficient leaching of ions, problems often exacerbated by the use of brackish irrigation water in areas of intensive agriculture (Gucci and Tattini, 1997). Saline water can arise from either drainage effluent from irrigated land or from contamination of fresh ground water supply by seawater in coastal areas. Salinity due to the geological origin of soils occurs less frequently. Salinity is one of the main factors limiting productivity in fruit trees (Flowers and Yeo, 1995).

Salt accumulation in soils impairs plant function and soil structure

The high Na^+ concentrations of sodic soils not only injure plants directly but also degrade the soil structure, deceasing porosity and water permeability. The higher the salt concentration of water, the greater is EC and lower is its osmotic potential. The quality of irrigation water is poor in semi-arid and arid regions.

The properties of good quality irrigation water (Concentration of ions in mM) :

Na^+	< 2.0
K^+	< 1.0
Ca^+	0.5-2.5
Mg^{2+}	0.25-1.0
CL-	<2.0
SO_4^{2-}	0.25-2.5
HCO_3^-	<1.5
Osmotic potential (MPa)	-0.039
Total dissolved salts (ppm)	500

Dissolved solutes in the rooting zone generate a low osmotic potential that lowers the soil water potential. The general water balance of plants is thus affected, because leaves need to develop an even lower water potential to maintain a "downhill" gradient of water potential between the soil and the leaves.

Under salt stress conditions, plant cells develop strategies to cope with Na^+ and Cl^-, including exclusion and compartmentalization, induction of antioxidant enzymatic systems and compatible solutes accumulation, such as proline. The precise function of this osmolyte still remains unclear. Proline may act on osmotic adjustment, as a free radical scavenger, protecting enzymes and avoiding DNA damages. It has been also suggested the role of proline in prevention of lipid peroxidation and as a signalling/regulatory molecule. A salt-sensitive *Citrus sinensis* 'Valencia late' cell line has a smaller growth rate and accumulates proline in the presence of NaCl (>200 mM). The addition of external proline to this cell line was evaluated in terms of cell metabolism. A positive influence on the relieve of salt stress symptoms due to the presence of exogenous proline 5 mM and 100 mM NaCl was obtained, with increased growth of this salt sensitive citrus cell line (LimaCosta *et al.*, 2010).

Salinity tolerance in Olive

Olive is more salt and drought tolerant than other temperate fruit trees (Rugini and Fedeli, 1990), and it is considered less demanding in terms of nutrients and energy inputs than other fruit crops (Bongi and Palliotti, 1994).

Typical symptoms of salt stress in olive plants are reduced growth, leaf tip burn, leaf chlorosis, leaf rolling, wilting of flowers, root necrosis, shoot dieback, and defoliation. Necrotic areas develop first at the distal end of mature leaves and then expand to the rest of the leaf. Bernstain (1964) reported a 10% reduction in growth when the EC of the soil solution was 4.6 mmho cm^{-1} and Bartolini *et al.* (1991) a 30% reduction at 90 meq L^{-1} NaCl. Shoot growth is completely inhibited at NaCl concentrations higher than 200 mM (Tattini *et al.*, 1995). Bartolini *et al.* (1991) reported that mortality of young plants ('Maurino') increased from 4 to 53% after one year of salinization at NaCl concentration of 40 and 90 meq L^{-1}, respectively. Salinity reduces viability and germinability of pollen, mean number of perfect flowers per inflorescence, and fruit set (Therios and Misopolinos, 1988; Cresti *et al.*, 1994) but does not significantly affect fresh weight, size, and drop of the fruit at moderate concentrations (Klein *et al.*, 1994).

Table 3: Classification of olive cultivars according to their relative salinity tolerance

Resistance	Cultivar	Reference
Tolerant	Megaritiki	Therios and Misopolinos, 1988
	Frantoio	Tattini *et al.* 1994
	Arbequina	Marin *et al.* 1995
	Picual	Marin *et al.* 1995
Intermediate	Carolea	Briccoli Bati *et al.* 1994
	Coratina	Tattini *et al.* 1994
	Moraiolo	Tattini *et al.* 1994
	Manzanillo	Klein *et al.* 1994
Sensitive	Leccino	Tattini *et al.* 1994
	Pajarero	Marin *et al.* 1995

The genes *SOS1*, *SOS2* and *SOS3* are postulated to encode regulatory components for salt tolerance. Do such genes exist in olives? Evidence indicates a critical role for K in salt tolerance (Zhu *et al.*, 1998). Furthermore, high salinity affects cellular and molecular responses (Hasegawa *et al.*, 2000).

Flooding stress

In flooded or waterlogged soils with poor drainage, the pores are filled by water and diffusion of O_2 is blocked. When temperatures are higher (>20°C) oxygen consumption by plant roots, soil fauna, and soil micro-organisms can totally deplete the oxygen from the bulk of the soil water in as little as 24 hours. Thus anaerobic condition sets in and the growth and survival of many plant species are greatly depressed under such conditions, and crop yields can be severely reduced.

Roots are injured in anaerobic soil water

In the absence of O_2 TCA cannot operate and ATP is produced only by fermentation. Thus roots begin to ferment pyruvate to lactate through the activity of Lactate Dehydrogenase. The lactate fermentation is transient because the accumulation of lactic acid lowers the cellular pH. At acidic pH Lactate Dehydrogenase is inactivated and Pyruvate Decarboxylase is activated which leads to production of ethanol from lactic acid.

The net yield in case of lactic acid fermentation is 2 moles of ATP as against 36 moles of ATP under aerobic respiration per mole of hexose sugar respired. Thus injury to root metabolism by O_2 deficiency originates in part from a lack of ATP to drive essential metabolic processes.

Under anaerobic conditions

$NO_3^- \rightarrow NO_2^- \rightarrow N_2O \rightarrow$ and $\uparrow N_2 \uparrow$

Severe anaerobic conditions

$Fe^{3+} \rightarrow Fe^{2+}$ (More soluble)

$SO_4^{2-} \rightarrow H_2S$ (Respiratory poison)

The failure of O_2 deficient roots to function injures shoots

Hypoxic roots lack sufficient energy to support physiological processes on which shoots depend. The failure of the roots to absorb nutrient ions and transport them to the xylem (and from there to the shoots) leads to quick shortage of ions within developing and expanding tissues. Older leaves senesce prematurely because of re-allocation of phloem mobile elements (N,P,K) to younger leaves. Hypoxia also accelerates production of the ethylene precursor, 1-amino cyclopropane-1-carboxylic acid (ACC), in roots. ACC travels via the xylem sap to the shoot, where, in contact with O_2, it is converted by ACC oxidase to ethylene.

Responses of fruit crops to flooding and its management

Plants vary considerably in their ability to tolerate waterlogged soil. Poor soil aeration associated with flooding results in changes that are usually deleterious to growth and development of many plant species (Schaffer *et al.* 1992). Flood responses among fruit crops are directly related to the dramatic changes in O_2 availability and the chemical and physical states of soil that occur after flooding. These changes have profound effects on plant growth and development.

Plant tolerance to waterlogged soil conditions may be influenced by a number of factors: (1) Soil type, porosity, and chemistry; (2) degree and duration of anaerobiosis; (3) soil microbe and pathogen status; (4) vapor pressure deficits and root zone and air temperatures; (5) plant are, stage of development, or season of the year; and (6) plant preconditioning responses as a result of prior climatic and edaphic conditions. The importance of each of these variables may differ with plant species. The tolerance of fruit trees to waterlogged conditions is mainly determined by the rootstock and not the scion (Anderson *et al.* 1984a, 1984b), although specific foliar symptoms of flooding injury may vary with the scion (Anderson *et al.* 1984b).

Table 4: Tolerance of citrus rootstocks to flooded soil conditions (Schaffer, *et al.* 1992)

Species	Common name	Tolerance level
Poncirus trifoliata	Trifoliate orange	Moderately tolerant
Citrus jambhiri	Rough lemon	Moderately tolerant
C. sinensis x C. reticulata	Carrizo citrance	Moderately sensitive
C. reticulata	Cleopatra	Moderately sensitive
C. sinensis	Sweet orange	Moderately sensitive
C. aurantium	Sour orange	Sensitive
C. auratifolia	Rangpur lime	Sensitive

Table 5: Relative tolerance of *Prunus* species to flooded soil conditions (Schaffer, *et al.* 1992)

Species	Cultivar	Tolerance level
Prunus japonica		Moderately tolerant
P. cerasifera (Marianna)	S2544-2	Moderately tolerant
P. cerasifera	GF 8-1	Moderately tolerant
P. domestica	Damas GF 1869	Moderately tolerant
P. domestica	Brompton, St. Julien A	Moderately sensitive
P. salicina	S37, S300, S2540	Sensitive
P. cerasus	Stockton Morello	Sensitive
P. avium	Mazzard	Very sensitive
P. mahaleb	Mahaleb	Extremely sensitive
P. persica	Lovell, Nemagard, Siberian C	Extremely sensitive
P. armeniaca		Most sensitive

Heat stress

Photosynthesis is inhibited before respiration

Both photosynthesis and respiration are inhibited at higher temperatures, but as temperature increases, photosynthetic rates decrease before respiration do. The temperature at which amount of CO_2 released by respiration in a given time is called temperature compensation point. Above this temperature photosynthesis cannot replace CO_2 used as substrate for respiration. As a result, carbohydrate reserves decline, and fruits lose sweetness. This imbalance between photosynthesis and respiration is one of the main cause of deleterious effects of high temperatures.

High temperature impairs the thermal stability of membranes and proteins

The stability of various cellular membranes is very important during high temperature stress, just as it is during chilling and freezing. Excessive fluidity of lipids at high temperatures is correlated with the loss of physiological function. At high temperatures the strength between hydrogen bonds and the electrostatic interactions between polar groups with the aqueous phase of the membrane decrease. Thus, integral membrane proteins (which associate with both hydrophilic and lipid region of the membrane) tend to associate more strongly with lipid phase. High temperatures thus modify membrane composition and structure and can cause leakage of ions.

Response of fruit crops to heat stress and its management

Exposure of the fruit to intense sunlight can cause sunburn damage in the form of large black spots on the fruit skin, which render the fruit unmarketable. The fruits with terminal bearing habit are more prone to sunburn.

Several approaches could be taken to reduce sunburn incidence. Different plant cultivars can have more leaf surface, therefore providing better shading over the fruits, or may have fruits that are more resistant to sunburn. An improved fertilization and irrigation regime could increase vegetative development and thereby improve protection of the fruits from direct sunlight. Or, shades or screens could be erected to shelter trees and fruits from exposure to direct sunlight.

An alternative means of protecting fruit from sunburn is to use overhead irrigation to cool the canopy and fruit by evaporative cooling (Evans *et al.*, 1995). This is seldom an option in dry arid climates, because of scarcity of water and its high salt content that could damage the canopy by salinity. Surround® WP, has been reported to reduce sunburn damage in apples (Glenn *et al.*, 2002) and in pomegranate (Melgarejo *et al*., 2004). In pomegranate the entire tree is white coated. Apart from its effect of reducing heat stress in plants, Surround WP was reported also to control a few insect and mite pests efficiently (Glenn *et al*., 1999).

Salicylic acid (SA) is a phenolic derivative found in a wide range of plant species. SA is involved in physiological processes of plant growth, development and flowering. In the field of stress physiology, salicylic acid was first demonstrated to play a role in responses to biotic stress. Nowadays more and more reports indicated that salicylic acid also plays a role in plant responses to abiotic stress. There are many reports on the application of salicylic acid which enhanced heat tolerance in vegetables, fruit trees, and ornamental plants. Possible mechanisms of salicylic acid involvement in the plant responses to heat stress were also discussed (Lin and Chang, 2009).

Reactive oxygen species (ROS) are induced during both biotic and abiotic stress, either as signaling molecules or as result of stress injury. ROS are highly destructive to cell components and the injury resulting from these compounds is referred to as oxidative stress. Antioxidant enzymes, such as superoxide dismutase (SOD), scavenge oxygen radicals preventing the injury resulting from oxidative stress. Artlip *et* al., (2009) conducted a research to produce transgenic apple plants (*Malus* × *domestica* 'Royal Gala') with enhanced production of a cytosolic SOD. A full-length SOD cDNA was isolated from spinach by a combination of RT-PCR and conventional plaque lift screening of a pea cDNA library. The SOD gene was mobilized into a binary vector consisting of pBINPLUSARS and pRTL2 for *Agrobacterium*-mediated transformation of apple. The resulting SOD-over expression (SOD-OX), blank-cassette, and un-transformed lines were evaluated for resistance to acute and prolonged exposure to high temperature, and freezing temperatures in non-acclimated and acclimated plants, by ion leakage assays of leaves and bark from one-year-old trees. Results indicated that SOD-OX leaves exhibited improved resistance to both acute (30 min) and longer-term exposure (2 h to 24 h) to elevated temperatures compared to the non SOD-OX lines. Cold tolerance of non-cold-acclimated SOD-OX tissues did not differ from the control plants. Cold acclimated (2 weeks exposure to a short day photoperiod at 4°C) leaves of SOD-OX trees, however, were more cold tolerant compared to the other lines, while bark was not. The over expression of antioxidant enzymes is believed to help cells recover from post-injury ROS rather than directly increase stress tolerance. Therefore, the differences observed in increased stress tolerance in the transgenic apple plants may be a reflection of the type and extent of injury caused by heat vs. freezing stress.

Bruising injury

Mechanical injury, such as bruises, cuts and abrasions cause marked deterioration in fruit quality. The bruising is initiated by the breakage of cell membranes, allowing cytoplasmic enzymes to act on sequestered substrates (Rouet-Mayer *et al*., 1990). The resultant browning is caused by the enzyme action on phenolic substrates.

Polyphenol oxidase (PPO) is well known enzyme responsible for tissue browning in mechanically injured fruits (Nicolas, *et al*., 1994). PPO catalyzes the oxidation of phenolic compounds to o-quinones which subsequently polymerize to form dark-coloured pigments (Rouet-Mayer *et al*., 1990). However, Lee *et al*. (2005) reported that polyphenol oxidase is not the only factor influencing the deterioration associated with bruising. Cell wall hydrolases are currently being assayed to determine if they also contribute the deterioration following bruising.

Conclusions

Under both natural and agricultural conditions plants are frequently exposed to stress. Some environmental factors (such as air temperature) can become stressful in just a few minutes; others may take days to become stressful or even months (some mineral nutrients) to become stressful. Evaluation of crop season in soil moisture variations based on water budgeting for selecting suitable crops and varieties and evolving site specific land use planning strategies. Adaptation of appropriate soil moisture conservation techniques helpful in overcome the water stress. Management of cultural practices in conjuction with use of plant bio-regulators and chemicals go a long way in management of various abiotic stresses. Over the past two decades, significant advances have been made in understanding the genetic regulation of cold hardiness, as well as resistance to other abiotic stresses. Recent genomic and proteomic research on stress response in fruit trees is also being used to develop a more comprehensive understanding of environmental stress resistance.

References

Anbu, S., Balasubramanyan, S., Venkatesan, K., Selvarajan, M. and Duarisingh, R. 2009. Evaluation of varieties and standardization of production technologies in ber (*Ziziphus mauritiana*) under rainfed vertisols. *Acta Horticulturae*, 840: 55-60

Andersen, P.C.; Lombard, P.B. and Westwood, M.N. 1984[a]. Effect of root anaerobiosis on the water relations of several *Pyrus* species. *Physiol. Plant*, 62: 245-252.

Andersen, P.C.; Lombard, P.B. and Westwood, M.N. 1984[b]. Leaf conductance, growth and survival of willow and deciduous fruit tree species under flooded soil conditions. *J. Am. Soc. Hort. Sci*. 109: 132-138.

Artlip, T. S., Wisniewski, M. E., Macarisin, D., and Norelli, J. L. 2009. Ectopic expression of a spinach *SOD* gene in young apple trees enhances abiotic stress resistance. *Acta Horticulturae*, 839: 645-650.

Ballard, J.K. and Proebsting, E.L. 1978. Frost and frost control in Washington orchards. *Wash. State Univ. Coop. Ext. Ser. Bull.* 634.

Bartolini, G., Mazuelos, C. and Troncoso, A. 1991. Influence of Na_2 SO_4 and NaCl salts on survival, growth and mineral composition of young olive plants in inert sand culture. *Adv. Hort. Sci.* 5: 73-76.

Bartolozzi, F., and Fontanazza, G.1999. Assessment of frost tolerance in Olive (*Olea europaea* L.). *Scientia Horticulturae*, 81: 309-319.

Bernstein, L. 1964. Effects of salinity on mineral composition and growth of plants. *Proc. 4th Int. Coll. Plant Analysis and Fertilizer Problems*, Bruxelles, Belgium, 4: 25-45.

Blanc, M.L., Geslin, H., Holzberg, I.A. and Mason, B. 1963. Protection against frost damage. *World Meteorol. Org. Tech. Note*. 51.

Bongi, G. and Palliotti, A. 1994. Olive. P. 165-187. In: B. Schaffer and P.c. Andersen (eds.) Handbook environmental physiology of fruit crops. CRC Press, Boca Raton, FL.

Brewer, R.F.; Burns, R.M. and Opitz, K.W. 1973. Evaluation of man made fog for frost protection. *Citrograph*, 59: 9-10.

Briccolli Bati, C.; Basta, P.; Tocci, C. and Turco, D. 1994. Influenza dell' irrigazione con acqua salmastra su giovani piante di olivo. *Olivae*, 53: 35-38.

Browne, L.T.; Leavitt, G. and Gerdts, M. 1978. Delaying almond bloom with ethephon. *Calif. Agr.* 32(3): 6-7.

Burns, R.M. 1970. Testing foliar sprays for frost protection of young citrus. *Proc. Fla.State Hort. Soc.* 83: 92-95.

Cresti, M.; Ciampolini, F.; Tattini, M. and Cimato, A. 1994. Effect of salinity on productivity and oil quality of olive (*Olea europeae* L.) plants. *Adv. Hort. Sci.* 8: 211-214.

Chartzoulakis, K., Patakas, A., Kofidis, G., Bosabalidis, A. and Nastou, A. 2002. Water stress affects leaf anatomy, gas exchange, water relations and growth of two avocado cultivars. *Scientia Horticulturae*, 95: 39-50.

Close, T.J.1997. Dehydrin: a commonality in the response of plants to dehydration and low temperature. *Physiologia Plantarum*, 100: 291-296.

Corgan, J.N. and Widmoyer, F.B. 1971. The effects of gibberellic acid on flower initiation,date of bloom and flower hardiness of peach. *J. Am. Soc. Hoert. Sci.* 96: 54-57.

Coston, D.C. 1985. Delay of peach flowering. *Proc.5th Intl. Symp. Growth Regulators Fruit Production.* Bologna-Rimini, Italy. P.16 [Abstr.]

Davies, F.S. Buchanan, D.W. and Anderson, J.A. 1981. Water stress and cold hardiness in field grown citrus. *J. Am. Soc. Hort. Sci.* 106: 197-200.

Dennis, F.G. 1976. Trials of ethephon and other growth regulators for delaying bloom in tree fruits. *J. Am. Soc. Hort. Sci.*101: 241-245.

Dennis, F.G., Crews, C.E. and Buchanan, D.W. 1977. Bloom delay in stone fruits and apple with rhizobitoxine analogue. *HortScience*, 12: 386. [Abstr.]

Edgerton, L.J. 1966. Some rffects of gibberellin and growth retardants on bud development and cold hardiness of peach. *Proc. Am. Soc. Hort. Sci.* 88: 197-203.

Evans, R.G.; Kroeger, M.W.; and Mhan, M.O. 1995. Evaporative cooling of apples by overtree sprinkling. *Appl. Eng. Agric.* 11: 93-99.

Evenari, M.; Shanan, L. and Tadmor, N.H. 1968. Runoff farming in the desert. I. Experimental Layout. *Agron. J.* 60: 29-32.

Finch, A.H. 1977. Use of artificial fog for low temperature control and its possible value in climate control to improve citrus fruiting in a desert climate. *Proc. Intl.Soc. Citriculture*, 1: 209-210.

Flore, J.A. and Howell, G.S. 1987. Environmental and physiological factors that influence cold hardiness and frost resistance in perennial crops, p. 139-150. In: F. Prodi, F. Rossi, G. Cristoferi (eds.). *Intl. Conf. on Agrometeorology*. Comune di Cesena, Cesena, Italy.

Flowers, T.J. and Yeo, A.R. 1995. Breeding for salinity resistance in crop plants: Where next? *Austral. J. Plant Physiol.* 22: 875-884.

Freeman, J. A. and Carne, I.C. 1970. Use of succinic acid 2,2-dimethylhydrazide (Alar) to reduce winter injury in strawberries. *Can. J. Plant Sci.* 50: 189-190.

Funt, R.C. and Ferree, D.C. 1985. Influence of ethephon on bloom delay, hardiness and yield of Redhaven peaches, Ohio, USA. *Proc. 5th Intl. Symp.* Growth Regulators Fruit Production, Bologna-Rimini, Italy, p.17 [Abstr.]

Glenn, D.M. Puterka, G.J. Vanderzwet, T. Byer, R.E. and Feldhake, C. 1999. Hydrophobic particle films: a new paradigm for suppression of arthropod pests and plant diseases. *J. Econ. Entomol.* 92: 759-771.

Glenn, D.M. Prado, E. Erez, A. Mc Ferson, J. and Puterka, G.J. 2002. A reflective processed - kaolin particle film affects fruit temperature, radiation, reflection and solar injury in apple. *J. Am. Soc. Hort. Sci.* 127: 188-193.

Gucci, R. and Tattini, M. 1997. Salinity tolerance in Olive. *Horticultural Reviews*, 21: 177-214.

Hasegawa, P.M., Bressan, S.A., Zhu, J.K. and Bohnert, H.J. 2000. Plant cellular and molecular responses to high salinity. *Annual Review of Plant Physiology/Plant and Molecular Biology*, 51: 463-499.

Hendershott, C.H. 1965. Chemical induction of dormancy and cold resistance in citrus. *Am. Meteorol. Soc. Monogr*. 6(28): 87-89.

Hicks, J.R. and Crane , J.C. 1968. The effect of gibberellin on almond flower bud growth, time of bloom and yield. *Proc. Am. Soc. Hort. Sci*. 92: 1-6.

Hitchcock, A.E. and Zimmerman, P.W. 1943. Summer sprays with potassium a-napthalene acetate retard bud opening of buds of fruit trees. *Proc. Am. Soc. Hort. Sci.* 42: 141-145.

Holmberg N and Bülow L. (1998) Improving stress tolerance in plants by gene transfer. *Trends in Plant Science* 3:61–66.

Holubowicz, T. and Boe, A..A. 1969. Development of cold hardiness in apple seedlings treated with gibberellic acid and abscisic acid. *J. Am. Soc.Hort. Sci.* 94: 661-664.

Howell, G.S. and Dennis, F.G. 1981. Cultural management of perennial plants to maximize resistance to cold stress, p. 175-204. In: C.R. Olein and M.N. Smith (eds.), Analysis and improvement of plant cold hrdiness. CRC Press, Boca Raton.

Huschke, R.E. (ed.). 1959. Glossary of Meteorology. *Am. Meteorol. Soc.*, Boston, Mass. 233, 239, 281.

Jaglo-Ottosen, K.R., Gilmour, G.S.J., Zarka, D.G., Schabenberger, O. and Thomashow, M.F. 1998. *Arabidopsis* CBF1 overexpression induces *Cor* genes and enhances freezing tolerance. *Science*, 280: 104-106.

Jeyakumar, P. Kavino M. Kumar, N. and Soorianathasundaram, K. 2007. Physiological performance of papaya cultivars under abiotic stress conditions. *Acta Horticulturae* 740: 450-58.

Kang, B.G., Kim, W.T., Yun, H.S., Chang, S.C. 2010. Use of plant growth-promoting rhizobacteria to control stress responses of plant roots. *Plant Biotechnology Reports,* 4(3):179-183

Karlidag, H., E. Yildirim and M. Turan. 2009. Salicylic acid ameliorates the adverse effect of salt stress on strawberry. *J. Agric. Sci.*, 66: 271-278.

Klein, I. Ben-Tal, Y. Lavee, S. DeMalach, Y. And David, I. 1994. Saline irrigation of cv. Manzanillo and Uovo di Piccione trees. *Acta Hort.* 356: 176-180.

Koo, R.J.C. 1981. The effect of fall irrigation of freeze damage to citrus. *Proc. Fla. State. Hort. Soc.* 94: 37-39.

Koshita, Y. and Takahara, T. 2004. Effect of water stress on flower bud formation and plant hormone content of Satsuma mandarin (*Citrus unshiu* Marc.). *Scientia Horticulturae*, 99: 301-307.

Kuiper, P.J.C. 1964. Inducing resistance to freezing and dessication in plants by decenylsuccinic acid. *Science*, 146: 544 – 546.

Lee, S.H. Choi, J.H. Kim, W.S. Han, T.H. Park, Y.S. and Gemma, H. 2006. Effect of soil water stress on the development of stone cells in pear (*Pyrus pyrifolia* cv. 'Niitaka') flesh. *Scientia Horticulturae*, 110: 247-253.

Lee, H.J.; Kim, T.C; Kim, S.J.; and Park, S.J. 2005. Bruising injury of persimmon (*Diospyros kaki* cv. Fuyu) fruits. *Scientia Horticulturae*, 103: 179-185.

LimaCosta, M. E., Ferreira,S., Duarte,A., Ferreira, A. L. 2010. Alleviation of salt stress using exogenous proline on a citrus cell line. *Acta Horticulturae,* 868:109-112.

Lin, L.N. and Chang, Y.S. 2009. Effects of salicylic acid on the plant responses to heat stress. *Journal of the Taiwan Society for Horticultural Science*, 55(4):193-206

Marin, L.; Benlloch, M. and Fernandez-Escobar, R. 1995. Screening for olive cultivars for s a l t tolerance. *Scientia Hort.* 64: 113-116.

Melgarejo, P., Martinez, J.J., Hernandez, F., Font, R.M., Barrows, P. and Erez, A. 2004. Kaolin treatment to reduce pomegranate sunburn. *Scientia Horticulturae*, 100: 349-353.

Mitchell, P.D.; Van Den Ende, B.; Jerie, P.H. and Chalmers, D.J. 1989. Response of Bartlett pear to withholding irrigation, regulated deficit irrigation and tree spacing. *J.Am.Soc. Hort. Sci.* 114: 15-19.

Modlibowska, I. 1968. Effects of some growth regulators on frost damage. *Cryobiology*, 5:175-187.

Naor, A.; Stern, R.A., Gal, Y., Peres, M. and Flashman, M. 2005. Timing and severity of ost-harvest water stress affect following year productivity and fruit quality of field grown 'Snow Queen' nectarine. *J.Am.Soc. Hort. Sci.* 130: 806-812.

Nath, V., Das, B., Dey, P., Rai M., Kumar, M. and Rai, A. 2005. Orchard management practices for litchi in Eastern India, *in-situ* water harvesting and moisture conservation for improving fruit yield and quality. *Progressive Horticulture*, 37(1): 56-62.

Nicolas, J.L., Richard-Forget, F.C., Goupy, P.M., Amiot, M.J., and Aubert, S.Y. 1994. Enzymatic browning reactions in apple and apple products. *CRC Crit. Rev. Food Sci. Nutr.* 34: 109-157.

Pareek, O.P., Vishal Nath and Mishra, A.L. 1996. Water harvesting and rain water management in fruit crops. *Proceedings of Silver Jublee National Symposium on Arid Horticulture*, December 5-6, CCSHAU, Hisar: 49-61.

Pellet, H.M. and Carter, J.V. 1981. Effect of nutritional factors on cold hardiness of plants. *Hort. Rev.* 3: 144-171.

Perry, K.B., Martsolf, J.D. and Norman, J.M. 1977. Radiant output from orchard heaters. *J. Am. Soc. Hort. Sci.* 102: 105-109.

Pretorius, J.J.B. and Wand, S.J.E. 2003. Late season stomatal sensitivity to microclimate is influenced by sink strength and soil moisture stress in 'Braester' apple trees in South Africa. *Scientia Horticulturae*, 98: 157-171.

Proebsting, E.L. and Mills, H.H. 1973. Bloom delay and frost survival in ethephon-treated sweet cherry. *HortScience*, 8: 46-47.

Proebsting, E.L. and Mills, H.H. 1974. Time of gibberellin application determines hardiness response of 'Bing' cherry buds and wood. *J. Am. Soc. Hort. Sci.* 99: 464-466.

Proebsting, E.L. and Mills, H.H. 1985. Cold resistance in peach, apricot and cherry as influenced by soil-applied paclobutrazol. *HortScience*, 20: 88-90.

Raese, J.T. 1977. Induction of hardiness in apple tree shoots with ethephon, NAA, and growth retardants. *J. Am. Soc. Hort. Sci.* 102: 789-792.

Raese, J.T. 1983. Conductivity tests to screen fall applied growth regulators to induce cold hardiness in young 'Delicious' apple trees. *J. Am. Soc. Hort. Sci.* 108: 172-176.

Reilly, C.C., Nyczepir, A.P., Sharpe, R.R., Okie, W.R. and Pusey, P.L. 1986. Shortlife of each trees as related to tree physiology, environment, pathogens and cultural practices. *Plant Dis.* 70: 538-541.

Rieger, M. 1989. Freeze protection of horticultural crops. *Horticultural Reviews*, 11: 45-110.

Rouet-Mayer, M.; Ralambosoa, J.; and Phillippon, J. 1990. Roles of 0-quinones and their polymers in the enymic browning of apples. *Phytochemistry*, 29: 435-440.

RuizSanchez, M.C., Domingo, R., Save, R., Biel, C. and Torrecillas, A. 1997. Effects of water stress and rewatering on leaf water relations of lemon plant. *Biol. Plantarum*, 39: 623-631.

Satisha, J., Ramteke, S.D. & Karibasappa, G.S., 2008. Physiological and biochemical characterization of grape rootstocks. S. Afr. J. Enol. Vitic. 28, 163-168.

Satisha, J., Somkuwar, R.G., Sharma, J., Upadhyay, A.K. and Adsule, V. 2010. Influence of Rootstocks on Growth Yield and Fruit Composition of Thompson Seedless Grapes Grown in the Pune Region of India. *S. Afr. J. Enol. Vitic.*, 31(1): 1-8

Savage, E.F. 1970. Cold injury as related to cultural management and possible protective devices for dormant peach trees. *HortScience*, 5: 425-431.

Save, R., Peiuelas, J., Marfa, O., Serrano, L. 1993. Changes in leaf osmotic and elastic properties and canopy architecture of strawberries under mild water stress. *HortScience*, 28: 925-927.

Schaffer, B. Andersen, P.C. and Ploetz, R.C. 1992. Responses of fruit crops to flooding. *Horticultural Reviews*, 13: 257-313.

Stewart, I. and Leonard, C.D. 1960. Increased winter hardiness in citrus from maleic hydrazide sprays. *Proc. Am. Soc. Hort. Sci.* 75: 253-256.

Southwick, S.M., and Devenport, T.L. 1986. Characterization of water stress and low temperature effects on flower induction in citrus. *Plant Physiol.* 81: 26-29.

Sullivan, D.T. and Widmoyer, F.B. 1970. Effects of succinic acid 2,2- dimethylhydrazide (Alar) on bloom delay and fruit development of Delicious apples. *Hort. Science*: 91-92.

Tattini, M. Gucci, R. Coradeschi, M.A. Ponzio, C. And Everard, J.D. 1995. Growth, gas exchange and ion content in *Olea europeae* plants during salinity stress and subsequent relief. *Physiol. Plant.* 95: 203-210.

Tattini, M. Ponzio, C. Coradeschi, M.A. Tafani, R. and Traversi, M.L. 1994.Mechanisms of salt tolerance in Olive plants. *Acta Hort.* 356: 181-184.

Therios, I.N. and Misopolinos, N.D. 1988. Genotypic response to sodium chloride salinity of four major olive cultivars (*Olea europeae* L.). *Plant Soil*, 106: 105-111.

Thomashow, M.F.1998. Role of cold responsive genes in plant freezing tolerance. *Plant Physiology*, 118: 1-8.

Thomashow, M. F. 1999. Plant cold acclimation: freezing tolerance genes and regulatory mechanisms. *Annual Review of Plant Physiology and Plant Molecular Biology*, 50: 571-599.

Van Camp, W., Willines, H., Bowler, C., Van Montagu, M., Inze, M., Reupold-Popp, P., Sandermann, H. and Langebartels,C.1993. Elevated levels of superoxide dismutase protect transgenic plants against ozone damage. *Biotechnology*, 12: 165-168.

Webster, A.D. 1985. Delaying flowering and improving yield of 'Victoria' plum with ethephon and gibberellic acid sprays. *Proc. 5th Intl. Symp. Growth Regulators Fruit Production.* Bologna-Rimini, Italy. P.22 [Abstr.]

Weiser, G.J. 1970. Cold resistance and acclimation in woody plants. *HortScience*, 5: 403- 410.

Westwood, M.N. 1970. Rootstock-scion relationship in hardiness of deciduous fruit trees. *Hort Science*, 5: 418-421.

Westwood, M.N. and Bjornstad, H.O. 1981. Winter injury to apple cultivars as affected by growth regulators, weed control method, and rootstocks. *J. Am. Soc. Hort. Sci.* 106: 430-432.

White, D.G. and Kennard, W.C. 1955. A preliminary report on the use of malic hydrazide to delay blossoming of fruits. *Proc. Am. Soc. Hort. Sci.* 50: 147-151.

Wisniewski, M., Bassett, C., Norelli, J. L., Artlip, T., and Renaut, J. 2007. Using biotechnology to improve resistance to environmental stress in fruit crops: the importance of understanding physiology. *Acta Horticulturae*, 738: 145-156.

Xiong, L., Schumaker, K.S., and Zhu, J.K. 2002. Cell signalling during cold, drought and salt stress. *Plant cell*, 14(1): 165-183.

Young, R. 1969. Cold hardening in Redblush grapefruit as related to sugars and water soluble proteins. *J. Am. Soc. Hort Sci.* 94: 252-254.

Young, R. 1971. Effect of growth regulators on citrus seedling cold hardiness. *J. Am. Soc. Hort. Sci.* 96: 708-710.

Young, R. and Olson, E.O. 1963. Freeze injury to citrus trees on various rootstocks in the lower Rio Grande Valley of Texas. *Proc. Am. Soc. Hort. Sci.* 83: 337-343.

Zhang, Y. H. and Shih, D. S. 2007. Isolation of an osmotin-like protein gene from strawberry and analysis of the response of this gene to abiotic stresses. *Journal of Plant Physiology*, 164(1): 68-77.

Zhu, J.K., Liu, J., and Xiong, L.1998. Genetic analysis of salt tolerance in *Arabidopsis*: evidence for critical role of potassium nutrition. *Plant cell*, 10: 1181-1191.

Zhu, L.H.; Peppel, A. Van de; Li, X.Y.and Welander, M. 2004. Changes of leaf water potential and endogenous cytokinins in young apple treees treated with or without paclobutrazol under drought conditions. *Scientia Horticulturae*, 99: 133-141.

11

Impact of Abiotic Stress on Vegetable Crops and its Possible Management

V C Dhyani and S K Maurya

India is the second largest producer of vegetables after China with 14 % contribution in total World vegetable production (1160 million tonnes). After the advent of green revolution, more emphasis is laid on the quality of the agricultural product along with the quantity to meet the ever-growing food and nutritional requirements. Both these demands can be met when the environment for the plant growth is suitably controlled. Olericulture a vital component of Indian Horticulture, which includes science and management of vegetables, has made a rapid stride during the last decade, recording appreciable growth in production (162.90 million tonnes against 81.89 million tonnes in 2000-01) and productivity (17.4 t/ha against 12.2 t/ha in 2000-01), availability (210 g/capita/day against recommendations of 300 g/capita/day) and export (36,94,860 tonnes worth Rs. 14,36,487 lakhs.) during 2013-14. Vegetables play a major role in Indian agriculture by providing food, nutritional and economic security and more importantly, producing higher returns per unit area and time (Srivastava *et al.* 2013). In addition, vegetables have higher productivity, shorter maturity cycle, high value and provide greater income leading to improved livelihoods. The total production of vegetables during 2013-14 was 162.90 million tonnes compared to 21.19 million tonnes in 1960-61, 81.89 million tonnes in 2000-01 and 146.55 million tonnes in 2010-11. Although, olericulture has exhibited leadership role but challenges are much greater. Looking to the requirement, which is estimated to be 225 million tonnes by 2020 and 350 million tonnes by 2030, which has to be produced from declining land and water resources and in the scenario of climate change, the task to meet the needs of growing population would be difficult but not impossible (Singh *et al.* 2014).

Improvement of vegetable crops has traditionally focused on enhancing a plant's ability to resist diseases or insect-pests. That is evidenced by the large number of disease or insect-resistant cultivars or germplasm released and used. Research on crop resistance or tolerance to abiotic stresses (heat, cold, drought, flood, salt, pH, etc.) has not received much attention. However, that is changing as a result of the research and publicity of global warming. "Adaptive research" aiming at adapting the vegetable industry to climate changes is becoming popular and is now one of the emphasis areas of funding agencies (Mou, 2011). Plants in nature may be exposed, during their ontogeny, to a wide variety of favorable or disadvantageous biotic and abiotic factors or stressors. Living being affecting the crop plants adversely are called biotic stressors and non living as abiotic stressors. Goel and Madan (2014) have reported that abiotic stresses cause more than 50 per cent reduction in crop yield. Here in this chapter emphasis thus has been laid to give information about abiotic stress impacts on vegetable crops and possible management option.

Environmental stresses represent the most limiting conditions for vegetable productivity and plant exploitation worldwide. Important factors among those are water, temperature, nutrition, light, oxygen availability, metal ion concentration, and pathogens (Schwarz *et al.* 2010). Fruits and vegetables are affected by abiotic stresses (Table 1) during their growth in field but these have potential to significantly impact on quality and nutritional status of fresh cut fruits and vegetables (Hodges &Toivonen, 2008).

Table 1: Sources of environmental (abiotic) stress in plants

Physical	Chemical	Biotic
Drought	Air pollution	Competition
Temperature	Heavy metals	Allelopath
Radiation	Pesticides	Herbivory
Flooding	Toxins	Diseases
Wind	Soil pH	Pathogenic fungi
Magnetic field	Salinity	Viruses
UV- β irradiation	Alkalinity	

Plants response to environmental stresses

Plant responses to abiotic stresses are dynamic and complex (Cramer, 2010); they are both elastic (reversible) and plastic (irreversible). The plant responses to stress are dependent on the tissue or organ affected by the stress. For example, transcriptional responses to stress are tissue or cell specific in roots and are quite different depending on the stress involved. In addition, the level and duration of stress (acute vs chronic) can have a significant effect on the complexity of the response.

In recent years, research has mainly concentrated on understanding plant responses to individual abiotic or biotic stresses, although the response to simultaneous stresses is bound to lead to a much more complex scenario (Rejeb *et al*, 2014). Also different environmental stresses to a plant may result in similar responses at the cellular and molecular level (Beck *et al.* 2007). In umpteen numbers of studies similar plant response due to stresses are being reported. For example water deficiency is being experienced by the plants in drought, salinity and cold temperature stress. Heat stress too is more severe when water stress is experienced by the plant. In general, plants tend to maintain stable tissue water status regardless of temperature when moisture is ample; however, high temperatures severely impair this tendency when water is limiting (Machado and Paulsen, 2001). Water availability in some cases however, might not offset the ill effects of heat stress though. In sugarcane though leaf water potential and its components were changed upon exposure to heat stress even though the soil water supply and relative humidity conditions were optimal, implying an effect of heat stress on root hydraulic conductance.

In short following morpho - anatomical responses of plants to different abiotic stresses are given. In heat stress, scorching of leaves and twigs, sunburns on leaves, branches and stems, leaf senescence and abscission, shoot and root growth inhibition, fruit discoloration and damage, and reduced yield. Major impact of high temperatures on shoot growth is a severe reduction in the first internode length resulting in premature death of plants (Reviewed by Wahid *et al.* 2007). Vegetative and reproductive processes in tomatoes are strongly modified by temperature alone or in conjunction with other environmental factors (Abdalla &Verkerk 1968). Long-term effects of heat stress on developing seeds may include delayed germination or loss of vigor, ultimately leading to reduced emergence and seedling establishment. High temperatures can cause significant losses in tomato productivity due to reduced fruit set, and smaller and lower quality fruits (Stevens & Rudich 1978). Hazra *et al.* (2007) summarized the symptoms causing fruit set failure at high temperatures in tomato; this includes bud drop, abnormal flower development, poor pollen production, dehiscence, and viability, ovule abortion and poor viability, reduced carbohydrate availability, and other reproductive abnormalities.

High salt content, especially chloride and sodium sulphates, affects plant growth by modifying their morphological, anatomical and physiological traits. Such growth impairment is due to osmotic effects and ionic imbalances affecting plant metabolis (Céccoli *et al.* 2011). These adverse effects of salt stress appears on whole plant level at almost all growth stages including germination, seedling, vegetative and maturity stages (Nawaz *et al.* 2010). Salinity has been known to affect time of germination, the size of the plants, branching and leaf size, and overall plant anatomy (Poljakoff-Mayber, 1975). According to the United States Department of Agriculture (USDA), onions are sensitive to saline soils, while cucumbers, eggplants, peppers, and tomatoes, are moderately sensitive (Peña & Hughes, 2007).

Drought stress progressively decreases CO_2 assimilation rates due to reduced stomatal conductance. It reduces leaf size, stems extension and root proliferation, disturbs plant water relations and reduces water-use efficiency. It disrupts photosynthetic pigments and reduces the gas exchange leading to a reduction in plant growth and productivity (Anjum, 2011).Vegetables, being succulent products by definition, generally consist of greater than 90% water (AVRDC 1990).

Under waterlogged conditions, all pores in the soil or soilless mixture are filled with water; so the oxygen supply is almost completely deprived. As a result, plant roots cannot obtain oxygen for respiration to maintain their activities for nutrient and water uptake. This is followed by loss of chlorophyll of the lower leaves, arrest of crop growth and proliferation of surface root growth with the retreat of water level. Various morpho-physiological, biochemical and anatomical changes induced in root system during flooding, for example reduction in the shoot-root relative growth rate. Plant transpiration is affected under anaerobic conditions and extended water logging results in root death due to inadequate oxygen supply. In general, damage to vegetables by flooding is due to the reduction of oxygen in the root zone which inhibits aerobic processes. Flooded tomato plants accumulate endogenous ethylene that causes damage to the plants (Drew, 1979). In vegetables waterlogging caused a marked reduction in stomata conductance of bitter melon.

Apart from these there is other common physio-biochemical and molecular responses are also seen in plants. There is production of reactive oxygen species (ROS) in all the above cited abiotic stresses. According to Gill *et al.*, 2010, the ROS comprises both free radical (O_2^{-}, superoxide radicals; .OH hydroxyl radical; O2H˙, perhydroxy radical and RO, alkoxy radicals) and non-radical (molecular) forms (H_2O_2, hydrogen peroxide and $^1O_2$, singlet oxygen). Stomatal closure, photosynthesis due to disruption of photosynthetic pigments as well as apparatus, increased respiration, etc. are also common physiological responses seen due to abiotic stresses. Photosynthesis, the most fundamental and intricate physiological process in all green plants, is severely affected in all its phases by abiotic stresses. Since the mechanism of photosynthesis involves various components, including photosynthetic pigments and photosystems, the electron transport system, and CO_2 reduction pathways, any damage at any level caused by a stress may reduce the overall photosynthetic capacity of a green plant (Ashraf, 2013). However, Rejeb *et al*, 2014 reported that, "a crucial step in plant defense is the timely perception of the stress in order to respond in a rapid and efficient manner. After recognition, the plants' constitutive basal defense mechanisms lead to an activation of complex signaling cascades of defense varying from one stress to another. All these responses are also considered as the plants' adaptive mechanism to cope up abiotic stresses. For example in drought stresses, curling of leaf is seen, which in turn reduces transpiration by the plant and saves plant from the catastrophic damages. There is also surge in production of low molecular weight osmolytes in the event abiotic stresses. The accumulation of low molecular weight water-soluble compounds known as "compatible solutes" or "osmolytes" is the common strategy adopted by many organisms to combat the environmental stresses. The most common compatible solutes are betaines, sugars (mannitol, sorbitol, and trehalose), polyols, polyamines, and amino acid (proline). Their accumulation is favored under water-deficit or salt stress as they provide stress tolerance to cell without interfering cellular machinery (Giri, 2011). Plants' antioxidant defense system also works to scavenge the ROS. The antioxidant defense machinery protects plants against oxidative stress damages. Plants possess very efficient enzymatic (superoxide dismutase, SOD; catalase, CAT; ascorbate peroxidase, APX; glutathione reductase, GR; monodehydroascorbatereductase, MDHAR; dehydroascorbatereductase, DHAR; glutathione peroxidase, GPX; guaicol peroxidase, GOPX and glutathione-*S*- transferase, GST) and non-enzymatic (ascorbic acid, ASH; glutathione, GSH; phenolic compounds, alkaloids, *non-protein amino acids and á-tocopherols) antioxidant defense systems which work in concert to control the cascades of uncontrolled oxidation and protect plant cells from oxidative damage by scavenging of ROS (Gill et al. 2011).*

Following exposure to abiotic and/or biotic stress, specific ion channels and kinase cascades are activated, phytohormones like abscisic acid, salicylic acid, jasmonic acid, and ethylene accumulate, and a reprogramming of the genetic machinery results in adequate defense reactions and an increase in plant tolerance in order to minimize the biological damage caused by the stress (Rejeb *et al.* 2014). Plant hormones play central roles in the ability of plants to adapt to changing environments, by mediating growth, development, nutrient allocation, and source/sink transitions. Although ABA is the most studied stress-responsive hormone, the role of cytokinins, brassinosteroids, and auxins during environmental stress is emerging. Recent evidence indicated that plant hormones are involved in multiple processes. Cross-talk between the different plant hormones results in synergetic or antagonic interactions that play crucial roles in response of plants to abiotic stress (Peleg *et al.* 2011).

Abiotic stresses influence the growth of a plant and secondary metabolite production. Plant secondary metabolites are often referred to as compounds that have no fundamental role in the maintenance of life processes in the plants, but they are important for the plant to interact with its environment for adaptation and defense. Accumulation of secondary metabolites often occurs in plants subjected to stresses including various elicitors or signal molecules. Secondary metabolites play a major role in the adaptation of plants to the environment and in overcoming stress conditions. (Ramakrishna and Ravishankar, 2011).Secondary metabolites like phenyl amides, polyamines, anthocyanin, flavonoids and phenolic acids, phenylalanine ammonia-lyase (PAL), shikonin,digitalinbetalains, betacyanins, lepidine, taxol, spermidine are produced in plants as response to various abiotic stresses (reviewed by Ramakrishna and Ravishankar, 2011).

Plants will counter abiotic stresses by adaptation or mitigation strategies of their cellular biochemical and molecular mechanisms to complete their life cycle. In this chapter, will discuss agronomic as well as breeding approaches for mitigate the effect of abiotic stresses.

Agronomic options to mitigate ill effects of abiotic stresses

There could be several agronomic optionsfor mitigation of heat stress, selection of heat tolerant varieties, use of nutrients, organic materials as source of materials and for moisture conservation (FYM, vermicompost, compost, use of sludge, industrial organic byproducts etc.,) foliar feeding, use of plant hormones at the critical stages etc. The agronomic options can help mitigate these abiotic stresses. As individual stresses come not in isolation, the effect of agronomic interventions too seen simultaneously on several kinds of stresses. Mulches conserve moisture in soil, reduce evaporative loss without decreasing transpiration. Thus more water is diverted for transpiration cooling.

Selection of crop varieties

In the event of heat stress one should choose heat tolerant material. There is availability of varieties which are heat tolerant (table 2):

Table 2: Heat tolerant vegetable varieties

Crop	Variety	Remarks
Tomato	Pusa Hybrid-1	High temperature
	Pusa Sheetal	Low temperature
	Pusa Sadabahar	Both high and low temperature
Bottle gourd	Pusa Santusti	Both high and low temperature
Carrot	PusaVrishti	Tolerant to heat
Cowpea	Arka Garima	Tolerant to heat and low moisture stress
Potato	KufriLauvkar	Tolerant to heat
	Kufri Surya	Tolerant to heat
Cabbage	Pusa Ageti	Tropical type

Use of balanced nutrients

Proper plant nutrition is one of the good strategies to alleviate the temperature stress and in crop plants (Waraich *et al.* 2012). Healthy plants can survive better in any of the stress, so emphasis should be given on nutrient application as per recommendation. Adequate nutrition is essential for the integrity of plant structure and key physiological processes such as nitrogen and magnesium is structural part of chlorophyll needed for photosynthesis, phosphorus is needed for energy production and storage, is a structural part of nucleic acids, potassium is needed for osmotic regulation and activation of enzymes (Waraich *et al.* 2011). Uchida *et al.* 2002 reported that northern blot analysis demonstrated that NO protected the chloroplast against oxidative damage under heat stress by inducing expression of gene encoding small heat shock protein 26 (HSP26). Apart from these primary nutrients, secondary and micro nutrients to impart tolerance to crop plant to with stand adverse effect of various abiotic stresses. Calcium has been found to alleviate adverse effect of heat stress in several vegetables. Some detrimental effects of heat stress on plant growth and stomatal function may be alleviated by Ca and N application during heat stress (Tawfic *et al.* 1996). Under heat stress, Ca^{2+} is required for maintenance of antioxidant activity and not for osmotic adjustment in some cool season grasses (Jiang and Haung, 2001). Ca^{2+} requirement for growth is high to mitigate adverse effects of the stress (Kleinhenz and Palta, 2002). Adequate zinc nutrition too has shown to increase thermo tolerance in wheat in zinc deficient soils (Peck and McDonald, 2010).

Use of Farm Yard Manure (FYM) compost

One strategy to strengthen the resilience of farming systems to abiotic stresses is to enhance the organic matter levels in soils for better soil moisture retention and water infiltration. The preservation and increased application of farm yard manure, which is organic matter prepared from various kinds of locally available animal excreta mixed with other organic materials is a suitable technology to augment the organic matter content in soils. Farm yard manure/compost is also a good source of nutrient, release nutrient slowly, nutrients remain available for longer time and more importantly it store moisture for longer period of time so will be good option against the heat stress too. Purchase of compost from outside vegetable farm will increase cost of production; in situ production of compost/vermicompost from farm waste should be encouraged. Under saline environment use of FYM is recommended specially while using saline water for irrigation. The synergy of FYM @ 20 t/ha and gypsum @ 100 per cent neutralization of RSC could be used to cultivate vegetable crops with alkali water.

Use of mulches

Mulches can also help greatly. One can increase reflection and dissipation of radiative heat using reflective mulches or use low density, organic mulches such as straw to reduce surface radiation and conserve moisture.There are variety of mulches available which are as follows

Organic mulch

Organic mulch is made up of natural substance such as bark, wood chips, pine needles, dry grasses, paddy straw, dry leaves, saw dust, grass clipping, etc. But organic mulch attracts insects, slugs and the cutworms that eat them. They get decomposed easily and need frequent replacements.

Inorganic mulch

Material Gravel, Pebbles and Crushed stones: These materials are used for perennial crops. Small rock Layer of 3-4 cm provides good weed control. But they reflect solar radiation and can create a very hot soil environment during summer. The use of organic and inorganic mulches has been observed to have important role in vegetable production, particularly under rainfed situations. These not only conserve the soil moisture, but along with regulation of soil temperature also suppress the weed population which ultimately helps in increasing the total and early yield. The yield increase in squash for example has been found to the tune of 30 % (Annda *et al.* 2008 and Sari *et al.* 1994).

Plastic mulch

Both, black and transparent films are generally used for mulching. Advancement in plastic chemistry has resulted in development of films with optical properties that are ideal for a specific crop in a given location. Plastic mulching helps to conserve moisture and reduce the need of irrigation, since there is no surface evaporation of water and it helps to stabilize soil moisture level for uptake by potato as compared to no mulching (reviewed by Singh and Ahmad, 2008). Under cold deserts with low relative humidity, mulching had significant influence on potato growth and yield. Emergence, plant height, and number of stems improved with black polythene mulching. Maximum tuber yield (35.2 t/ha) was recorded with black polythene mulching followed by white polythene (Singh and Ahmad, 2008). Plastic mulch had better growth, yield and economic returns in squash when compared with inorganic, organic mulch under the conditions of low temperature in initial growth phase followed by moisture stress and temperature fluctuation problems in later parts of crop growth (Bhatt *et al*, 2011).

Use of shade cloth

High temperature, clear skies and high irradiance cause sunburn in fruits and vegetables. There are three types of sunburns are seen on fruits and vegetables. First sunburn necrosis where skin, peel or tissue dies upon exposure to sunlight, second sunburn browning where there is loss of pigmentation but not death of tissue and third is photo oxidative sunburn where fruits become bleached due to sudden pruning. Shade cloths which provide 10-30 per cent can be use in such circumstances but where large acreage is planted this is not a practical solution. For large farms or acreage kaolin clay based solution can be sprayed which make a thin film outside the fruits or vegetables. An Iranian study on pomegranate reported that spray of Kaolin clay (5 %) reduced sunburn by close to 40 per cent. Spray of kaolin (5%) on pomegranate trees was no ill-effect on leaves chlorophyll and photosynthesis (Farazmand, 2013).

Frequent and supplemental irrigation

Provision of sufficient water during times of high heat is useful in dealing with heat stress. Light irrigation with more frequency is required. However, with conventional irrigation system it is not possible. It is best to use a drip irrigation system and check this regularly to be sure that all plants are being fed an adequate supply of water. High-frequency water management by drip irrigation provides daily requirement of water to a portion of the root zone of each plant, and maintains a high soil matric potential in the rhizosphere to reduce plant water stress (Nakayama and Bucks, 1986). Now a day there is availability of low cost drip irrigation which ensured high water as well as nutrient use efficiency too.The payback period of the investment in net house cultivation of tomato using drip irrigation was found out to be one and a half years (three seasons) by which time the system became beneficial (Dunage *et al.* 2009). Adequate watering ensured transpiration at high rate keeps the vegetable canopy cool, thus alleviate adverse effect of heat stress.The major method to reduce heat stress is by overhead watering, sprinkling, and misting for improved water supply, reduction of tissue temperature, and lessening of the water vapor pressure deficit.

There is enough water to be harvested in rainfed areas. According to one estimate about 114 billion cubic meters (BCM) of runoff is generated from this 28 M. ha of rainfed areas. In rainfed conditions application of harvested water at the most critical stage has shown to increase the yields about 50 % under farmer's fields (www.crida.in).

Plant hormones

Plant hormones play important roles in the ability of plants to adapt to changing environments, by mediating growth, development, nutrient allocation, and source/ sink transitions. Abscisic acid (ABA) and ethylene (C_2H_4), as stress hormones, are involved in the regulation of many physiological properties by acting as signal molecules. Different environmental stresses, including high temperature, result in increased levels of ABA (Wahid *et al.* 2007). Although ABA is the most studied stress-responsive hormone, the role of cytokinins, brassinosteroids, and auxins during environmental stress is emerging (Peleg, 2011). Among other hormones, salicylic acid (SA) has been suggested to be involved in heat-stress responses elicited by plants. SA is an important component of signaling pathways in response to systemic acquired resistance (SAR) and the hypersensitive response (HR) (reviewed by Wahid *et al.* 2007). Foliar application of SA increased the IAA content in broad bean leaves (Xin *et al.* 2000). Foliar application of SA exerted a significant effect on plant growth metabolism when applied at physiological concentration, and thus acted as one of the plant growth regulating substances (Kalarani *et al.* 2002). Ethylene also plays an important role in stress condition. It has variety of roles. It hastens the senescence and also causes epinasty of plant leaves. Ethylene action inhibitor (1-MCP or 1-methyl cyclo propane) was tested to have positive effect in wheat's tolerance to heat stress as sprayed plants did not exhibit kernel abortion and reduction in kernel weight as compared to control (Hays *et al.* 2007). Several naturally occurring plant extract too have plant growth regulators and thus cause enhancement in growth and productivity of crops. *Moringa oleifera* leaf extract (MOLE) is found to have higher phenolic antioxidants and other biochemical parameters in spinach compared to control (Aslam, 2015).

Use of microorganisms

Microorganisms could play an important role in adaptation strategies and increase of tolerance to abiotic stresses in agricultural plants. Plant-growth promoting rhizobacteria (PGPR) are associated with plant roots and mitigate most effectively the impact of abiotic stresses (drought, low temperature, salinity, metal toxicity, and high temperatures) on plants through the production of exopolysaccharates and biofilm formation. When plants are expos*ed to stress conditions, rhizospheric microorganisms affect plant cells by different mechanisms like induction of osmoprotectors and heat shock proteins (Milosevię, 2012). Under stress, most of the rhizobacteria produce osmoprotectants (K^+*, glutamate, trehalose, proline, glycine beatine, proline betaine and ectoine etc.) to modulate their cytoplasmic osmolarity (Blanco and Bernard, 1994 and Magdy *et al.* 1990). Endophytic fungus also imparts drought tolerance at least in rice plant. An endophytic fungus Trichoderma-colonized rice seedlings wilted slowly in response to drought due to late disruption of stomatal conductance, net photosynthesis and leaf greenness. The primary direct effect of Trichoderma colonization was promotion of root growth, regardless of water status, which caused delay in the drought responses of rice plants (Shukla *et al.* 2012).

Breeding for tolerance to abiotic stresses in vegetable crops

A drastic reduction in the yield is observed under several abiotic stress factors, i.e. high and low temperature, drought, excessive moisture, salinity and atmospheric pollutants. There is a need to search for genotypes that can tolerate abiotic stress conditions. A better understanding of genetics and mechanism of stress tolerance will enable the development of suitable varieties for stress conditions.

Characters affected adversely by abiotic stresses in vegetable crops

- Germination
- Root development
- Shoot development
- Yield related characteristic mostly reduction in fruit weight, number of clusters/plant

Success in any breeding process is based on the followings:

i. Recognition, as precisely as possible, of the characteristic to be improved.

ii. The existence of variability for the characteristic within the same species or alternatively, in related species, together with high expression of the characteristic in one or more accessions (s) and

iii. High heritability for the characteristic (which is the same as expression of the characteristic with little influence of the environment).

Vegetable crops are cropped for the immature pant part which is harvested at horticultural maturity stage than physiological maturity. The abiotic stress tolerance of a vegetable cultivar is then characterized by the response of its yield to abiotic stress conditions. The final yield, after reduction by abiotic stress, must still be sufficient not only to cover the cropping expense but also to provide profit for the producer. Yield should therefore be the leading characteristic in any breeding programme and that by which final success of the process is evaluated. However, abiotic stress produces so many disturbances to plant morphology and physiology that the only way to achieve profitable yields under abiotic stress conditions might be by combining in one cultivar morphological and physiological characteristics each of which alone only improves a particular deleterious effect of abiotic stress altogether the combination makes the cultivar close to the ideotype (Maurya *et al.* 2013).

A rather large number of characteristic suitable for use in breeding for abiotic stress have emerged. These are:

i. At germination: percentage of germination, speed of germination etc.

ii. In the root: root dry weight, number of feeder roots, root/shoot dry weight etc.

iii. In the shoot: vigour, shoot dry weight, stem growth, leaf area, leaf growth rate, leaf dry weight, succulence, water-use efficiency etc.

iv. At fruiting: fruit size, number of fruits, pollen quantity, blossom end rot etc.

v. At the whole plant level: survival, yield, models that relate EC with yield, plant dry weight etc.

Screening genotypes in natural abiotic stress condition is not a general practice that can be recommended due to the variability in abiotic stress conditions over the locations in a region and in these conditions plants exhibits differential responses. Hence, screening in artificial condition is one of the better options particularly for abiotic stress like salt tolerance etc. The abiotic stress at which genotypes are to be evaluated must be carefully chosen as the type of gene action active for characteristic like yield and number of fruits changes depending on the intensity of stress although, for other characteristics like plant height and mean fruit weight or speed of germination similar or identical genes are involved at different stress levels. The intensity of stress to be used in the evaluation should be similar to that of condition available for field cultivation.

Heritabilities found in studies for most of the characteristic involved in abiotic stresses suggest that those characteristic can be improved by selection. Yield and fruit set have the lowest heritabilities found, but the former in principle, would not be the object of direct breeding as with the strategy proposed here yield would be indirectly improved by enhancing characteristic related with plant growth and fruit set with higher heritabilities under stress condition. Heritabilities of the characteristic should then be estimated with the wider range of stress in which selections have been recommended. On the other hand, dominance and even heterosis have been found for characteristic related with vigour ('relative leaf area' and 'stem growth') and succulence which could be very useful for breeding new cultivars of different vegetables for the fresh market, where mainly hybrids are employed.

Indirect selection by morphological and/or molecular genetic makers could be extremely helpful for characteristic that are so affected by the environment as those related with abiotic stress tolerance. Unfortunately, the genetics of these characteristic in most of the vegetables is still poorly known and no genetic marker closely linked to any of them has been found to-date. Some work has been done in this direction that clearly emphasizes the possibilities of markers in breeding programme between cultivated species and their related species under abiotic stress conditions of some of the vegetables like tomato, cucumber etc.

Undoubtedly, the necessary effort to find new markers more closely linked to the different quantitative trait loci (QTL) that build the complex 'abiotic stress tolerant' will produce most future advances towards breeding of cultivars tolerant to abiotic stress.

An additional obstacle to breeding for abiotic stress in most of the vegetable crops is that all the above mentioned traits related with a particular abiotic stress are not combined together in a single donor but in several genotypes. Some of those genotypes may even show phenotypes sensitive to abiotic stress but with a high expression of a particular characteristic that combined with other positive traits would give a tolerant phenotype. A number of donors should then be employed in the breeding programme for pyramiding all those characteristic in a single cultivar which would exhibit a tolerance surpassing that of any existent cultivar against a particular abiotic stress. This idea is already being put into practice in rice with promising results. Vegetable breeding should also resort to pyramiding characteristic since no described trait alone is likely to produce genotype.

Sources of genetic variation for vegetable improvement

Success in any breeding process is based on existence of variability for the characteristic within the same species or alternatively, in related species, together with high expression of the characteristic in one or more accessions (s). A diverse germplasm collection is the backbone or bloodline of any crop improvement program. Germplasm may have genes or traits we need, or they can be recombined to generate novel or enhanced traits in the battle against abiotic/biotic stresses (Mou, 2011).

- Breeders produce plants with improved combinations of genes by crossing (hybridization) and selection within the primary gene pool, which is comprised of a crop species and its closest related wild species.
- Tissue culture methods such as embryo culture are commonly used to enable genes from the secondary gene pool to be transferred into the cultivated species.
- Other methods such as somatic hybridization sometimes allow genes from the tertiary gene pool of more distantly related species to be transferred into crop plants.

- The immense gene resources of the quaternary gene pool (essentially all other organisms) can be used for crop improvement only via transgenic methods.

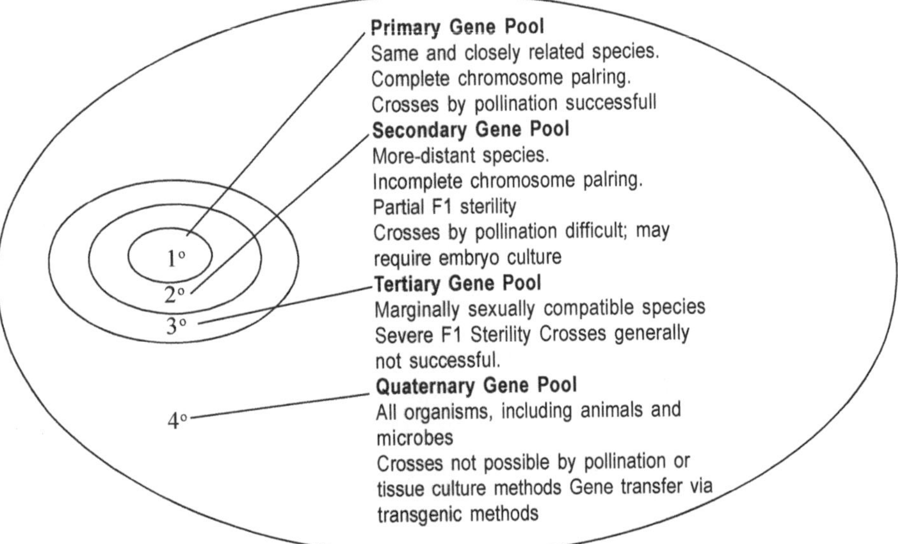

Fig. 1: Source of Genetic Variation

Common breeding and selection schemes for abiotic stresses in vegetable crops

In most of the commercial vegetable crop hybrids have monopolized the market for fresh produce for many years and new hybrids for quality traits are progressively replacing the traditional open-pollinated varieties. Current hybrids for the fresh produce possess many traits such as high productivity, adaptation to intensive cultivation, uniformity in shape and size, excellent colour, long shelf life, resistances to many diseases and insect-pests, which are the result of many years of breeding efforts. Characteristic involved in abiotic stress should then be pyramided step by step within the parents of current hybrids in such a way that they acquire tolerance to abiotic stress and at the same time maintain all the traits that make a current hybrid competitive. Some of the common breeding and selection schemes for abiotic stresses in vegetable crops are follows:

Backcrossing

The breeding method would be backcrossing, if the considered characteristic had a monogenic control. A donor line (blue bar) featuring a specific gene of interest (red) is crossed to an elite line targeted for improvement (white bar), with progeny repeatedly backcrossed to the elite line. Each backcross cycle involves selection for the gene of interest and recovery of increased proportion of elite line genome.

Introduction of the characteristic related to stress tolerance in parents of current cultivar should require separate breeding programmes for each trait.

The following breeding methods will be used with low selection pressure, if the abiotic stress is polygenic in nature.

Gene pyramiding

Genes/QTLs associated with different beneficial traits (blue, red, orange, green) are combined into the same genotype via crossing and selection.

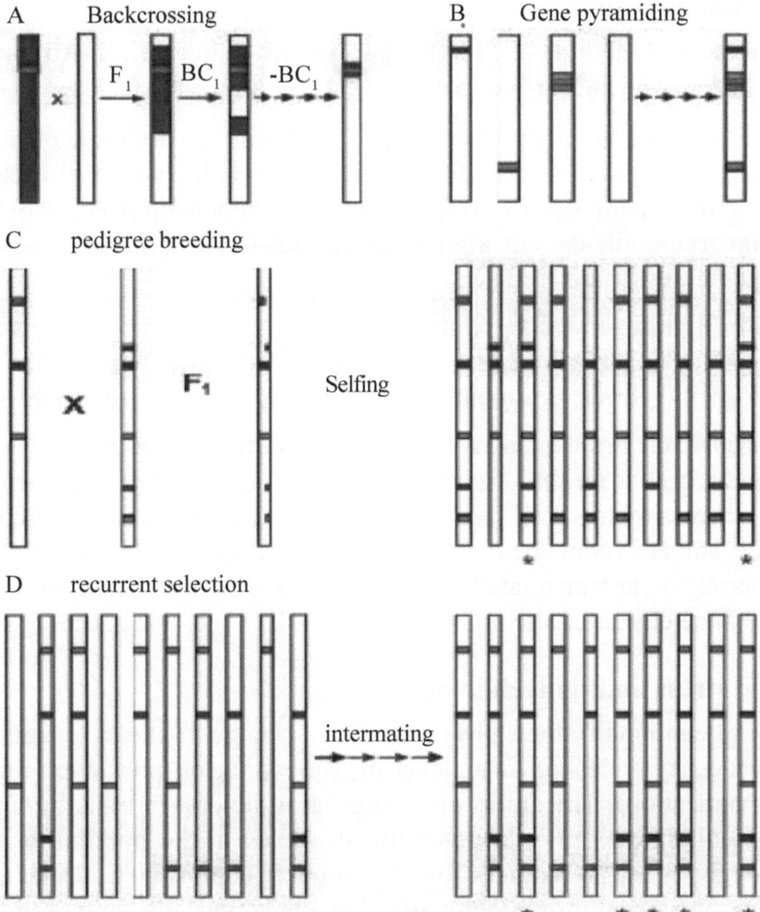

Fig. 2: Gene Pyramiding

- Each vertical bar is a graphical representation of the genome for an individual within a breeding population, with coloured segments indicating genes and/or QTLs that influence traits under selection.
- Genes associated with different traits are shown in different colours (e.g. red, blue).
- "X" indicates a cross between parents, and arrows depict successive crosses of the same type. Asterisk below an individual signifies a desirable genotype.

Pedigree breeding

Two individuals with desirable and complementary phenotypes are crossed; F_1 progeny are self-pollinated to fix new, improved genotype combinations.

Recurrent selection

A population of individuals (10 in this example) segregate for two traits (red, blue), each of which is influenced by two major favorable QTLs.

- Intermating among individuals and selection for desirable phenotypes/genotypes increases the frequencies of favorable alleles at each locus. For this example, no individual in the initial population had all of the five favorable alleles, but after recurrent selection half of the population possesses the desired genotype.
- For hybridized crops, recurrent selection can be performed in parallel within two complementary populations to derive lines that are then crossed to form hybrids; this method is called reciprocal recurrent selection.

Nevertheless, taking advantage of the fact that grafting in some of the vegetables like tomato, brinjal, melons, cucumber etc. is a technique available commercially, the time necessary for developing cultivars tolerant to moderate stress condition (chilling, salt etc.) could be reduced by grafting a cultivar developed for those tolerance characteristic related to shoot performance onto a cultivar, in which, the tolerance characteristic related to the root had been introduced.

Abiotic stress and biotechnology

Organism can modify their physiology by a change in alleles (regulation by allele frequency = adaption) or a change in gene expression (phenotypic regulation = acclimatization). The ability to change stress tolerance traits by means of allele frequency requires a diversity of alleles in the population. Infact, heterozygosity is a prerequisite for adaptation. In contrast, if a species changes its stress tolerance traits by acclimatization, phenotypic plasticity is important. Phenotypic plasticity is the phenotypic response of an organism to change in its environment (Schlichting and Levin, 1986). The advantage of extensive genetic heterogeneity and/or effective plasticity depends on the scale and pattern and spatial stress variation.

The potential success of genetic engineering in a stress physiology context is independent on the ability to isolate particular genes that confer stress tolerance. The complexity inherent in the physiology of plant adaptation to environmental stress makes it difficult to globally identify particular genes that are individually responsible for a particular stress tolerance trait. There are several basic strategies for identifying stress tolerance mechanisms at the molecular level after determining the basic mechanism by which stress tolerance traits are genetically regulated.

- **Discovery strategy** - a native protein is discovered in the test species and shown to regulate stress tolerance.
- **Inference strategy** - foreign proteins are introduced in to the test species and the consequence of that introduction is evaluated.
- **Antisense strategy**- is verified that a particular gene confers stress tolerance. In this strategy a DNA sequence is synthesized that will code for an mRNA fragment that is complementary to the mRNA produced by the putative stress tolerance gene. This is termed as antisense construct. In the presence of antisense construct, no putative stress tolerance gene product is produced. If the tolerant line loses tolerance when transformed with an antisense construct, the gene probably confers stress tolerance.
- **Mutant strategy** - is to induce mutations in the organisms and search the mutations for one that has specifically lost the putative stress tolerance protein. The mutant can be screened for stress tolerance capacity.

Methods of protein separation and identification

Cellular proteins can be isolated and separated by a number of different techniques including two dimensional gel electrophoresis and column chromatography. In two dimensional gel electrophoresis, commonly used to identify stress induced proteins, protein extract are separated by isoelectric focusing in one direction and SDS polyacrylamide electrophoresis in the other direction. Western blotting utilizes a dual antibody technique to locate specific proteins in a gel.

Gene isolation

Restriction endonucleases are bacterial enzymes that cut double stranded DNA at specific nucleotide sequences called restriction sites. Restriction endonucleases are used to separate desired piece of DNA from the large background of DNA in the cell. Cloning is a mechanism of multiplying restriction fragments in a biological vector usually a bacteria or virus. Restriction fragments are inserted in to bacterial plasmids that contain an antibiotic screening device and a polylinker, i.e., multiple restriction site sequence. In a genomic clone, *E. coli* are transformed with plasmids containg restriction fragments, resulting in a large numbers of bacterial clone, each with a unique restriction fragment. In partial genomic cloning, mRNA is extracted from the cell induced to form complementary DNA (cDNA) by reverse transcriptase. The cDNA is incorporated into a bacterial vector in order to multiply each cDNA fragment. It is easier to screen cDNA clone for desired gene because (I) extraneous introns are removed by the cell during RNA processing before the mRNA was extracted (ii) extracted mRNA is likely to contain a large quantity of the desired mRNA and (iii) there are fewer members of cDNA clone than of a genomic clone. A probe is needed to screen a genomic or partial genomic library. The probe can be a small piece of radiolabelled DNA that codes for a six to eight amino-acid-long segment of the putative stress tolerance protein. Southern blotting is used to select the clone with the correct DNA or cDNA.

Mechanism of gene transfer

There are two mechanisms of gene transfer given hereunder.

Indirect mechanism

Gene transfer systems are inefficient due to the low number of competent cells in plant. The ability to transfer gene into plants is dependent on the evolutionary history of the species. The major vector used to transfer genes in to plant cells is the plasmid (Ti) from *Agrobacterium tumefaciens*. T-DNA is the portion of a Ti plasmid that is responsible for incorporating plasmid DNA in to the host cells. A binary plasmid construct can be made containing a restriction fragment from another species and T-DNA from the Ti plasmid. The binary plasmid can be used to transfer genes in to host cells in a number of ways. A viral vector can also be used to transfer DNA in to a host cell, but the DNA will not be transferred in to the genome or the germ cell. Thus, a viral vector cannot be used to develop a true breeding line of transgenic organisms.

Direct mechanism

Electroporation becoming an important technique for transforming plants and is found to be effective with several species. In this technique, genes are directly transferred in to protoplasts by providing an electrical shock. Further, as long as the genetic material is protect to some degree from the strong chemical treatment, chemical shock induction of gene transfer is also possible. Microprojectile bombardment is a technique in which genetic material is coated on a bead projectile and bombarded in to leaf discs, or callus. Genetic material on the beads is incorporated in to the genomes of some cells with a low frequency, although this technique is preferred in soybean and many cereals.

Bioengineering of stress tolerance traits in plants

Thermally induced proteins

Heat Shock Proteins (HSP) were the first stress induced proteins discovered. HSP induction is based on a molecular regulation system involving heat shock factors and heat stock elements. Both translation and transcription regulation are involved in heat shock induction. In plants, HSP 21 and HSP 24 may be important to protecting PSII activity during high temperature conditions. In some cases, over expression of HSPs has resulted in thermo tolerance, but not in other cases. In contrast, chilling stress can result in the down regulation of enzyme genes, such as that of the low molecular weight fragment of rubisco. Other enzymes, such as GAPT, can be up-regulated by cold shock. Over expression of COR genes have resulted in increased protection from chilling in transgenic plants.

Oxidation induced proteins

Many null mutants for important antioxidant enzymes such as the SOD enzymes have been isolated. These mutants are very sensitive to stress conditions that increase oxidative compounds in cells. Two different bacterial antioxidant genes have been transferred in to sensitive mutants. The result is an increase in antioxidant capacity and an increased tolerance to stress. In plants systems, an over expression of ascorbate peroxidase provides oxidant protection in transgenic plants.

Anaerobically induced protein

In non-wetland species, a group of about 20 anaerobically stimulated proteins (ASPs) are induced. In general, anaerobic conditions induce an array of enzymes associated with glycolysis. Introduction begins after about 90 min of anaerobic treatment and 70 % of all protein synthesize after 5h are ASPs.

Alcohol dehydrogenase (ADH) for example is an important ASP. ADH induction is maximum under hypoxic condition, but minimum under anoxic conditions. The promoter region of ADH genes may be an inducible element in much the same manner as heat shock proteins. Up regulations of ADH with a surely anaerobic promoter may induce increased tolerance of anaerobic conditions.

Cellular dehydration caused by drought or salinity is a widespread abiotic stress, constituting one of the most stringent factors limiting plant growth and productivity (Boyer, 1982). Drought tolerance includes three major components: osmotic adjustment, antioxidant capacity and desiccation tolerance *per se* (Zhang *et al.* 1996). May different genes, encompassing many classes of gene product are induced by drought stress. These genes can be divided into different classes based on their DNA sequence, expression characteristics and/or predicted functions. The classification used for drought induced genes which have been identified in vegetable crops are grouped under following three heads and presented in the table (i) genes expressed in vegetative tissues which are homologous to late embryogenesis abundant (LEA) genes Table (3), (ii) enzymes, proteins with non-enzymatic function (Table 4) and (iii) genes with no sequence identity in the similar amino acid sequence (Table 5).

Table 3: Genes encoding late embryogenesis abundant proteins (LEA) and induced by drought (D), salt stress (S), low temperature (L) or high temperature (H) in vegetables; if the genes are not induced by these stress a_sign follows the symbol

Gene/cDNA	Species	Stress induction	ABA regulated	References
Genes encoding late embryo genesis abundant proteins (LEA)				
Group 2				
TAS14	Tomato	DSL_	+	Godoy *et. al.*, 1990
le4	Tomato	DSL_H_	+*	Cohen *et. al.*,1990
Group 4				
le25	Tomato	DSL_H_	+*	Cohen *et. al.*,1990
D113	Tomato	Preserve membrane structure		Cohen *et. al.*, 1990
le25	Tomato	Preserve membrane structure		Cohen *et. al.*,1990

• Regulated by endogenous ABA

Table 4: Genes encoding enzyme activities and regulated by abiotic factors, i.e. drought (D), salt stress (S), low temperature (L) or high temperature (H) in vegetables

Gene/cDNA	Species	Stress induction	ABA regulated	References
Genes encoding enzyme activities				
Lipoxygenase				
*lox*A*	Soybean	D	- #	Bell & Mullet, 1991
*lox*B*	Soybean	D	- #	Bell & Mullet, 1991
*lox Pl**	Pea	D	-	Bell & Mullet, 1991
Acid Phosphate				
*vsp*A	Soybean	D	-	De Wald *et. al.* 1992
*vsp*B	Soybean	D	-	Guerrero *et. al.* 1990
Thiol protease				
*15*a	Pea	D	-	

Contd.

Gene/cDNA	Species	Stress induction	ABA regulated	References
Δ-Pyrroline-5-carboxylate reductase				
*p*Pro *CI*	Soybean	S	?	Delauney &Verma, 1990
Genes encoding water channel proteins				
7a	Pea	D	+	Guerrero *et. al.* 1990
Aldehyde dehydrogenase				
*26*g*	Pea	D	-	Guerrero *et. al.* 1990
Cytosolic Glutathione				
cDNA	Pea			Creissen *et. al.* 1994
Cytosolic Super oxide dismutase				
*Sod*2	Pea			White & Zilinskas, 1991
Ascorbate peroxidase				
Apx I	Pea			Mittler & Zilinskas, 1992

+ induce; - inhibits; - # also induced by methyl jasmonate; ? probable effect but questionable.
* Function has not been determined by enzyme assay, predicted based on amino acid sequence identity with other genes in the gene banks.

Table 5: Genes encoding protein with non-enzymatic function and genes with no sequence identity in the similar amino acid sequence as regulated by drought (D) salt stress (S) low temperature (L) or high temperature (H) in vegetables

Gene/cDNA	Species	Stress induction	ABA regulated	References
Lipid transfer protein				
*Le*16	Tomato	DSLH	+*	Plant *et al.* 1991
*TSW*12	Tomato	HS	+	Torres-Schumann *et.al.* 1992
HI Histone				
*le*20	Tomato	DS	+*	Bray (1991)
*PGE*16	Soybean	D	+	Creelman &Mullet, 1991
*PGE*95	Soybean	D	-	Creelman &Mullet, 1991

• +induce; * regulated by endogenous ABA.

Genes corresponding to desirable traits

These are several methods such as map-based cloning and transposon tagging which are being employed to isolate gene(s) corresponding to desirable traits. There are many steps involved for getting success in map based cloning (Natalya *et al.* 1996). First, the evaluation of target phenotype should be easy and reproductive, second, the population size should be large enough to detect recombination between the target gene and DNA markers Third, a wide repertoire of jobs ranging from the handling of megabased - sized genomic DNA to DNA. And fourth, transformation of plants is to be planned using most effective and time saving method. Samples of isolate gene are presented in Table 6.

Table 6: List of genes cloned in tomato plants

Gene	Motif/coding for the gene	Method	References
Pto	Ser/Thr protein kinase	Map-based	Martin *et. al.* 1993
Fen	Ser/Thr protein kinase	Map-based	Martin *et. al.* 1994
Ptil	Ser/Thr protein kinase	Two-hybrid	Zhou *et. al.* 1995
Prf	LZ, NBS, LRR	Map-based	Salmeron *et. al.* 1996
cf-9	LRR	Transposon tagging	Jones *et. al.* 1994
cf-2	LRR	Map-based cloning	Dixon *et. al.* 1996

LZ: leuinezipper; NBS: nucleotide binding site; LRR: leneine rich repeat

Conclusions

Adapting olericulture to future conditions is essential to meet the need of the growing population and increasing demand for vegetables, and other horticultural products. This enormous and difficult task requires tremendous research efforts from multiple disciplines. Stress physiology studies identify mechanisms of stress tolerance and provide an approach, method, and traits for screening stress-resistant genotypes. Molecular biology and genomic investigations lead to a better understanding of the structural organization and functional properties of genetic variation for stress-related traits, allow gene-based selection through identification of molecular markers and high-throughput genotyping techniques, and increase the gene pool available, including new sources of stress-tolerant traits or transgenes. Vegetable breeders needs to translate these findings into stress-tolerant crop varieties by using all tools available that include germplasm screening, marker-assisted selection, plant transformation, and conventional breeding methods.

Our ability to breed new varieties for adaptation to future environments is undermined by the systematic erosion of the biological basis of olericulture—the genetic diversity of plants, which is in turn threatened by climate changes. Present-day vegetable crops are facing a narrowing genetic base as monoculture dominates throughout the country. Global warming accelerates the loss of diversity because organisms may not be able to adapt to environmental changes and extinguish. Plant diversity is the key for the future vegetable industry to buffer attacks from diseases, insects, and environmental adversities. We should devote necessary effort and funding to safeguard, preserve, manage, collect, maintain, characterize, and use plant germplasm resources for us and our children. Indeed, climate change probably would affect future generations more than our generation.

Transgenic plants have shown great promise in tolerance to abiotic stresses such as heat, cold, drought, and salt. Despite demonstrated benefits, the commercialization of transgenic fruits, vegetables, nuts, and ornamentals has largely lagged behind the agronomic crops, although considerable research is still being conducted on these vegetable crops. Use of transgenic vegetable crops is hindered by many hurdles, including consumer acceptance, technical difficulties with the transformation and expression of transgenes in certain crops, intellectual property right, market and economic issues, and regulatory problems. Still, it is critical to continue working on these challenges to provide growers and consumers options and alternatives in solving abiotic stress problems. Because most of the vegetable produce is for direct human consumption, it is particularly important to perform science based research to assess the impact of genetic engineering on biodiversity, environmental safety, and human health.

The changing environments pose serious and imminent threats to global agriculture and place unprecedented pressures on the sustainability of horticulture industry. On one hand, the climate change makes crop production more difficult. On the other hand, population growth and health-conscious consumers demand more and better vegetable produce. The challenges and opportunities coexist for our dynamic and resilient industry. In addition to curbing carbon emission and conserving resources, we should mitigate abiotic stresses and adapt to the warming planet.

References

Abdalla, A.A. and Verderk, K.1968. Growth, flowering and fruit set of tomato at high temperature. *The Neth J Agric Sci*. 16:71-76.

Ahn S. J., Im, Y.J., Chung, G.C., Cho, B. H. and Suh, S. R. 1999. Physiological responses of grafted-cucumber leaves and rootstock roots affected by low root temperature. *Sci. Hortic.* 81: 397–408.

Allen, D. J. and Ort, D. R. 2001. Impacts of chilling temperatures on photosynthesis in warm-climate plants. *Trends Plant Sci.* 6: 36–42.

Annda, A., Venkatesha, M. M., Kiran Kumar, K. C., Krishnamurthy, N. and BhanuPrakash, V. H. 2008. Mulching: an ideal approach for soil and water conservation in dry lands. *Rashtriya Krishi*. 3(2): 128-129.

Arnon, I. 1980. Breeding for higher yields. In: Physiological Aspects of Crop productivity. Proc 15th Colloquium of the International Potash Institute, Wageningen, *The Netherlands*. pp 77-81.

Ashraf, M., Harris P.J.C. 2013. Photosynthesis under stressful environments: an overview. *Photosynthetica.* 51(2):163–190.

Aslam, M., Sultana, B., Anwar, F. and Munir, H. 2015. Foliar spray of selected plant growth regulators affected the biochemical and antioxidant attributes of spinach in a field experiment Turkish *Journal of Agriculture and Forestry*. doi:10.3906/tar-1412-56.

AVRDC. 1990. Vegetable Production Training Manual.Asian Vegetable Research and Training Center. Shanhua, Tainan. pp 447.

Bell, E. and Mullet, J.E. 1991. Lipoxygenase gene expression is modulated in plants by water deficit, wounding and methyl jasmonate. *Mole Gen Genet.* 230:456-462.

Bhatt, L., Rana, R., Uniyal, S.P. and Singh, V.P. 2011. Effect of mulch materials on vegetative characters, yield and economics of summer squash (*Cucurbita pepo*) under rainfed mid-hill condition of Uttarakhand. *Veg. Sci.* 38:165-168.

Blum, A. 1988. Plant breeding for Stress Environments. CRC Press, Inc, Boca Raton, Florida, USA.

Boyer, J.S. 1982. Plant productivity and environment. *Science*. 218: 443-448.

Bradford, K.J. and Yang, S.F. 1981. Physiological responses of plants to water logging. *Hortscience*. 16:25.

Bray, E.A. 1991. Regualtion of gene expression by endogenous ABA during drought stress. In: Davies WJ and Jones HG (eds), Abscisic Acid: Physiology and biochemistry. *BIOS Scientific Publishers Oxford*, pp. 81-99.

Cohen, A., Plant, A.L., Moses, M.S. and Bray, E.A. 1991. Organ specific and environmentally regulated expression of two abscisic acid induced genes of tomato. *Plant Physiol*. 97:1367-1374.

Cramer, G.R. 2010. Abiotic stress & plant responses from the whole vine to the genes. *Aust J Grape Wine Res*. 16:86-93.

Creelman, R.A. and Mullet, J.E. 1991. Water deficit modulates gene expression in growing zones of soybean seedlings. Analysis of differentially expressed cDNAs, a new â-tubuline gene and expression of genes encoding cell wall proteins. *Plant Mol Biol*. 17:591-608.

Wald, D.B., mason, H.S. and Mullet, J.E. 1992. The soybean vegetative storage proteins VSP or VSPß are acid phosphatase active on polyphosphates. *J Biol Chem*. 267:15958-15964.

Delauney, A.J. and Verma, D.P.S. 1990. A Soybean gene encoding Al-pyrroline-5- carboxylate reductase was isolated by functional complementation is Esherichia coli and is found to be osmoregulated. *Mole Gen Genet*. 221:229-305.

Dixon, M.S., Jones, D.A., Keddie, J.S., Thomas, C.M., Harrison, K. and Jones, J.D.G. 1996. The tomato Cf-2 disease resistance locus compriss two functional genes encoding leucine-rice repeat proteins. *Cell*. 84: 951-59.

Dunage, V.S., Balakrishnan, P. and Patil, M.G. 2009. Water use efficiency and economics of tomato using drip irrigation under net house conditions. *Karnataka J. Agric. Sci*. 22(1): 133-136.

Farazmand H (2013). Effect of kaolin clay on pomegranate fruits sunburn. *Applied Entomology & Phytopathology*. 80 (2-95):173-183.

Giri, J. 2011. Glycine betaine and abiotic stress tolerance in plants.*Plant Signal Behav*. 6: 1746–1751.

Godoy JA, Pando JM and Pintor-Toro JA (1990). A tomato cDNA inducible by salt stress and abscisic acid: nucleotide sequence and expression pattern. *Plant Mol. Biol*. 15:685-705.

Goel, S., Madan, B. 2014. Genetic engineering of crop plants for abiotic stress tolerance IN: Ahmed P, Rasool S (Ed): Emerging Technologies and Management of Crop Stress Tolerance, *Volume 1*. Elsevier, Amsterdam.

Guerrero, F.D., Jones, J.T. and Mullet, J.E. 1990. Turgor responsive gene transcription and RNA levels increase rapidly when pea shoots are wilted: sequences and expression of three inducible genes. *Plant Mol. Biol*. 15: 11-26.

Hodges, D. M. and Toivonen, P. M. A. 2008. Quality of fresh-cut fruits and vegetables as affected by exposure to abiotic stress. *Postharvest Biology and Technology*. 48(2): 155–162.

Rejeb, I.B., Pastor, V. and Mauch-Mani, B. 2014. Plant Responses to Simultaneous Biotic and Abiotic Stress: *Molecular Mechanisms Plants*. 3: 458-475.

Jiang, Y. and Haung, B. 2001. Plants and their environment. Effects of calcium on antioxidant activities and water relations associated with heat tolerance in two cool-season grasses. *J. Exp. Bot.* 52: 341-349.

Kalarani, M.K.M., Thangaraj, R., Sivakumar and V, Mallika. 2002. Effect of salicylic acid on tomato (*Lycopersion esculentum*) *productivity. Crop Res.* 23: 486-492.

Kalloo, G. 1998. Breeding vegetable crops for tolerance to stress environments. In: Vegetable Breeding Vol II. *CRC Press .Inc, Boca Raton Fla.* pp 169-202.

Kleinhenz, M.D., Palta, J.P. 2002. Root zone calcium modulates the response of potato plants to heat stress. *Physiol. Plant.* 115: 111-118.

Kramer, U., Coller-Howells, J.D., Chasnock, J.M., Baker, A.J.M. and Smith, A.C. 1996. Free histidine as a metal chelator in plants that accumulate nickel. *Nature*. 379: 635- 638.

Martin, G.B., Brommonschenkel, S.H., Chunwaongse, J., Frasy, A., Ganal, M.W., Spirey, R., Wu, T., Earle, E.D. and Tanksley, S.D. 1993. Map-based cloning of a protein uinase gene conferring disease resistance in tomato. *Science*. 262:1432-1436.

Martin, G.B., Fray, A., Wu, T., Brommonschenkel, S.H., Chunwaongse, J., Earle, E.D. and Tanksley, S.D. 1994. A member of the tomato *Pto* gene family confers sensitivity to fenthion resulting in rapid cell death. *plant Cell*. 6: 1543-1552.

Maurya, S. K., Singh, A. and Rawat, M. 2013. Plant Genetic Resources Conservation, Evaluation, Characterization and Utilization in Vegetable Crop Improvement. *In* : Prasad, B. and Kumar, S. eds. Reshaping Technology for Agricultural Development. *Satish Serial Publishing House, Delhi.* pp. 55-88.

Miloseviæ, N., Marinkoviæ, J. B. and Tintor, B. 2012. Mitigating abiotic stress in crop plants by microorganisms. *Zbornik Matice Srpske Za Prirodne Nauke*. 123: 17–26.

Misra, S. and Gedami, L. 1989. Heavy metal tolerant transgenic *Brassica napus* L. and *Nicotiana tobacum* L. plants. *Theor Appl Genet*. 78:161-168.

Mittler, R. and Zilinskas, B.A. 1992. Molecular clonning and characterization of a gene encoding pea cytosolic ascorbate peroxidase. *J Biol Chem*. 267:21802-21807.

Mou, B. 2011. Improvement of Horticultural Crops for Abiotic Stress Tolerance: *An Introduction. Hort Science*. 46 (8) : 1068-1069.

Nakayama, F.S. and Bucks, D.A. 1986. Trickle Irrigation for Crop Production, Design, Operation and Management. *Elsevier Scientific Publishers, Netherlands*. p.376.

Natalya, Y., klueva, Zthang j. and Nguyen, H.T. 1996. Molecular strategies for managing environment stress. *In :Proc of the 2nd Int Crop Sci Cong. New Delhi, India.* pp 501-524.

Niklas, K.J. 1994. Plant Allometry. The scaling of form and process. *The University of Chicago Press Chicago.*

Nilsen, E.T. and Orcutt, D.H. 1996. Physiology of Plants under Stress-Abiotic Factors. *John Wiley & Sons. Inc NY, USA.*

Opena, R.T., Yang, C.Y., Lo, S.H. and Lai, S.H. 1993. The breeding of vegetables adapted to the low land tropics Tech Bull Food Fert Tech Center Taiwan, No. 77:16.

Osmond, C.B., Austin, M.P., Berry, J.A., Billing, W.D., Boyer, J.S., Dacey, J.W.H., Nobel, P.S., Smith, S.D. and Winner, W.E. 1987. Stress physiology and the distribution of plants. *Bio. Science*. 37:37-48.

Pan, N., Yang, M., Tie, F., Li, L., Chen, Z. and Ru, B. 1994. Expression of mouse metallothionein-I gene confers cadmium resistance in transgenic tobacco plants. *Plant Hort Biol*. 24:341-351.

Peck, A.W. and McDonald, G.K. 2010. Adequate zinc nutrition alleviates the adverse effects of heat stress in bread wheat. *Plant Soil*. 337:355-374.

Peleg, Z., Blumwald, E. 2011. Hormone balance and abiotic stress tolerance in crop plants. *Curr Op Plant Biol*. 14:290–295.

Plant, A.L., Cohen, A., Moses, M.S. and Bray, E.A. 1991. Nucleotide sequence and spatial expression pattern of a drought and absicisic acid induced gene in tomato. *Plant physiol.* 97:900-906.

Ponnamperuma, F.N. 1972. The chemistry of submerged soils. *Adv Agron.* 24: 29-96.

Ramakrishna, A. and Ravishankar, G.A. 2011. Influence of abiotic stress signals on secondary metabolites in plants. *Plant Signaling & Behavior.* 6(11): 1720–1731.

Rehigan, J.B., Villareal, R.L. and Lai, S.H. 1977. Reaction of three tomato cultivar to heavy rainfall and excessive soil moisture. *Philipp j Crop Sci.* 2:221.

Rugh, C.L., Wilde, H.D., Staak, N.M., Thompso, D.M., Summers, A.O. and Meagher, R.B. 1996. Mercuric ion reduction and resistance in transgenic *Arabidopsis thaliana* plants expressing a modified bacterial mer A gene. *Proc Natl Acad Sci USA.* 93:3182-3187.

Rush, D.W. and Epstein, E. 1976. Genotypic responses to salinity: differences between salt sensitive and salt tolerant genotypes of tomato. *Plant Physiol.* 57:162.

Rush, D.W. and Epstein, E. 1978. Salt tolerance in tomatoes: feasibility of using wild germplasm to increase salinity tolerance in domestic species. *Plant Physiol.* 61:94.

Rush, D.W. and Epstein, E. 1981. breeding and selection for salt tolerance by the incorporation of wild germplasm into a domestic tomato. *J. Am. Soc. Hortic. Sci.* 106:699.

Rush, D.W. and Epstein, E. 1981a. Comparative studies on the sodium, potassium and chloride relation of a wild halophytic and domestic salt sensitive tomato specie4s. *Plant physiol.* 68:1308.

Salmeron, J.M., Oldroyd, E.D.G., Remmens, C.M.T., Scoofield, S.R., Kim, H.S., Lavelle, D.T., Dahlbeck, D. and Staskawicz, B.J. 1996. tomato *Prf* is a member of the leucine-rich repeat class of plant disease resistance genes and lies embeded within the *Pto* kinase gene cluster. *Cell.* 86:123-133.

Samaras, Y., Bresson, R.A., Csonka, L.N., Garcia-Rios, M., PainaD'Urzo, M., Rhodes, D. 1995. Proline accumulation during water deficit. In: Sambrook N (ed) Environment and plant metabolism flexibility and acclimation. *BioScientific publishers, Oxford.*

Sari, N., Guler, H.Y., Abak, K. and Pakyurek, Y. 1994. Effect of mulch and tunnel on the yield and harvesting period of cucumber and squash. *Acta Horticulturae.* 371: 305-310.

Schlicting, S.D. and Levin, D.A. 1986. Phenotypic plasticity: an evolving plant character. *Biol. J. Linnean Soc.* 29:37-47.

Schwarz, D., Rouphael, Y., Colla, G. and Venema, J.H. 2010. Grafting as a tool to improve tolerance of vegetables to abiotic stresses: thermal stress, water stress and organic pollutants. *Sci. Hortic.* 127:162–171.

Shukla, N., Awasthi, R.P., Rawat, L. and Kumar, J. 2012. Biochemical and physiological responses of rice (Oryza sativa L.) as influenced by Trichoderma harzianum under drought stress. *Plant Physiol Biochem.* 54:78–88.

Silverira, J., Viegas, R.D., Da Rocha, I.M., Moreira, A.C., Moreira, R.D. and Oliveira, J.T. 2003. Proline accumulation and glutamine synthetase activity are increased by salt induced proteolysis in cashew leaves. *J. Plant Physiol.* 160:115–123.

Singh, N. and Ahmad, Z. (2008). Effect of mulching on potato production in high altitude cold arid zone of Ladakh. *potato J.* 35 (3-4):118-121.

Singh, S., Maurya, S. K. and Gangola, P. 2013. Management of Soil Fertility for Sustainable Vegetable Production. *In* : Kumar, S. and Prasad, B. eds. Modern Technology for Sustainable Agriculture. New India Publishing Agency, New Delhi. pp. 33-54.

Skirycz, A. and Inze, D. 2010. More from less: plant growth under limited water. *Curr. Opin.Biotechnol.* 21(2):197-203. 9.

Srivastava, A. P., Rama Rao, D., Basade, Y., Singh, A. K., Sikarwar, M. and Ashar, N. 2013. Livelihood Enhancement through Improved Vegetable Cultivation in Backward Districts of India Published by : National Director National Agricultural Innovation Project Indian Council of Agricultural Research KrishiAnusandhanBhawan – II, Pusa, New Delhi – 110012.

Stevens, M.A. and Rudich, J. 1978. Genetic potential for overcoming physiological limitations on adaptability, yield, and quality in tomato. *Hort. Science.*13:673-678.

Tawfik, A.A., Kleinhenz, M.D. and Palta, J.P. 1996. Application of calcium and nitrogen for mitigating heat stress effects on potatoes. Amer. *Potato J.* 73:261-273.

Thomas, C.H., Hammond-Kosalk, K.E., Balint-Kurti, P.J. and Jones, J.D.G. 1994.

Isolation of the tomato *cf*-9 gene for resistance to *cladosporium fuluum* by transposon tagging. *Science.* 266:789-793.

Toivonen, P.M.A., Hampson, C. and Stan, S. 2005. Apoplastic Levels of Hydroxyl Radicals in Four Different Apple Cultivars are Associated with Severity of Cut-Edge Browning. *Acta. Horticulturae.* 682:1819-1824.

Torres-Schumann, S., Godoy, T.A. and Pintor-Toro, J.A. 1992. A Probable lipid transfer protein is induced by NaCl in stems of tomato plants. *Plant Mol. Biol.* 18:749-757.

Uchida, A.T., Jagendorf, T. and Hibino. 2002. Effects of hydrogen peroxide and nitric oxide on both salt and heat stress tolerance in rice. *Plant Sci.* 163: 515-523.

Venema, J.H., Linger, P., Van Heusden, A.W., Van Hasselt, P.R. and Brüggemann, W. 2005. The inheritance of chilling tolerance in tomato (*Lycopersicon* spp.). *Plant Biol.* 7: 118–130.

Wahid, A., Gelani, S., Ashraf, M. and Foolad, M.R. 2007. Heat tolerance in plants: an overview. *Environ. Exp Bot.* 61:199–223.

White, D.A. and Zilinskas, B.A. 1991. Nucleotide sequence of a complementary DNA endocing pea cyltosolic copper/fine superoxidase dis mutase. *Plant Physiol.* 96:1391-1392.

Xin, L., Yan, L. and Qiu, Z.S. 2000. Effect of salicyclic acid on growth and content of IAA and IBA of broadbean seedlings. *Plant Physiol.* 2: 50-52.

Zhang, J., Klueva, N. and Nguyen, H.T. 1996. Plant adaptation and crop improvement for arid and semi arid environments. In: Proc Fifth Int Conf on Desert Development, Aug. 12-17, Lubbock, Texas.

Zhou, J., Loh, Y.T., Bressan, R.Y. and Martin, G.B. 1995. The tomato gene *Ptil* encodes a serine/ threonine protein kinase that is phosphorylated by *pto* and is involved in the hypersensitive response. *Cell.* 83:925-935.

12

Abiotic Stress: Impact and Management in Ornamental Crops

Nomita Laishram and Arvinder Singh

Plants are exposed to a wide range of environmental stresses and have to adapt physiologically to these as the local environment changes. Unfavorable soil properties, fertility imbalances, moisture extremes, temperature extremes, chemical toxicity, physical injuries, and other problems are some of the most discussed examples of abiotic disorders that can hamper plant growth which ultimately lead to death of the plants under extreme conditions. Furthermore, many of these abiotic stresses can predispose plants to diseases caused by infectious microbes. These stresses in plants also lead to a series of physiological, biochemical and molecular changes. Impact of biotic stresses on agricultural plants such as wheat, rice, maize, groundnut and many others have been top listed area of interest for researchers but until now, ornamental plants have not been a major object of such studies even though these plants constitute a major part of horticultural production and play an important role in everyday human life.

Ornamental plants are an integral part of the urban public places and private gardens. However, these plants are highly sensitive to abiotic stresses, which are known as the most harmful factor concerning the growth and development of these plants. Plant stress implies some adverse effect on the physiology of a plant induced upon a sudden transition from some optimal environmental condition.

Abiotic stresses are at their most harmful stage when they occur together, in combinations of two or more abiotic stress factors (Mittler, 2006). Although some biotic stresses can be controlled by germicide, pesticides, antibiotic treatment and/or plant metabolism, it is urgent to control abiotic stress, which can possibly affect response to biotic stress control with detrimental effect on plant growth and yield (Josine *et al.*, 2011). Salinity and drought are the most commonly observed abiotic stress; they significantly reduce yield and affect almost every aspect of the physiology and biochemistry of plants. Great efforts have been devoted to understanding the physiological aspects of response in plants to salinity and drought-tolerant genotypes (Cuartero *et al.*, 2006). Management strategies to fight against abiotic stresses include selection of better adaptable genotypes, genetic manipulation to overcome extreme climatic stresses, measures to improve water and nutrient-use efficiency. In this chapter, the impact of abiotic stress on ornamental plant growth and yield is discussed and strategies to overcome the harmful consequences are outlined.

How do plants respond to stress?

Abiotic stresses like extreme temperatures (low/high), soil salinity, drought and floods are detrimental to growth of ornamental plants. Some plants may be injured by a stress, which means that they exhibit one or more metabolic dysfunctions. If the stress is moderate and short term, the injury may be temporary and the plant may recover when the stress is removed. If the stress is severe enough, it may prevent flowering, seed formation, and induce senescence that leads to plant death. Such plants are considered to be susceptible. Some plants escape the stress altogether, such as ephemeral, or short-lived, desert plants. Ephemeral plants germinate, grow, and flower very quickly following seasonal rains. They thus complete their life cycle during a period of adequate moisture and form dormant seeds before the onset of the dry season. In a similar manner, many arctic annuals rapidly complete their life cycle during the short arctic summer and survive over winter in the form of seeds. Because ephemeral plants never really experience the stress of drought or low temperature, these plants survive the environmental stress by stress avoidance. Avoidance mechanisms reduce the impact of a stress, even though the stress is present in the environment. Many plants have the capacity to tolerate a particular stress and hence are considered to be stress resistant. Stress resistance requires that the organism exhibit the capacity to adjust or to acclimate to the stress.

Types of abiotic stress

Abiotic stress is the negative impact of physical factors (e.g., light, temperature) that the environment may impose on a plant. Abiotic stress hampers growth and development in most of the ornamental plants worldwide. Plants may respond similarly to avoid one or more stresses through morphological or biochemical mechanisms (Capiati *et al.*, 2006). Environmental interactions may cause stress response of plants more complex. Measures to adapt to these stresses are critical for sustainable flower production. There is a need to do more research on how ornamental crops are likely to be affected by the increased abiotic stresses. Some of the important abiotic stresses which affect ornamental plants have been reviewed below.

High temperature stress

A constantly high temperature causes an array of morpho-anatomical changes in plant which affect the seed germination, plant growth, flower shedding, pollen viability, gametic fertilization, fruit setting etc. These problems can be minimized by improvement in the cultural practices and breeding approaches. Protected structures can play important role to minimize the impact of temperature fluctuation, over/under precipitation, fluctuating sun shine hour and infestation of disease and pest (Singh and Satpathy, 2005).

Maya and Matsubara in 2014 reported that association of AM fungi had the ability to enhance the growth of cyclamen under heat stress condition and could decrease the anthracnose incidence in cyclamen production. Non-acclimation to heat stress causes more susceptibility of plants to pathogen. For example, ornamental plant roots directly exposed to 45°C soil temperatures increased severity of *Phytophthora infestans* (causal agent of root rot in ornamentals).

Mesophytic plants (terrestrial plants adapted to temperate environments that are neither excessively wet nor dry) have a relatively narrow temperature range of about 10°C for optimal growth and development. Outside of this range, varying amounts of damage occur, depending on the magnitude and duration of the temperature fluctuation. Most actively growing tissues of higher plants are tillable to survive extended exposure to temperatures above 45°C or even short exposure to temperatures of 55°C or above. However, non growing cells or dehydrated tissues (e.g., seeds and pollen) remain viable at much higher temperatures. Pollen grains of some species can survive 70°C and some dry seeds can tolerate temperatures as high as 120°C.

Most plants with access to abundant water are able to maintain leaf temperatures below 45°C by evaporative cooling, even at elevated ambient temperatures. However, high leaf temperatures combined with minimal evaporative cooling causes heat stress. Leaf temperatures can rise to 4 to 5°C above ambient air temperature in bright sunlight near midday, when soil water deficit causes partial stomatal closure or when high relative humidity reduces the gradient driving evaporative cooling. Increases in leaf temperature during the day can be more pronounced in plants experiencing drought and high irradiance from direct sunlight.

High temperature stress can result in damaged membranes and enzymes. Plant membranes consist of a lipid bilayer interspersed with proteins and sterols, and any abiotic factor that alters membrane properties can disrupt cellular processes. The physical properties of the lipids greatly influence the activities of the integral membrane proteins, including H+-pumping ATPases, carriers, and channel-forming proteins that regulate the transport of ions and other solutes. High temperatures cause an increase in the fluidity of membrane lipids and a decrease in the strength of hydrogen bonds and electrostatic interactions between polar groups of proteins within the aqueous phase of the membrane. High temperatures thus modify membrane composition and structure, and can cause leakage of ions. High temperatures can also lead to a loss of the three-dimensional structure required for correct function of enzymes or structural cellular components, thereby leading to loss of proper enzyme structure and activity. Misfolded proteins often aggregate and precipitate, creating serious problems within the cell.

Temperature stress can inhibit photosynthesisand respiration.Typically, photosynthetic rates are inhibited by high temperatures to a greater extent than respiratory rates. Although chloroplast enzymes such as rubisco, rubisco activase, NADP-G3P dehydrogenase, and PEP carboxylase become unstable at high temperatures, the temperatures at which these enzymes began to denature and lose activity are distinctly higher than the temperatures at which photosynthetic rates begin to decline. This would indicate that the early stages of heat injury to photosynthesis are more directly related to changes in membrane properties and to uncoupling of the energy transfer mechanisms in chloroplasts.

This imbalance between photosynthesis and respiration is one of the main reasons for the deleterious effects of high temperatures. On an individual plant, leaves growing in the shade have a lower temperature compensation point than leaves that are exposed to the sun (and heat). Reduced photosynthate production may also result from stress-induced stomatal closure, reduction in leaf canopy area, and regulation of assimilate partitioning.

Cold stress

Cold stress is one of the main abiotic stresses that limit crop productivity by affecting their quality and post-harvest life. Because plants are immobile, they must modify their metabolism to survive such stress. Most temperate plants acquire chilling and freezing tolerance upon their exposure to sub-lethal cold stress, a process called cold acclimation. Cold stress affects virtually every aspects of cellular function in plants. The cold stress signal is transduced through several components of signal transduction pathways. Major components are Ca^{2+}, ROS, protein kinase, protein phosphatase and lipid signaling cascades. ABA also mediates the response of cold stress. Different plant species tolerate cold stress to a different way, which depends on the concern gene expression to modify their physiology, metabolism, and growth. The cold response mechanism may be related to various changes like, the expression of kinases related to signal transduction, accumulation of osmolytes and membrane lipid composition.

Freezing temperatures cause ice crystal formation and dehydration. Freezing temperatures result in intra- and extracellular ice crystal formation. Intracellular ice formation physically shears membranes and organelles. Extracellular ice crystals, which usually form before the cell contents freeze, may not cause immediate physical damage to cells, but they do cause cellular dehydration. Low temperature probably alters chemical changes associated with respiration and the plant dies due to freezing as a result of ice formation. Ice formation in the cell protoplasm disrupts the colloidal system and destroys the normal properties of the cell. In the beginning ice formation occurs in the intercellular spaces because of which solution of this area becomes concentrated. This increased concentration pulls the water from cell interior and consequently the cells become dehydrated. The volume of protoplasm inside the cell is reduced and the cell wall contracts to a possible limit beyond which it can rupture. Besides, concentration of salt increases which denatures the number of proteins and enzyme molecules.

Table1: Categories for freeze hardiness of various annual flowers

Hardy	Tolerant	Tender	Sensitve
Cornflower	Bells of Ireland (*Moluccella*)	Aster	*Ageratum*
Ornamental cabbage	*Coreopsis*	*Nicotiana*	Balsam
Pansy	Pinks (*Dianthus*)	Petunia	Begonia
Primrose	Pot Marigold (*Calendula*)	*Scabiosa*	Cockscomb
Violet	Snapdragon	Statice	*Impatiens*
	Stock (*Matthiola incana*)	Sweet alyssum	Lobelia
	Sweet pea	Verbena	Marigold
	Torenia		Moss rose (*Portulaca*)
			Periwinkle (*Vinca*)
			Phlox, annual
			Salpiglossis
			Salvia
			Zinnia

Source: Based on Purdue University publication HO-14, as cited by Caplan, 1988

Table 2: Frost tolerance, light preference, and relative drought tolerance of bedding plants (*Armitage, 1988*)

Bedding plant	Frost tolerance	Light preference	Relative drought tolerance
Ageratum	Tender	Full sun	Moderate
Alyssum	Tolerant	Full sun to partial shade	Low
Aster	Moderate tolerance	Full sun to partial shade	Low
Begonia	Tender	Full sun to heavy shade	Low
Gaillardia	Tolerant	Full sun	Moderate
Browallia	Moderate tolerance	Full sun to partial shade	Low
Candytuft	Tolerant	Full sun to partial shade	Low
Calendula	Tolerant	Full sun to partial shade	Low
Celosia	Tender	Full sun	Moderate
Coleus	Moderate tolerance	Partial to heavy shade	Low
Cornflower	Moderate tolerance	Full sun to partial shade	Low
Cosmos	Moderate tolerance	Full sun	Moderate
Dahlia	Tender	Full sun	Low
Dianthus	Very tolerant	Full sun to partial shade	Low
Senecio	Tolerant	Full sun to partial shade	Moderate
Geranium	Tender	Full sun	Low
Globe Amaranth	Moderate tolerance	Full sun	High
Rudbeckia	Tolerant	Full sun to partial shade	Moderate
Impatiens	Tender	Full sun to heavy shade	Low
Lisianthus	Tender	Full sun	High
Lobelia	Tolerant	Full sun to partial shade	Low
Marigold	Moderate tolerance	Full sun	Moderate
Ornamental Pepper	Tender	Full sun	Low
Pansies	Very tolerant	Full sun to partial shade	Low
Petunia	Tolerant	Full sun to partial shade	Moderate
Phlox	Tolerant	Full sun to partial shade	Low
Portulaca	Tender	Full sun	High
Salvia	Tender	Full sun	Low
Snapdragon	Very tolerant	Full sun to partial shade	Low
Cleome	Tender	Full sun	Moderate
Gazania	Tolerant	Full sun	High
Verbena	Moderate tolerance	Full sun to partial shade	Moderate
Vinca	Tender	Full sun to partial shade	Moderate
Zinnia	Tender	Full sun	Moderate

Drought stress

Drought Stress is the deficiency or dearth of water severe enough to check the plant growth. Drought injury occurs primarily due to deficiency of soil moisture but, atmospheric factors such as high temperature, low humidity, fast wind, etc. aggravate the adverse effects of it. Drought resistance is the capacity of the plants to endure drought and to recover rapidly after the onset of permanent wilting with minimum damage to the plant itself. Drought in plants adversely affects the functioning of stomata, carbohydrate metabolism in green leaves, reduces the photosynthetic activity in green cells, osmotic pressure of the plant cell increases and the permeability increases specially to water, urea and glycerine. Water deficiency adversely affects protein synthesis. Plants exposed to stress are found to have higher levels of ABA and ethylene is also produced in response to various kinds of stress. Lin and Kao (1998) suggested that the accumulation of ammonium, which is toxic to plants, may be a factor that further aggravates the stress. Chlorophylls are essential pigments of the higher plant assimilatory tissues, responsible for light absorbing and proper functioning of the photosynthetic apparatus. According to Ueda *et al.* (2003), the chlorophyll a+b content in the leaves can be indicative of stressful conditions, such as water or salt stress. Proper levels of chlorophylls also have a strictly ornamental value as chlorotic, yellowing plants lose their aesthetic appeal.

Classification of plants in drought prone areas

a. Ephemerals: These are short lived plants and they complete their life cycle within a smaller favourable period during the rainy season. They pass dry periods in the form of seeds. e.g. *Solanum xanthocarpum, Argemone Mexicana.*

b. Succulent plants accumulate large quantities of water and use it slowly during dry periods. Such plants develop several morphological adaptations for reducing transpiration such as thick cuticle, reduced surface area, sunken stomata, etc. Examples are Agave, Euphorbia, Opuntia, etc.

c. Non succulent plants are the real drought enduring plants (euxerophytes). They tolerate drought without adapting any mechanism to ensure continued supply of water. They develop a greyish colour, reflecting surfaces, smaller leaves, extensive root system, leaf fall during dry season, sunken stomata and thick cuticle. The stomata remains closed mostly in dry periods.

According to Dr. Elwynn Taylor, Professor of agricultural meteorology at Iowa State University, for every 10^0 increase in temperature, the water requirement by a crop would increase by 50% (Anonymous, 2010). Water is a essential resource which is required in more amounts by horticultural crops typically than agronomic crops. For overcoming the problems of drought stress in ornamental plants, drought-tolerant cultivars are needed to improve water use efficiency, enhance water conservation, reduce irrigation costs, and maintain or increase the acreage of horticultural crops. For certain areas or specific seasons that are prone to flooding, new varieties should be developed to adapt to the wet soil. Plants have evolved a number of adaptations for surviving and thriving within the water-limited environments.

Water availability is highly sensitive in flower cultivation and severe water-stress conditions will affect crop productivity. In combination with elevated temperatures, decreased precipitation could cause reduction in availability of irrigation water and increase in evapo-transpiration, leading to severe crop water-stress conditions (IPCC, 2001). Drought-stress causes an increase in solute concentration in the environment (soil), leading to an osmotic flow of water out of plant cells. This leads to an increase in the solute concentration in plant cells, thereby lowering the water potential and disrupting membranes and cell processes such as photosynthesis. Precise irrigation requirements can be predicted based on crop water-use and effective precipitation values. Lack of water influences the crop growth in many ways and the effect depends on the severity, duration, and time of stress in relation to the stage of growth. Nearly all ornamental crops are sensitive to drought during flowering period.

Comparative drought tolerance of foliage of several ornamental crops was studied by Robert *et al.* in 2003. They reported that the most dehydration tolerant of the four genera studied was Salvia, as characterized by lethal leaf water potential, and showed the highest osmotic adjustment. Dahlia and the two Impatiens cultivars had similar water relations at the lethal point. Foliage of woody species tended to be more tolerant of dehydration than foliage of herbaceous species. Dehydration tolerance is a measure of tissue capacity for withstanding desiccation and has been defined for several agronomic species as the water potential (C) of the last surviving leaves (called the lethal value) on a plant subjected to a slow, continuous soil drying episode (Ludlow, 1989).

Foliar dehydration tolerance of several woody and herbaceous ornamental species, ranked by lethal leaf water potential was given by Robert *et al.* 2003

Table 3: Lethal leaf water potential of ornamental speecies

Species	Lethal leaf water potential (MPa)
Pelargonium hortorum	-2.01
Impatiens wallerana	-2.06
Dahlia hybrida	-2.29
New Guinea impatiens	-2.37
Helianthus angustifolia	-2.58
Monarda didyma	-3.02
Rudbeckia fulgida	-3.56
Echinacea purpurea	-3.77
Oxydendrum arboreum	-3.98
Cornus kousa	-4.01
Rosa hybrida	-4.16
Acer ginnala	-4.19
Pentas lanceolata	-4.27
Acer rubrum	-4.43
Quercus alba	-4.60
Salvia splendens	-5.16
Hypericum patulum	-6.67

Cicevan *et al.* (2016) evaluated twelve cultivars of three ornamental *Tagetes* species (*T. patula*, *T. Tenuifolia* and *T. erecta*) for Drought tolerance and reported considerable differences in the evaluated traits among the control and drought-stressed plants. Drought stress generally caused a marked reduction in plant growth and carotenoid pigments, and an increase insoluble solutes and oxidative stress. For most cultivars, proline levels in stressed plants increased between 30 and 70-fold compared to the corresponding controls. A considerable variation in the tolerance to drought was found within each species, on average *T. erecta* proved to be more tolerant to drought than *T. patula* and *T. tenuifolia.*

Franco (2011) reported a decrease in shoot: root ratio, a common observation under drought-stress, which results either from an increase in root growth or from a relatively larger decrease in shoot growth than in root growth. It was also reported in *Lonicera implexa* (Navarro *et al.*, 2008), *Lotus creticus* (Franco *et al.*, 2001; Banon *et al.*, 2004), *Myrtus communis* (Banon *et al.*, 2002), *Nerium oleander* (Banon *et al.*, 2006), *Rosmarinus officinalis* (Sanchez-Blanco *et al.*, 2004) and *Silene vulgaris* (Arreola *et al.*, 2006; Franco *et al.*, 2008).

Response of two bedding annuals viz. impatiens (*Impatiens walleriana* Hook) and geranium (*Pelargonium hortorum* L. H. Bailey) to water stress was studied by Chylinski *et al.* (2007). In both species roots were significantly longer in plants grown at 30% SWC as compared to 80% SWC while plant height and flower number were reduced by drought only in impatiens. Ammonium content in leaves of both species increased significantly under stress but the ranges of increase were different in both species. There was a significant increase in the free amino acids content in leaves of impatiens as compared to geranium but this rise was more time than drought dependent. The reduction in the *a* + *b* chlorophyll concentration in leaves of impatiens was significantly time and stress dependent while no reaction in geranium was observed.

Jaleel (2007) investigated whether $CaCl_2$ increases *Catharanthus roseus* drought tolerance and if such tolerance is correlated with changes in oxidative stress, osmoregulation and indole alkaloid accumulation. *Catharanthus roseus* plants were grown under water deficit environments with or without $CaCl_2$. Drought induced oxidative stress was measured in terms of lipid peroxidation (LPO) and H_2O_2 contents, osmolyte concentration, proline (PRO) metabolizing enzymes and indole alkaloid accumulation. The plants under pot culture were subjected to 10, 15 and 20 days interval drought (DID) stress and drought stress with 5 mM $CaCl_2$ and 5 mM $CaCl_2$ alone from 30 days after planting (DAP) and regular irrigation was kept as control. The plants were uprooted on 41 DAS (10 DID), 46 DAS (15 DID) and 51 DAS (20 DID). Drought stressed plants showed increased LPO, H_2O_2, glycine betaine *(GB) and PRO contents and decreased proline oxidase (PROX) activity and increased γ-glutamyl kinase (γ-GK) activity when compared to control. Addition of* $CaCl_2$ to drought stressed plants lowered the PRO concentration by increasing the level of PRO.

Water management

The quality and efficiency of water management determine the quality of ornamental plants. Too much or too little water causes abnormal plant growth, predisposes plants to infection by pathogens, and causes nutritional disorders. If water is scarce and supplies are erratic or variable, then timely irrigation and conservation of soil moisture reserves are the most important agronomic interventions to maintain yields during drought stress. There are several methods of applying irrigation water and the choice depends on the crop, water supply, soil characteristics and topography. Surface irrigation methods are utilized in more than 80% of the world's irrigated lands, yet its field level application efficiency is often 40-50% (Von *et al.*, 2004).

Drip irrigation minimizes water losses due to run-off and deep percolation and water savings of 50-80% are achieved when compared to most traditional surface irrigation methods. Crop production per unit of water consumed by plant evapo-transpiration is typically increased by 10-50%. Thus, more plants can be irrigated per unit of water by drip irrigation, and with less labour.

Water harvesting for dry land is also another traditional water management technology to ease future water scarcity in many arid and semi arid regions of the world. Rainwater and flood water harvesting have the potential to increase the productivity of arable land by increasing the yields and reducing the risk of crop failure under stress situation.

Flooding stress

When there is excess amount of water than its optimum requirement is known as flooding/ water logging. It leads to replacement of gaseous phase by liquid phase. Production is often limited during the rainy season due to excessive moisture brought about by heavy rains. Most of the ornamental crops are highly sensitive to water logging and genetic variation with respect to this character is limited. In general, the damage to plants by flooding is due to reduction of oxygen in the root zone, which inhibits aerobic processes.

Salinity stress

The term salinity is used to describe excessive accumulation of salt in the soil solution. Salinity stress has two components: nonspecific osmotic stress that causes water deficits, and specific ion effects resulting from the accumulation of toxic ions, which disturb nutrient acquisition and result in cytotoxicity. Salt-tolerant plants genetically adapted to salinity are termed *halophytes*, while less salt-tolerant plants that are not adapted to salinity are termed *glycophytes*. 20% of the world's irrigated lands are affected by salinity (Zhu, 2001), a situation worsened by climate changes. High salinity limits crop production in 30% of the irrigated land in the United States (Kumar *et al*., 2004). Several anomalies associated with the elemental composition of soils can result in plant stress, including high concentrations of salts (e.g., Na^+ and Cl^-) and toxic ions (e.g., As and Cd), and low concentrations of essential mineral nutrients, such as $Ca^{2+,}$ Mg^{2+}, N, and P. Saline soils are often associated with high concentrations of NaCl, but in some areas $Ca^{2+,}$ Mg^{2+}, and SO^{4-} are also present in high concentrations in saline soils. High Na+ concentrations that occur in sodic soils (soils in which Na^+ occupies 10% of the cation exchange capacity) not only injure plants but also degrade the soil structure, decreasing porosity and water permeability. Salt incursion into the soil solution causes water deficits in leaves and inhibits plant growth and metabolism.

Salinity is a serious problem that reduces growth and development of ornamental crops in many salt-affected areas. It is estimated that about 20% of cultivated lands and 33% of irrigated agricultural lands worldwide are afflicted by high salinity (Foolad, 2004). In addition, the salinized areas are increasing at a rate of 10% annually. Low precipitation, high surface evaporation, weathering of native rocks, irrigation with saline water, and poor cultural practices are the major contributors to the increasing soil salinity. Soil salinity poses a serious threat to floriculture crops as it reduces the crop yield in the affected areas. Salinity stress limits crop growth and yield in different ways. Salt puts two primary effects on plants: osmotic stress and ionic toxicity.

Under normal condition the osmotic pressure in plant cells is higher than that in soil solution. Plant cells use this higher osmotic pressure to take up water and essential minerals from soil solution in to the root cells. Under salt stress the osmotic pressure in the soil solution exceeds the osmotic pressure in plant cells due to the presence of more salt, and thus, limits the ability of plants to take up water and minerals like K^+ and Ca^{2+} meanwhile Na^+ and Cl^- ions can enter into the cells and have direct toxic effects on cell membranes, as well as on metabolic activities in the cytosol. During salt stress the key mechanisms against ionic stress include the reduced uptake of toxic ions such as Na^+ and Cl^- into the cytosol, and also sequestration of these toxic ions either into the vacuole or into the apoplast.

Ion toxicity, water deficit, nutritional imbalance and high salinity in the root area sternly inhibits normal plant growth and development, resulting in reduced crop productivity or total crop failure (Ghassemi *et al.*, 1995). Young seedlings and plants at anthesis appear to be more sensitive to salinity stress than at the mature stages (Lutts *et al.*, 1995). One of the most effective ways to overcome salinity problems is the use of tolerant species and varieties. The response of plants to increasing salt application may differ significantly among plant species as a function of their genetic tolerance.

Few studies on salinity stress have dealt specifically with ornamental plants used in landscapes, despite the fact that salt stress causes serious damage in these species (Cassaniti *et al.*, 2009; Marosz, 2004). Salinity is of rising importance in landscaping because of the increase of green areas in the urban environment where the scarcity of water has led to the reuse of wastewaters for irrigation (McCammon *et al.*, 2009; Navarro *et al.*, 2008). Salinity is also a reality in coastal gardens and landscapes, where plants are damaged by aerosols originating from the sea (Ferrante *et al.*, 2011) and in countries where large amounts of de-icing salts are applied to roadways during the winter months (Townsend and Kwolek, 1987).

Salt tolerance varies considerably among the different genotypes of ornamentals used in landscaping. Cassaniti *et al.* (2009b) showed that the decrease in shoot dry weight and leaf area were the first visible effects of salinity both in sensitive and tolerant species such as *Cotoneaster lacteus* and *Eugenia myrtifolia*, respectively. Another common response to high salt level is leaf thickening, which occurred in ornamental plants such as *Coleus blumei* and *Salvia splendens* (Ibrahim *et al.*, 1991).

Research programs conducted in Israel have revealed ornamental species suitable for saline environments or to be irrigated with salt waters (Forti, 1986). Niu and Rodriguez (2006) observed the response to salt stress on eight

Table 4: List of ornamental shrubs and its response to salt stress

Species	Rating	Salt response	Salinity threshold	References
Bougainvillea spectabilis, Lantana camara var. *aculeate*	Tolerant	Maintains a high visual quality	1.94 dS m^{-1}	Devitt *et al.*, 2005
Euonymus japonica, Fraxinus pennsylvanica var. *lanceolata, Taxus cuspidata, Tilia europaea*	Sensitive	Low rank score in visual quality	2.1 dS m^{-1}	Quist *et al.*, 1999
Crataegus opaca	Sensitive	Reduction in relative growth rate (RGR)	3.15 dS m^{-1}	Picchioni and Graham, 2001
Lantana ×hybrida 'New Gold', *Lonicera japonica* 'Halliana', *Rosmarinus officinalis* 'Huntington Carpet'	Tolerant	Little reduction in growth, good aesthetic appearance	5.4 dS m^{-1}	Niu *et al.*, 2007
Lantana montevidensis	Sensitive	Reduction in growth index, low aesthetic appearance	5.4 dS m^{-1}	Niu *et al.*, 2007
Potentilla fruticosa 'Longacre', *Cotoneaster horizontalis*	Tolerant	No growth reduction and visible effects	12 dS m^{-1}	Marosz, 2004

herbaceous perennials (*Penstemon eatonii, P. pseudospectabilis, P. strictus, Ceratostigma plumbaginoides, Delosperma cooperi, Lavandula angustifolia, Teucrium chamaedrys, Gazania rigens*). The relative water content significantly declined as salinity increased in *C. plumbaginoides* and in *G. rigens* at the highest salt level; *D. cooperi* showed the highest water potential due to increased succulence of the leaves, a common mechanism of salt tolerance (Kozlowski, 1997). Plants of *L. angustifolia* and most of the species of *Penstemon* showed symptoms of necrosis and eventually died with an EC more than 3.2 dS m^{-1}.

A study conducted in California (Wu *et al.*, 2001) on ten ornamentals used in the landscape (*Pistacia chinensis, Nerium oleander, Pinus cembroides, Buxus microphylla, Liquidambar styraciflua, Bignonia violacea, Ceanothus thyrsiflorus, Nandina domestica, Rosa sp., Jasminum polyantum*) confirmed that the species showed a higher sensitivity when irrigated with sprinkler than drip irrigation. In studies conducted on native species of coastal areas other authors observed a large variation in foliage damage (with no symptoms to severe injury) in species that showed similar salinity tolerance in the roots (Cartica and Quinn, 1980; Sykes and Wilson, 1988), confirming that plants can evolve resistance to saline aerosols.

Table 5: List of salt tolerance of 38 landscape woody plant species grown under sprinkler irrigation with two NaCl concentrations (Source: Wu *et al.*, 2001).

Scientific name	Tolerance to NaCl	
	500 mg L^{-1}	1500 mg L^{-1}
Abelia ×grandiflora	Low	Low
Albizia julibrissin	Moderate	Low
Buddleja davidii	Low	Low
Cedrus deodara	High	High
Cotoneaster microphyllus	Moderate	Low
Forsythia ×intermedia	High	Moderate
Jasminum polyanthum	High	Moderate
Juniperus virginiana	High	High
Koelreuteria paniculata	Moderate	Low
Nandina domestica	Moderate	Low
Nerium oleander	High	High
Olea europaea	High	High
Plumbago auriculata	High	High
Sambucus nigra	Moderate	Low

As for all species, ornamentals differ in their tolerance to stress. For example, *Rudbeckia hirta* 'Becky Orange' and *Phlox paniculata* 'John Fanick' accumulated large quantities of Cl^- in the leaves which led to dry weight reduction of about 25%, while *Lantana ×hybrida* 'New Gold' and *Cuphea hyssopifolia*

'Allyson' tolerated salinity extremely well showing the low Cl accumulation (Cabrera *et al*., 2006).

Shade conditions can also influence the effects of salinity on plants. Devitt *et al.* (2005) quantified the foliar damage and flower production of 19 flowering landscape plants, sprinkle irrigated with reuse water and reuse water plus a period of shade (24% reduction in solar radiation). Results indicated that about half of the treated species had an acceptable levels of foliar damage at 1.94 dS m^{-1} in both of the reuse treatments (*Asteriscus maritimus* 'Gold Coin', *Centaurea cineraria*, *Lantana camara* var. *aculeate*, *Bougainvillea spectabilis*, *Gazania* spp., *Hemerocallis fulva*, *Gaillardia aristata*, *Mesembryanthemum crystallinum*).

Air Pollutants

Some landscape sites, especially in highly urbanized areas, are subjected to significant levels of air pollution. The most damaging of these pollutants are sulfur dioxide (SO_2), ozone (O_3), and peroxyacetyl nitrate (PAN). Symptoms of SO_2 injury include necrotic (dead) spots between the major veins, where the tissue turns light tan and papery in texture. The most common symptom of exposure to O_3 is the formation of tiny, light-colored flecks or spots on the upper surfaces of affected leaves, similar to spider mite damage. PAN injury is expressed as silvering, glazing, bronzing, and sometimes death of the lower leaf surfaces. Bedding plants do exhibit relative sensitivity and tolerance to these materials (Table 6), and if pollutants are a problem, plants should be selected accordingly.

Table 6: Bedding plant sensitivity to air pollutants.

Sulfur Dioxide		
Sensitive	Intermediate	Resistant
Begonia	Dianthus	Chrysanthemum
Cosmos	Nasturtium	
Coleus	Zinnia	
Aster		
Centaurea		
Geranium		
China Aster		
Marigold		
Poppy		
Ozone		
Ageratum	Impatiens	China Aster
Fuchsia	Verbena	Chrysanthemum
Aster		Geranium
Marigold		Lobelia

Contd.

Begonia	Ornamental Pepper
Pansy	
Petunia	
Salvia	
Dahlia	
Peroxyacetyl Nitrate (PAN)	
Aster	Begonia
Ornamental Pepper	Calendula
Dahlia	Chrysanthemum
Petunia	Coleus
Fuchsia	Gaillardia
Salvia	Pansy
Snapdragon	Periwinkle
Impatiens	

Adapted from Rogers, M.N. 1976. Air pollution, p. 441-481.

Role of phytohormones in plant response to abiotic stresses

Plant response to abiotic stresses depends on various factors, phytohormones are considered the most important endogenous substances for modulating physiological and molecular responses, a critical requirement for plant survival as sessile organisms.

Abscisic Acid (ABA)

Abscisic acid (ABA) is perhaps the most studied phytohormone for its response and distinct role in plant adaptation to abiotic stresses, and is accordingly termed a "stress hormone." Abscisic acid is considered an essential messenger in the adaptive response of plants to abiotic stress. In response to environmental stresses, endogenous abscisic acid levels increase rapidly, activating specific signaling pathways and modifying gene expression levels (O'Brien and Benkova, 2013). Nemhauser *et al.* (2006) have reported that abscisic acid transcriptionally regulates up to 10% of protein-encoding genes. Abscisic acid also acts as an internal signal enabling plants to survive under adverse environmental conditions (Keskin *et al.*, 2010). Under water-deficit conditions, abscisic acid plays a vital role in providing plants the ability to signal to their shoots that they are experiencing stressful conditions around the roots, eventually resulting in water-saving antitranspirant activity, notably stomatal closure and reduced leaf expansion (Wilkinson *et al.*, 2012). Abscisic acid is also involved in robust root growth and other architectural modifications under drought stress (Giuliani *et al.*, 2005) and nitrogen deficiency (Zhang *et al.*, 2007). Abscisic acid regulates the expression of numerous stress-responsive genes and in the synthesis of LEA proteins, dehydrins, and other protective proteins (Verslues *et al.*, 2006) and (Sreenivasulu *et al.*, 2012). abscisic acid up regulates the processes involved in cell turgor maintenance and synthesis of osmoprotectants and antioxidant enzymes conferring desiccation tolerance (Chaves *et al*., 2003). Zhang *et al.* (2006) reported a proportional increase in abscisic acid concentration upon exposure of plants to salinity.

Auxins

Auxins is vital not only for plant growth and development but also for governing and/or coordinating plant growth under stress conditions (Kazan, 2013). There is growing evidence that IAA plays an integral part in plant adaptation to salinity stress (Fahad *et al*, 2015) and (Iqbal *et al*., 2014). It increases root and shoot growth of plants growing under salinity or heavy metal stresses (Sheng and Xia, 2006) and (Egamberdieva, 2009). Auxin is regarded as an influential constituent of defense responses via regulation of numerous genes and mediation of crosstalk between abiotic and biotic stress responses (Fahad *et al*., 2015).

Cytokinins (CKs)

Alteration of endogenous levels of cytokinins in response to stress indicates their involvement in abiotic stress (O'Brien and Benkova, 2013), including drought (Kang *et al*, 2012) and salinity (Nishiyama *et al*., 2012). Mutants and transgenic cells/tissues with altered activity of cytokinin metabolic enzymes or perception machinery points toward their crucial involvement in several crop traits including productivity and increased stress tolerance (Zalabak *et al*, 2013). cytokinins are often considered abscisic acid antagonists (Pospisilova, 2003). In water-stressed plants, decreased cytokinin content and accumulation of abscisic acid lead to an increased ABA/CK ratio. The reduced cytokinin levels enhance apical dominance, which, together with the abscisic acid regulation of stomatal aperture, aids in adaptation to drought stress.

Ethylene (ET)

Abiotic stresses including low temperature and salinity alter endogenous ethylene levels in plants. Enhanced tolerance was accordingly achieved with higher ethylene concentrations (Shi *et al*, 2012). Ethylene also plays a major role in the defense response of plants to heat stress (Larkindale *et al*, 2005). Environmental stress induces ethylene accumulation which increases plant survival chances under these adverse conditions (Gamalero and Glick, 2012). Ethylene has been proposed to function via modulation of gene expression considered as the effectors of ethylene signal (Klay *et al*, 2014).

Ethylene in combination with other phytohormones such as jasmonates and Salicylic Acid often acts cooperatively. These are considered the main players involved in regulating plant defense against pests and pathogens (Kazan, 2015). The biosynthesis, transport, and accumulation of these hormones trigger a cascade of signaling pathways involved in plant defense (Matilla-Vazquez and Matilla, 2014). As concluded by Yin *et al.* (2015), ethylene and abscisicacid seem to act synergistically or antagonistically to control plant growth and development.

Gibberellins (GA)

Recently, experiments have been performed to investigate the role of Gibberellins in osmotic stress response in *Arabidopsis thaliana* seedlings (Skirycz *et al*, 2011) and (Claeys *et al.*, 2012). Gibberellins are known to interact with all other phytohormones in numerous developmental and stimulus-response processes (Munteanu *et al.,* 2014). The interactions between Gibberellin and ethylene include both negative and positive mutual regulation depending on the tissue and signaling context (Munteanu *et al.,* 2014).

Jasmonates

Jasmonates activates plant defense responses to pathogenic attack as well as environmental stresses including drought, salinity, and low temperature (Pauwels *et al.*, 2009) and (Seo *et al.*, 2011). Jasmonates are vital signaling molecules induced by various environmental stresses including salinity (Pauwels *et al.*, 2009), drought (Seo *et al.*, 2011) and (Du *et al.*,2013), and UV irradiation (Demkura *et al.*, 2010). They have great potential to mitigate an array of threatening environmental stresses (Dar *et al.*, 2015). Remarkably, endogenous levels of Jasmonates increased in rice roots under salinity stress and reported to counteract the deleterious effects of salinity stress (Wang *et al.*, 2001). Jasmonates applications alleviate heavy metal stress in plants by activating the antioxidant machinery (Yan *et al.*, 2013). Jasmonates confers tolerance in *A. thaliana* plants against Cu and Cd stress via accumulation of phytochelatins (Maksymiec *et al.*, 2007).

Salicylic acid (SA)

Low concentrations of Salicylic acid enhance the antioxidant activity in plants, but high concentrations of Salicylic acid cause cell death or susceptibility to abiotic stresses (Jumali *et al.*, 2011). Most genes that respond positively to acute Salicylic acid treatment are associated with stress and signaling pathways that eventually led to cell death. Salicylic acid consists of genes encoding chaperones, heat shock proteins, antioxidants, and genes involved in the biosynthesis of secondary metabolites, such as sinapyl alcohol dehydrogenase, cinnamyl alcohol dehydrogenase, and cytochrome P450 (Jumali *et al.*, 2011).

Salicylic acid is involved in plant response to abiotic stresses such as drought (Miura *et al.*, 2013), salinity (Fahad and Bano, 2012) and (Khodary,

2004), chilling (Yang *et al*, 2012), and heat (Fayez and Bazaid, 2014). Salicylic acid along with abscisic acid is involved in the regulation of drought response (Miura and Tada, 2014). Drought stress induced a five-fold increase in the endogenous levels of Salicylic acid in *Phillyrea angustifolia* (Munne-Bosch and Penuelas, 2003).

Breeding approaches in abiotic stress management

Development of crop cultivars tolerant to abiotic stresses is an important goal of plant breeding. Both traditional plant breeding and transgenic technology are being employed to achieve this goal. New technologies being developed through plant stress physiology research can potentially contribute to mitigate threats from abiotic stress. Many institutes have been working on addressing the effect of environmental stress on ornamental plant production. Germplasm of major ornamental crops, which are tolerant to high temperatures, flooding and drought have been identified and advanced breeding lines are being developed in many institutions. *Rosa indica* var odorata has the abilities to withstand high level of pH an indication of excessive soil salinity. Such species can be used in breeding program to transfer the genes of interest. Similarly resistance to high level of salts is found in carnation species *Dianthus aydogduii* which can also be incorporated in breeding program.

Use of molecular technologies has revolutionized the process of traditional plant breeding. Combining of new knowledge from genomic research with traditional breeding methods has enhanced our ability to improve crop plants. The use of molecular markers as a selection tool provides the potential for increasing the efficiency of breeding programmes by reducing environmental variability, facilitating earlier selection, and reducing subsequent population sizes for field testing. Molecular markers facilitate efficient introgression of superior alleles from wild species into the breeding programmes and enable the pyramiding of genes controlling quantitative traits; thus, enhancing and accelerating the development of stress-tolerant and higher-yielding cultivars.

Genetic engineering offers powerful means to develop transgenic crop plants with improved tolerance to range of abiotic stresses. In addition it helps in elucidating the physiological function of different stress related genes and gaining valuable information towards a better understanding of the mechanisms governing stress tolerance. Transgenic approach is turning out to be a viable option to improve crop tolerance to various abiotic stresses by incorporating genes involved in stress protection from diverse sources, related or unrelated to important ornamental plants. Many genes have been identified which get induced upon exposure of plants to various abiotic stresses. Some of these useful genezs have been implicated in conferring stress tolerance to plants (Abdin *et al.* 2002, Tayal *et al.*, 2004)

Abiotic stress tolerance in transgenic plants has been engineered in most cases by deploying genes belonging to the following categories: a. genes for biosynthesis of organic compatible solutes (proline, sugars, sugar alchohols, polyols, betaine, ectoine etc., or osmoprotectants, b. genes encoding stress proteins such as chaperones, late embryogenesis abundant (LEA) proteins, and c. genes involved in antioxidant systems (enzymes and metabolites). Several traits of ornamental plants have already been modified including flower color, fragrance, flower shape, plant architecture, flowering time, postharvest life and resistance for both biotic and abiotic stresses. Currently, at least 50 ornamental plants can now be transformed (Chandler and Sanchez, 2012). Transgenic ornamentals have been produced by several different techniques, the most common techniques being *Agrobacterium*-mediated transformation and particle bombardment (Hammond, 2006). Ornamental Biosciences in Germany is now focusing on improved abiotic stress resistance, specifically frost tolerance. This would increase the range of environments in which ornamental plant could be grown (Chandler and Sanchez, 2012). Research on GM for improved abiotic stress resistance is being explored for pot plants by Ornamental Biosciences (Potera, 2007), utilizing genes known to be involved in drought tolerance. Frost tolerance in *Petunia × hybrida* (petunia) may be increased by transfer of the CBF3 gene from *Arabidopsis thaliana* (Warner, 2011) and this would potentially increase the range of environments in which this bedding plant could be grown. Yamada *et al.* (2005) reported that AtP5CS or OsP5CS gene transfer in petunia plant exhibit 1.5 to 2 fold higher proline accumulation as compared with wild type plants under stress conditions. Also the transgenic plants showed enhanced drought tolerance. Delayed senescence and enhanced drought tolerance was also observed in transgenic petunia plants using IPT (isopentenyl transferase) driven by senescence inducible promoter P-SAG12 (Clark *et al.*, 2004). Constitutive overexpression of ZPT2-3 in transgenic petunia plants also increased tolerance to dehydration (Sugano *et al.*, 2003).

Song *et al.*, 2014 worked on a chrysanthemum Heat Shock Protein that confers tolerance to abiotic Stress. They reported that when *CgHSP70* was stably over-expressed in chrysanthemum, the plants showed an increased peroxidase (POD) activity, higher proline content and inhibited malondialdehyde (MDA) content. After heat stress, drought or salinity the transgenic plants were better able to recover, demonstrating CgHSP70 positive effect. Hong *et al.*, 2009 found that over-expression of the *AtDREB1A* gene in chrysanthemums significantly improved heat tolerance, suggesting that over-expression of this transcriptional regulator can confer a broader range of abiotic stress tolerance in chrysanthemums. These results suggest that improvement of heat stress tolerance in transgenic chrysanthemum may be associated with enhanced tolerance of photosynthesis.

References

Abdin MZ, Rehman RU, Israr M, Srivastava PS, Bansal KC. 2002. Abiotic stress related genes and their role in conferring resistance to plants. *Indian Journal of Biotechnology*. 1: 225-244.

Anonymous. 2010. How real is global warming? Pioneer Growing Point Magazine January: 12–13.

Armitage A. 1988. heat tolerant annuals for the landscaper. Greenhouse Grower. 6(13):54-56.

Arreola J, Franco J A, Vicente M J and Martinez Sanchez J J. 2006. Effect of nursery irrigation regimes on vegetative growth and root development of *Silene vulgaris* after transplantation into semi-arid conditions. *Journal of Horticultural Science and Biotechnology.* 81: 583-592.

Banon S, Fernandez J A, Franco J A, Torrecillas A, Alarcon J J and Sanchez-Blanco M J. 2004. Effects of water stress and night temperature pre-conditioning on water relations and morphological and anatomical changes of *Lotus creticus* plants. *Scientia Horticulturae.* 101: 333-342.

Banon S, Ochoa J, Franco J A, Alarcon J J and Sanchez-Blanco M J. 2006. Hardening of oleander seedlings by deficit irrigation and low air humidity. *Environmental and Experimental Botany.* 56: 36-43.

Banon S, Ochoa J, Franco J A, Alarcon J J, Fernandez T and Sanchez Blanco M J. 2002.The influence of acclimation treatments on the morphology, water relations and survival of *Myrtus communis* L. plants. In: Sustainable Use and Management of Soils in Arid and Semiarid Regions. (Faz, A., Ortiz, R. and Mermut, A. R., Eds.). Quaderna Editorial,Murcia, Spain. 275-277.

Cabrera RI, Rahman L, Niu G, McKenney C and Mackay W. 2006. Salinity tolerance in herbaceous perennial. *Hort. Science.* 41: 1054.

Capiati DA, Pais SM and Tellez-Inon MT. 2006.Wounding increases salt tolerance in tomato plants: Evidence on the participation of calmodulin-like activities in cross-tolerance signaling. *J. Exp. Bot.* 57: 2391-2400.

Cartica RJ and Quinn JA. 1980. Responses of populations of *Solidago sempervirens* (Compositae) to salt spray across a barrier beach, *American Journal of Botany.* 67: 1236–1242.

Cassaniti C, Li Rosi A and Romano D. 2009. Salt tolerance of ornamental shrubs mainly used in the Mediterranean landscape. *Acta Horticulturae.* 807: 675-680.

Chandler C and Sanchez. 2012. Genetic modiûcation; the development of transgenic ornamental plant varieties. *S.F. Plant Biotechnology Journal*, pp. 891–903.

Chaves M, Manuela JP, Maroco JS. 2003. Pereira, Understanding plant responses to drought-from genes to the whole plant. *Funct. Plant Biol.* 30: 239–264.

Chylinski W K, Lukaszewska A J and Kutnik K. 2007. Drought response of two bedding plants. *Acta Physiol Plant*. 29: 399.

Cicevan R, Hassan M A, Sestras A F, Prohens J, Vicente O, Sestras R E and Boscaiu M. 2016. Screening for drought tolerance in cultivars of the ornamental genus Tagetes (Asteraceae). *PeerJ* DOI 10.7717/peerj.2133

Claeys H, Skirycz A, Maleux K, Inze D. 2012. DELLA signaling mediates stress-induced cell differentiation in *Arabidopsis* leaves through modulation of anaphase-promoting complex/cyclosome activity. *Plant Physiol.* 159: 739–747.

Clark DG, Dervinis C, Barret JE, Klee H, Jones M. 2004. Drought induced leaf senescence and horticultural performance of transgenic P-SAG12-IPT petunias. *Journal of the American Society of Horticultural Sciences.* 129: 93-99.

Cuartero J, Bolarin MC, Asins MJ, Moreno V. 2006. Increasing salt tolerance in the tomato. *Journal of Exp. Bot.* 5: 1045–1058.

Dar TA, Uddin M, Khan MMA, Hakeem KR, Jaleel H. 2015. Jasmonates counter plant stress: a review. *Environ. Exp. Bot.* 115: 49–57.

Demkura PV, Abdala G, Baldwin IT, Ballare CL. 2010. Jasmonate-dependent and-independent pathways mediate specific effects of solar ultraviolet B radiation on leaf phenolics and antiherbivore defense. *Plant Physiol.* 152:1084–1095.

Devitt DA, Morris RL and Fenstermaker LK. 2005. Foliar damage, spectral reflectance, and tissue ion concentrations of trees sprinkle irrigated with waters of similar salinity but different chemical composition, *Hort. Science* 40: 819-826.

Du H, Liu HB, Xiong LZ. 2013. Endogenous auxin and jasmonic acid levels are differentially modulated by abiotic stresses in rice. *Front. Plant Sci.* 4: 397.

Egamberdieva D. 2009. Alleviation of salt stress by plant growth regulators and IAA producing bacteria in wheat. *Acta Physiol. Plant.* 31: 861–864

Fahad S, Bano A. 2012. Effect of salicylic acid on physiological and biochemical characterization of maize grown in saline area. *Pak. J. Bot.* 44: 1433–1438.

Fahad S, Hussain S, Matloob A, Khan FA, Khaliq A, Saud S, Hassan S, Shan D, Khan F, Ullah N, Faiq M, Khan MR, Tareen AK, Khan A, Ullah A, Ullah N, Huang JL. 2015. Phytohormones and plant responses to salinity stress: a review. *Plant Growth Regul.* 75:391-404.

Fayez KA, Bazaid SA. 2014. Improving drought and salinity tolerance in barley by application of salicylic acid and potassium nitrate. *J. Saudi Soc. Agric. Sci.* 13: 45–55.

Ferrante A, Trivellini A, Malorgio F, Carmassi G, Vernieri P and Serra G. 2011. Effect of seawater aerosol on leaves of six plant species potentially useful for ornamental purposes in coastal areas, *Scientia Horticulturae.* 128: 332–341.

Foolad MR. 2004. Recent advances in genetics of salt tolerance in tomato. *Plant Cell Tissue Organ Cul.* 76: 101-119.

Forti M. 1986. Salt tolerant and halophytic plants in Israel, *Reclamation and Revegetation Research* 5: 83-96.

Franco J A, Banon S, Fernandez J A and Leskovar D I. 2001. Effect of nursery regimes and establishment irrigation on root development of *Lotus creticus* seedlings following transplanting. *Journal of Horticultural Science and Biotechnology.* 76: 174-179.

Franco J S. 2011. Root development under drought stress. *Technology and Knowledge Transfer e-bulletin* **2**(6): 1-3.

Franco JA, Martinez-Sanchez JJ, Fernandez JA and Banon S. 2006. Selection and nursery production of ornamental plants for landscaping and xerogardening in semi-arid and environments, *Journal of Horticulture Science and Biotechnology* 81: 3-17.

Gamalero E and Glick BR. 2012. Ethylene and abiotic stress tolerance in plants. In: P. Ahmed, M.N.V. Prasad (Eds.), Environmental Adaptations and Stress Tolerance of Plants in the Era of Climate Change, Springer, New York, pp. 395–412.

Ghassemi F, Jakeman AJ and Nix HA. 1995. Salinisation of Land and Water Resources: HumanCauses, Extent Management and Case Studies. Canberra, Australia. The Australian National University, CAB International, Wallingford, England. 526 p

Giuliani S, Sanguineti MC, Tuberosa R, Bellotti M, Salvi S, Landi P. 2005.Root-ABA1 a major constitutive QTL affects maize root architecture and leaf ABA concentration at different water regimes. *J. Exp. Bot.* 56: 3061–3070.

Hammond J, Hsu H, Huang Q, Jordan R, Kamo K *and* Pooler M. 2006. Transgenic approaches to disease resistance in ornamental crops. *J. Crop Improvement,* 17: 155–210.

Hong B, Chao M, Yang Y, Wang T, Yamaguchi K, Shinozaki, Gao J. 2009. Over-expression of *AtDREB1A* in chrysanthemum enhances tolerance to heat stress. *Plant Molecular Biology.* 70(3):231–240.

Ibrahim KM, Collins JC and Collin HA. 1991. Effects of salinity on growth and ionic composition of *Coleus blumei* and *Salvia splendens*. *Journal of Horticulture Science.* 66: 215-222.

IPCC. 2001. Climate change 2001: Impacts, adaptation and vulnerability. *Intergovermental Panel on Climate Change*. New York, USA.

Iqbal N, Umar S, Khan NA, Khan MIR. 2014. A new perspective of phytohormones in salinity tolerance: regulation of proline metabolism. *Environ. Exp. Bot.* 100: 34–42.

Jaleel C A, Manivannan P, Sankar B, Kishorekumar A, Gopi R, Somasundaram R, Panneerselvam R. 2007. Water deficit stress mitigation by calcium chloride in *Catharanthus roseus*: Effects on oxidative stress, proline metabolism and indole alkaloid accumulation 60(1): 110-116.

Josine T L, Jing Ji, Gang Wang and Chun Feng Guan. Advances in genetic engineering for plants abiotic stress control. *African Journal of Biotechnology.* 10(28): 5402-5413.

Jumali SS, Said IM, Ismail I, Zainal Z. 2011. Genes induced by high concentration of salicylic acid in *Mitragyna speciosa. Aust. J. Crop. Sci.* 5: 296–303.

Kang NY, Cho C, Kim NY, Kim J. 2012. Cytokininreceptor-dependent and receptor-independent path ways in the dehydration response of *Arabidopsis thaliana. J. Plant Physiol.* 169: 1382-1391.

Kazan K. 2013. Auxin and the integration of environmental signals into plant root development *Ann. Bot.* 112: 1655–1665.

Kazan K. 2015. Diverse roles of jasmonates and ethylene in abiotic stress tolerance. *Trends Plant Sci.* 20: 219–229.

Keskin BC, Sarikaya AT, Yuksel B, Memon AR. 2010. Abscisic acid regulated gene expression in bread wheat. *Aust. J. Crop. Sci.* 4: 617–625

Khodary SEA. 2004. Effect of salicylic acid on growth, photosynthesis and carbohydrate metabolism in salt stressed maize plants. *Int. J. Agric. Biol.* 6: 5–8.

Klay I, Pirrello J, Riahi, Bernadac A, Cherif A, Bouzayen M, Bouzid S. 2014. Ethylene response factor *Sl-ERF.B.3* is responsive to abiotic stresses and mediates salt and cold stress response regulation in tomato. *Sci. World J.* p. 167681

Kozlowski TT. 1997. Responses of woody plants to flooding and salinity, *Tree Physiology Monograph N° 1*, Heron Publishing, Victoria, Canada.

Kumar S, Dhingra A, and Daniell H. 2004. Plastid-expressed betaine aldehyde dehydrogenase gene in carrot cultured cells, roots, and leaves confers enhanced salt tolerance. *Plant Physiol.* 136:2843–2854.

Larkindale J, Hall DJ, Knight MR, Vierling E. 2005. Heat stress phenotypes of Arabidopsis mutants implicate multiple signaling pathways in the acquisition of thermo-tolerance. *Plant Physiol.* 138: 882–897.

Lin J N and Kao CH. 1998. Water stress, ammonium and leaf senescence in detached rice leaves. *Plant Growth Regul.* 26: 165–169

Ludlow MM. 1989. Strategies in response to water stress. In: Kreeb HK., Richter H, Hinkley TM. (Eds.), Structural and Functional Response to Environmental Stresses: Water Shortage. SPB Academic Press, The Netherlands, pp. 269–281.

Lutts S, Kinet JM and Bouharmont J. 1995. Changes in plant response to NaCl during development of rice (*Oryzasativa* L.) varieties differing in salinity resistance. *J. Exp. Bot.*46: 1843-1852.

Maksymiec W, Wojcik M, Krupa Z. 2007. Variation in oxidative stress and photochemical activity in *Arabidopsis thaliana* leaves subjected to cadmium and excess copper in the presence or absence of jasmonate and ascorbate. *Chemosphere*. 66: 421–427.

Marosz A. 2004. Effect of soil salinity on nutrient uptake, growth and decorative value of four ground cover shrubs, *Journal of Plant Nutrition.* 27: 977-989.

Matilla Vazquez MA, Matilla AJ. 2014. Ethylene: Role in plants under environmental stress. In: P. Ahmad, M.R. Wani (Eds.), Physiological Mechanisms and Adaptation Strategies in Plants under Changing Environment, vol. 2, Springer Science + Business Media, New York. pp. 189–222

Maya MA, Ito M, Matsubara Y. 2014. Tolerance to heat stress and anthracnose in mycorrhizal Cyclamen *Acta Horticulturae* 1025: International Symposium on Orchids and Ornamental Plants.

McCammon TA, Marquart Pyatt ST and Kopp KL. 2009. Water-conserving landscapes: an evaluation of homeowner preference, *Journal of Extension* 47(2): 1-10.

Mittler R. 2006. Abiotic stress, the field environment and stress combination. *Trends In Plant Sci.* 11: 15-19.

Miura K, Okamoto H, Okuma E, Shiba H, Kamada H, Hasegawa PM, Murata Y. 2013. SIZ1 deficiency causes reduced stomatal aperture and enhanced drought tolerance via controlling salicylic acid-induced accumulation of reactive oxygen species in *Arabidopsis. Plant J.* 49: 79–90.

Miura K, Tada Y. 2014. Regulation of water, salinity, and cold stress responses by salicylic acid. *Front. Plant Sci.* 5:4.

Munne-Bosch S, Penuelas J. 2003. Photo and antioxidative protection and a role for salicylic acid during drought and recovery in field-grown *Phillyrea angustifolia* plants. *Planta.* 217: 758–766

Munteanu V, Gordeev V, Martea R, Duca M. 2014. Effect of gibberellin cross talk with other phytohormones on cellular growth and mitosis to endoreduplication transition. *Int. J. Adv. Res. Biol. Sci.* 1 (6): 136–153.

Navarro A, Vicente M J, Martínez-Sanchez J J, Franco J A, Fernandez J A and Banon S. 2008. Influence of deficit irrigation and paclobutrazol on plant growth and water status in *Lonicera implexa* seedlings. *Acta Horticulturae.* 782: 299-304

Nemhauser JL, Hong F, Chory J. 2006. Different plant hormones regulate similar processes through largely non overlapping transcriptional responses. *Cell.* 126:467–475

Nishiyama R, Watanabe Y, Fujita Y, Tien LD, Kojima M, Werner T, Vankova R, Yamaguchi-Shinozaki K, Shinozaki K, Kakimoto T, Sakakibara H, Schmuelling T, Lam-Son PT. 2011. Analysis of cytokinin mutants and regulation of cytokinin metabolic genes reveals important regulatory roles of cytokinins in drought, salt and abscisic acid responses, and abscisic acid biosynthesis. *Plant Cell.* 23: 2169–2183

Niu G and Rodriguez DS. 2006. Relative salt tolerance of five herbaceous perennials, *Hort. Science.* 41: 1493-1497.

O'Brien JA, Benkova E. 2013.Cytokinin cross-talking during biotic and abiotic stress responses. *Front. Plant Sci.* 4: 451.

Pauwels L, Inze D, Goossens A. 2009. Jasmonate-inducible gene: what does it mean? *Trends Plant Sci.* 14: 87–91.

Pospisilova J. 2003. Participation of phytohormones in the stomatal regulation of gas exchange during water stress. *Biol. Plant.* 46: 491–506.

Potera C. (2007) Blooming biotech. *Nat. Biotechnol.* 25: 963–965.

Robert M, Auge Ann, Stodola JW, Moore JL, William E, Klingeman and Xiangrong Duan. 2003. Comparative dehydration tolerance of foliage of several ornamental crops. *Scientia Horticulturae.* 98:511–516

Rogers MN. 1976. Air pollution. In: J. Mastalerz (Ed.). Bedding Plants, Prentice-Hall, Englewood Cliffs, N.J. p. 441-481.

Sanchez-Blanco M J, Ferrandez T, Navarro A, Banon S and Alarcon J J. 2004. Effects of irrigation and air humidity preconditioning on water relations, growth and survival of *Rosmarinus officinalis* plants during and after transplanting. *Journal of Plant Physiology* 161: 1133-1142.

Seo JS, Joo J, Kim MJ, Kim YK, Nahm BH, Song SI, Cheong JJ, Lee JS, Kim JK, Choi YD. 2011. OsbHLH148, a basic helix-loop-helix protein, interacts with OsJAZ proteins in a jasmonate signaling pathway leading to drought tolerance in rice. *Plant J.* 65: 907–92.

Sheng XF, Xia JJ. 2006. Improvement of rape (*Brassica napus*) plant growth and cadmium uptake by cadmium-resistant bacteria. *Chemosphere*. 64: 1036–1042

Shi Y, Tian S, Hou L, Huang X, Zhang X, Guo H, Yang S. 2012. Ethylene signaling negatively regulates freezing tolerance by repressing expression of *CBF* and Type-A *ARR* genes in Arabidopsis. *Plant Cell*, 24: 2578–2595.

Singh RK and Satpathy KK. 2005. Scope and Adoption of Plasticulture Technologies in North Eastern Hill Region. In: Agricultural Mechanization in North East India, published by ICAR,Research Complex for NEH Region, Barapani. p. 114-21.

Skirycz A, Claeys H, De Bodt S, Oikawa A, Shinoda S, Andriankaja M, Maleux K, Eloy NB, Coppens F, Yoo SD, Saito K, Inze D. 2011. Pause-and-stop: the effects of osmotic stress on cell proliferation during early leaf development in *Arabidopsis* and a role for ethylene signaling in cell cycle arrest. *Plant Cell*, 23: 1876–1888.

Song Aiping, Xirong Zhu, Fadi Chen, Haishun Gao, Jiafu Jiang and Sumei Chen A. 2014. Chrysanthemum heat shock protein confers tolerance to abiotic stress. *Int J Mol Sci.* 15(3): 5063-5078.

Sreenivasulu N, Harshavardhan VT, Govind G, Seiler C, Kohli A. 2012. Contrapuntal role of ABA: does it mediate stress tolerance or plant growth retardation under long-term drought stress? Gene, 506: 265–273.

Sugano S, KaminakaH, Rybka Z, Catala R, Salinas J, Matsui K, Ohme TM, Takatsuji H. 2003. Stress responsive zinc finger gene ZPT2-3 play a role in drought tolerance in Petunia. *The Plant Journal.* 36: 830-841.

Sykes MT and Wilson JB. 1988. An experimental investigation into the response of some New Zealand sand dune species to salt spray, *Annals of Botany* 62: 159–166.

Tayal D, Srivastava PS, Bansal KC. 2004. Transgenic crops for abiotic stress tolerance. In: Srivastava PS, Narula A, Srivastava S (eds) Plant Biotechnology and Molecular Markers, Kluwer Academic Publishers, The Netherlands, pp 346-365.

Townsend AM and Kwolek WF. 1987. Relative susceptibility of thirteen pine species to sodium chloride spray, *Journal of Arboriculture* 13: 225-227.

Ueda A, Kanechi M, Uno Y and Inagaki N. 2003: Photosynthetic limitations of a halophyte sea aster (*Aster tripolium* L.) under water stress and NaCl stress. *J. Plant Res.* 116: 63–68.

Verslues PE, Agarwal M, Katiyar-Agarwal S, Zhu J, Zhu JK. 2006. Methods and concepts in quantifying resistance to drought, salt and freezing, abiotic stresses that affect plant water status. *Plant J.* 45: 523–539.

Von WS and Chieng SS. 2004. A comparison between low-cost drip irrigation, conventional drip irrigation, and hand watering in Nepal. *Agric. Water Management.* 64: 143-160.

Wang Y, Mopper S, Hasentein KH. 2001. Effects of salinity on endogenous ABA, IAA, JA, and SA in *Iris hexagona. J. Chem. Ecol.* 27: 327–342.

Warner R. 2011. Genetic approaches to improve cold tolerance of petunia. *Floricult. Int.* June, 15–16.

Wilkinson S, Kudoyarova GR, Veselov DS, Arkhipova TN, Davies WJ. 2012. Plant hormone interactions: innovative targets for crop breeding and management. *J. Exp. Bot.* 63:3499–3509.

Wu L, Guo X, Hunter K, Zagory E, Waters R and Brown J. 2001. Studies of salt tolerance of landscape plant species and California native grasses for recycled water irrigation, *Slosson Report*.

Yamada M, Morishta H, Urano K, Shiozaki N, Yamaguchi SK, Shinozaki K, Yoshiba Y. 2005. Effects of free proline accumulation in petunias under drought stress. *Journal of Experimental Botany*. 56: 1975-1981.

Yan Z, Chen J, Li X. 2013. Methyl jasmonate as modulator of Cd toxicity in *Capsicum frutescens* var. *fasciculatum* seedlings. *Ecotoxicol. Environ. Saf.* 98: 203–209.

Yang Z, Cao S, Zheng Y, Jiang Y. 2012. Combined salicyclic acid and ultrasound treatments for reducing the chilling injury on peach fruit. *J. Agric. Food Chem*. 60: 1209–1212.

Yin CC, Ma B, Collinge DP, Pogson BJ, He SJ, Xiong Q, Zhang JS. 2015. Ethylene responses in rice roots and coleoptiles are differentially regulated by a carotenoid isomerase-mediated abscisic acid pathway. *Plant Cell*, 27: 1061–108.

Zalabak D, Pospisilova H, Smehilova M, Mrizova K, Frebort I, Galuszka P. 2013.Genetic engineering of cytokinin metabolism: prospective way to improve agricultural traits of crop plants. *Biotechnol. Adv*. 31:97–117.

Zhang J, Jia W, Yang J, Ismail AM. 2006. Role of ABA in integrating plant responses to drought and salt stresses. *Field Crops Res.* 97:111–119.

Zhang S, Hu J, Zhang Y, Xie XJ, Knapp A. 2007. Seed priming with brassinolide improves lucerne (*Medicago sativa* L.) seed germination and seedling growth in relation to physiological changes under salinity stress. *Aust. J. Agric. Res*. 58: 811–815.

Zhu J. K. 2001. Plant salt tolerance. *Trends In Plant Sci*. 6:66–71.